Water Management in Agriculture

Water Management in Agriculture

Editor: Davis Twomey

STATES
ACADEMIC PRESS
www.statesacademicpress.com

States Academic Press,
109 South 5th Street,
Brooklyn, NY 11249, USA

Visit us on the World Wide Web at:
www.statesacademicpress.com

ISBN: 978-1-63989-565-6 (Hardback)

Cataloging-in-Publication Data

Water management in agriculture / edited by Davis Twomey.
p. cm.
Includes bibliographical references and index.
ISBN 978-1-63989-565-6
1. Water in agriculture. 2. Water-supply, Agricultural. 3. Water resources development.
4. Irrigation. I. Twomey, Davis.
S494.5.W3 W38 2022
631.7--dc23

Table of Contents

Preface

Agriculture uses a large portion of fresh water in the world. Rising world population is increasing the need of food for consumption which is eventually leading to more use of water globally. This escalates the risk of water scarcity. Water management refers to the practice of planning, developing, allocating and controlling the use of water resources. It is an important aspect in reducing water scarcity. Water management in agriculture can be done by using various strategies such as making more rainwater available to the crops, using improved crop varieties, and using improved cropping systems and agronomics. It can also be achieved by increasing crop productivity through efficient methods of irrigation and by decreasing water losses through soil evaporation. This book unfolds the innovative aspects of water management in agriculture which will be crucial for the progress of this field in the future. It elucidates new techniques and their applications in a multidisciplinary manner. This book will serve as a reference to a broad spectrum of readers.

All of the data presented henceforth, was collaborated in the wake of recent advancements in the field. The aim of this book is to present the diversified developments from across the globe in a comprehensible manner. The opinions expressed in each chapter belong solely to the contributing authors. Their interpretations of the topics are the integral part of this book, which I have carefully compiled for a better understanding of the readers.

At the end, I would like to thank all those who dedicated their time and efforts for the successful completion of this book. I also wish to convey my gratitude towards my friends and family who supported me at every step.

Editor

1

Effect of under Irrigation Management on Potato Performance Components

Fakhrodin Ghasemi Sahebi[1], Masoud Hekmat[2*], Eman Pourkhiz[3]

Abstract

To evaluate the effect of water, tape drip and furrow irrigation methods on the performance components and features of Agria cultivar potato, an experiment was carried out in the agriculture faculty of Kermanshah. Repeated three times, this study was carried out in the form of full random blocks with the main factor of different amounts of irrigation water in three levels (50, 75, and100 percent of cumulative evaporation from an A- class evaporation pan) and the secondary factor of irrigation method (including drip and furrow irrigations). At 3 repetitions, results showed that the minimum performance (19.168 tons per hectare) was related to the drip irrigation method with 50 percent evaporation from the A-class pan and the maximum one (34.455 tons per hectare) was related to the drip irrigation method with 100 percent evaporation from the A-class pan. The rate of irrigation and effectiveness of method on the number of the main stem per square meter and number of the main stem in the bush were not significant. The minimum percent (number of tubers) of tubers smaller than 35mm (26.47percent) and the maximum tuber production (37.17 percent) in the size of food and market-friendly (tuber bigger than 55mm) were attained by drip irrigation method with 100 percent evaporation from the A-class pan. The impact of Irrigation method on and the specific gravity of tubers and starch (dry matter) was and was not significant, respectively.

Keywords:
Tape drip irrigation, Furrow irrigation, Potato, Dry matter

[1] *M.Sc., Irrigation and Drainage, Razi University, Kermanshah, Iran.*

[2] *Department of Agricultural Management, Shoushtar Branch, Islamic Azad University, Shoushtar, Iran.*

[3] *Young Researchers Club, Ghaemshahr Branch, Islamic Azad University, Iran.*

* *Corresponding author's email: Masoud.hekmat1@gmail.com*

INTRODUCTION

Potato is one of the glandular products having an important role in human and animal nutrition. It is cultured in different climates and is the fourth product after wheat, rice, and popcorn; in the subject of producing countries number, it is the second product after popcorn (Dokhani and Rabiee Motmaen, 2001).

Studies showed that potato is very sensitive to the aridity stress .moisture stress has a bad effect on the potato`s growth and performance. Moisture stress, at the chlorophyll growth stage and before the tubers formation stage, reduces the leaf area, number of root system`s secondary branches, bush height, and the overall growth of the plant`s green cover. The second stage of potato growth is the tuber formation stage in which aridity reduces one of the performance components which is the number tuber per each bush; the tubers` mean size and specific gravity is highly reduced, too. The third stage of the potato growth is the tubergrowth stage. At this stage, the product`s quality and performance is highly affected by the aridity stress. The last stage is the ripening stage. At this stage, the green cover of the plant senesces and the tuber skin becomes thick, the water need is reduced, and even the moisture stress, after harvesting, increases the tuber resistance to the pests (Mosavi Fazl and Faeznia, 2003).

Produced tubers` size and weight distribution depends on each gland`s growth and the competitive relation between the tubers. All tubers of a bush do not, simultaneously, start the dry matter aggregation. This differences are resulted from the tuber formation time and the location of the tuber on the bush causing the distribution of the gland`s size and weight (Elizabeth, 1992). It seems that despite the reduction of a bush`s mean weight of each tuber caused by the increase of the produced tubers number, a larger number of tubers gratifies the weight reduction and the weight of all tubers is more in such bushes; but the market-friendly of tubers with a lesser mean weight would be reduced (Wurr and Allen, 1994).

Yuan *et al.,* evaluated the effects of different irrigation regimes on the potato growth at the drip irrigation. Irrigation water quantity was considered as 125, 100, 75, 50, and 25 percent of evaporation from the water level in a ceramic evaporation pan (0.2 m diameter). Plant height, biomass amount, total fresh glandular products, and market-friendly tubers (more than 85 g) were increased by increasing the irrigation water. Plant height and the total glandular product at the treatment of 125 percent evaporation from the evaporation pan were close to the 100 percent one. Increasing the irrigation water not only have the number of tubers increased, but increased the mean weight of the tubers, too. Irrigation water increased the quantity of the tubers but reduced their quality (Yuan *et al.,* 2003).

Nadler and Heour (1995) believed that the specific gravity of the tubers is reduced and the reducing sugars rates are increased by reducing the irrigation water. This reduction in the gland`s specific gravity is caused by a reaction to the under irrigation during the long term of growth season.

This study is aimed at evaluating the effect of different irrigation water quantities and tape drip and furrow irrigation methods on the performance, performance components, and features of potato (dry matter and specific gravity).

MATERIALS AND METHODES

The experiment, to compare the performance of potato (agria cultivar) under tape drip and furrow irrigation methods, was carried out in the agriculture faculty`s research farm of Kermanshah in the spring of 1386. Farm`s soil texture was salty clay. Repeated three times, this study was carried out in the form of full random blocks plan.

The main factor was different amounts of irrigation water in three levels of 50, 75, and100 percent of cumulative evaporation from an A-class evaporation pan and the secondary factor was two irrigation methods including drip and furrow irrigations. Each check was 13 m length and 3m width in which 4 lines were cultivated. The distance of culture lines was 75cm and the bushes distance on the row was 30cm. volume contours were used to measure and control the amount of influent irrigation water of each plot in each treatment. In measuring the gross irrigation requirement performance efficiency for the furrow method was 40 percent because the water was running off from the end of check.

In the tape drip irrigation method volume contours 30 m length T tapes were used and regulating valves were used to measure and control the amount of irrigation water in each treatment. Application efficiency for the tape drip method

Table 1: Total quantity of irrigated water during the plant growth period in each method

treatment	100 percent of water requirement	75 percent of water requirement	50 percent of water requirement	Furrow irrigation
Total quantity of the irrigated water (m^3) in each hectare	9749.62	7312.215	4874.81	21357.207

was 90 percent. Data of evaporation from the A-class evaporation pan located in the weather station near the agriculture faculty was used to calculate the plant water requirement. The potential rate of evaporation and transpiration was calculated by the following equation:

$$ET = K_p . E_{pan} \quad (1)$$

In which

ET: evaporation and transpiration (in millimeter) between two irrigations

K_p: evaporation pan coefficient

E_{pan}: rate of evaporation (in millimeter) from the pan between two continuous irrigations

The amount of water for each treatment was then attained by multape driplying the coefficients related to the water level of treatment (50, 75, and 100 percent) by the evaporation and transpiration calculated by equation1.

Agricultural operations such as weeding, fertilizing, and spraying were equally carried out based on the customs and requirements of the farms in each experimental unit. At the end of the growth season, two centric 6 m adjacent lines were selected from each check and their potato tubers were harvested. Based on the size, tubers were divided in three groups of smaller than 35, 35 to 55, and bigger than 55 mm. samples of each treatment were prepared and sent to the laboratory to carry out the qualitative tests of product such as measuring the specific gravity and the percent of dry matter.

Measurement of the specific gravity

Specific gravity was attained from division of samples - weight and volume. Tubers were randomly selected, peeled, chopped to very little pieces, and mixed to determine the dry matter. About 5 g of this mixture was poured in the ceramic bushes which have reached the fixed weight at 105° C in the oven; after 6 hours in the oven with air flow, it reached the fixed weight. Then, the ceramic bushes were cooled down in the desiccators, weighted again, and the percentage of the dry matter was calculated (Dokhani and Rabiee Motmaen, 2001). Consumptive water volume in different irrigation water treatments is illustrated in table 1. SAS software was used for statistical analysis and was compared to Duncan. Measure and control the amount of irrigation water input to each plot in each treatment was done by volumetric meters.

RESULTS AND DISCUSSION

Tubers` performance

Table 2 below shows the effect of the irrigation water rate on the tuber performance was significant at a 1 percent level. The minimum performance related to the 50 percent treatment was 19.168 tons per hectare and the maximum one related to the 100 percent treatment was 34.455 tons/ha as demonstrated on table 3. The 75 percent treatment and the furrow irrigation

Table 2: Analysis of variance of potato tuber`s performance

Source changes	degree of freedom	Bush performance(kg)	performance ton per hectare
repetition	2	0.0156	6.1351
treatment	3	0.4251**	354.0292**
error	6	0.0368	27.6018
sum	11	0.4777	378.7662
CV(%)		8.85	8.16

ns, no significant; ** p≤0/01, * p≤0/05

Table 3: Mean Comparison of potato tuber`s performance

experimental treatment	Bush performance (kg)	performance ton per hectare
50 percent cumulative evaporation from the pan	0.62167 c	19.168 c
75percent cumulative evaporation from the pan	0.95967 b	25.938 b
100 percent cumulative evaporation from the pan	1.13667 a	34.455 a
Furrow irrigation	0.82600 b	25.582 b

Means followed by same letter (s) in each column are not significantly different ($p < 0.05$)

treatment, based on Duncan groups, classified in one group. This experiment showed that the more water is provided for the plant, the more performance of tubers is resulted.

Size of the tubers

Percent of tubers smaller than 35m

The effect of irrigation water on the percent of the smaller than 35mm tubers was not significant as illustrated on table 4. Minimum losses (small tubers, 26.47 percent) were resulted from drip 100 percent treatment and the maximum ones (37.39 percent) were resulted from the furrow irrigation treatment.

Percent of 35-55 mm tubers

The effect of irrigation water on the percent of 35-55mm tubers was not significant as showed on table 4. The size of 35-55 mm is considered as seed size in classifying the tubers size. If the objective is to produce the potato seed size, the maximum seed tuber (47.87 percent) can be produced by the 50 percent drip method.

Percent of bigger than 55 mm tubers

Table 4 shows the effect of irrigation water on the percent of bigger than 55 mm tubers was not significant. More than 55 mm tubers are considered as the food size and market-friendly ones.

Table 4. Analysis of variance of tuber percent in the groups of smaller than 35mm, 35-55mm, and bigger than 55mm (Mean Squares)

Number of the main stem

Table 4 illustrates the effect of irrigation water on the number of the main stem per meter square was not significant. The number main stem depends on the size, physiologic age of the seed gland, and the genetic characteristics of the cultivar - number of eyes in the seed tuber (Allen, 1992). Maximum number of the main stem per square meter in the 100 percent treatment was 25.08 as shown on table 6.

The number of tubers

Number of tubers per meter square

The effect of irrigation water on the number of tubers was not significant (table 6). The maximum tuber number per meter square in the 50 percent water treatment was 35086 and the minimum one in the 75 percent treatment was 34.06. Shock *et al.,* (1993) showed that there a great correlation between stolen and the number of tubers.

Percent of the dry matter

The effect of irrigation water on the dry matter was not significant shown on table 9 but the percent of the dry matter reduced by increasing the

Table 4: Analysis of variance of tuber percent in the groups of smaller than 35mm, 35-55mm, and bigger than 55mm (Mean Squares)

Source changes	degree of freedom	smaller than 35mm	35-55mm	bigger than 55mm
Repetition	2	356.94	7.875	326.44
treatment	3	270.86ns	211.13ns	370.14*
error	6	388.61	505.60	274.83
sum	11	1016.42	724.60	971.42
CV(%)		26.26	22.31	23.98

ns, no significant; ** p≤0/01, * p≤0/05 .ger than 55mm (Mean Squares)

Table 5: Mean comparison of in the groups of smaller than 35mm, 35-55mm, and bigger than 55mm

Source changes	smaller than 35mm	35-55 mm	bigger than 55mm
50 percent cumulative evaporation from the pan	26.427 a	47.877 a	25.700 ab
75percent cumulative evaporation from the pan	31.677 a	40.457 a	27.867 ab
100 percent cumulative evaporation from the pan	26.470 a	36.357 a	37.170 a
Furrow irrigation	37.933 a	39.857 a	22.153 b

Means followed by same letter (s) in each column are not significantly different ($p < 0.05$)

Table 6: Analysis of the variance of the number of main stem and tubers (Mean Squares)

Source of variation	degree of freedom	Number of stem in the bush	Number of stem per meter square	Number of tuber in the bush	Number of tuber per meter square
Repetition	2	2.50	54.62	2.2386	45.93
Treatment	3	2.56*	63.04^{ns}	1.86^{ns}	44.93^{ns}
Error	6	1.63	51.50	10.32	260.35
Sum	11	6.707	169.17	14.42	351.22
CV(%)		12.25	13.73	19.29	19.29

ns, no significant; ** p≤0/01, * p≤0/05

Table 7: Mean comparison of the main stem and tubers numbers

Experimental treatment	Number of stem in the bush	Number of stem per meter square	Number of tuber in the bush	Number of tuber per meter square
50 percent cumulative evaporation from the pan	4.0300 ab	19.753 a	7.097 a	35.863 a
75percent cumulative evaporation from the pan	4.1933 ab	21.297 a	6.193 a	31.060 a
100 percent cumulative evaporation from the pan	5.0267 a	25.080 a	6.713 a	33.903 a
Furrow irrigation	3.8033 b	19.217 a	7.197 a	35.710 a

Means followed by same letter (s) in each column are not significantly different ($p < 0.05$).

irrigation water quantity. It is believed that by increasing the quantity of irrigation, percent of the dry matter is increased (Mohammadi and Faeznia, 2001). If the rate of dry matter is very low oil absorption would be high and soft chips can be produced (Dokhani and Rabieae Motmaen, 2001).

Gland`s specific gravity

The effect of irrigation water on the Gland's specific gravity was significant at a 1 percent probability level. The maximum specific gravity (1.11 g/cm^3) was related to the tape drip 100 percent treatment. Increasing the irrigation water, the specific gravity of the tubers was also increased, which is ion accordance with the findings of Yuan *et al.,* (Yuan *et al.,* 2003).

RECOMMENDATION

1- Reciprocal effect of both chemical and organic fertilizers on the potato in different weathers of the province and at different soils should be evaluated.

2- Water stress should be evaluated at the three stages of germination, growth, and gland formation.

3- Effects of placing the tape at different depths on the water use efficiency and the potato product performance should be evaluated.

4- Drip tape drip irrigation for different varieties of potato should be analyzed.

5- Effect of different irrigation methods and regimes on the qualitative characteristics of potato should be evaluated.

6- Regimes (1.25 and 0.25) of 1 time cumula-

Table 8: Analysis of variance of specific gravity and arid matter (Mean Squares)

Source of the variations	degree of freedom	Specific gravity (g/cm^3)	dry matter (percent)
Repetition	2	0.00002	8.05
Treatment	3	0.00112*	13.78^{ns}
Error	6	0.00006	12.58
Sum	11	0.00130	34.43
CV(%)		0.29	6.974

ns, no significant; ** p≤0/01, * p≤0/05

Table 9: Mean comparison of specific gravity and dry matter

Experimental treatment	Specific gravity (g/cm^3)	Dry matter (percent)
50 percent cumulative evaporation from the pan	1.085433 d	19.168 c
75percent cumulative evaporation from the pan	1.102067 b	25.938 b
100 percent cumulative evaporation from the pan	1.1134 a	34.455 a
Furrow irrigation	1.093333 c	25.582 b

Means followed by same letter (s) in each column are not significantly different ($p < 0.05$)

tive evaporation from the evaporation pan should be evaluated.

REFERENCES

1- Allen, E. J. (1992). The Effect of Row width on the Yield of Three Potato Varieties, J. Agric. Sci. camb. 79: 315-321.

2- Almekinfers, C.J.M. (1991). Flowering and tTrue Seed Production in Potato (*Solanum tuberosum* L.): Effect of Stem Density and Pruning at Lateral Stems, J. Potato Res. 34: 379-388.

3- Dokhani, Sh., & Rabeae Motmaen, L. (2001). Evaluation of the Variation Rate of Glucose and Organic Acids of Potato Varieties (Moren, Mafona, and Agria) in Esfahan Province Dduring Storage with High Efficiency Chromatographic Method, journal of agriculture and natural resources sciences and technologies, 5th edition, 1:161-171.

4- Elizabet, G.C. (1992). Structure and Development of Potato Plant, PP. 65-146. In: Harris, P.M. (Ed), the potato crop: Scientific basis for improvement, Chapman and Hall, London.

5- Haverkort, A.J., DeWart, M.V., & Bodlaender, K.B.A. (1990). Inter-relationship of the Number of Initial Sprouts, Stems, Stolons and Tubers per Potato Plant, Potato Res. 33: 269-274.

6- Mohammadi, A., & Faeznia, F. (2001). Effect of Moisture Stress on the Growth and Performance of Two Potato Varieties, Experimental Report of Semnan Agriculture Research Center (Shahroud)

7- Mosavi Fazl, H., & Faeznia, F. (2003). Effect of Water and Azot Fertilizer Different Quantituies on the Qualitative and Quantitative Characteristics of Potato, Papers of the 11th Conger of the Iran National Committee of the Irrigation and Drainage.

8- Nadler, A., & Heuer, B. (1995). Effect of Saline Irrigation and Water Deficit on Tuber Quality, Potato Res. 38: 119-123.

9- Shock, C.C., Holmes, Z.A., Strieber, T.D., Eldredge, E.P., & Zhang, P. (1993). The Effect of Timed Water Stress on Quality, Total Solids and Reducing Sugar Content of Potatoes, Am. Potato J. 70(3): 227-241.

11- Van Heemst, H.D.J. (1996). The Distribution of Dry Matter During Growth of a Potato Crop, J. Potato Res. 28(1): 55-66.

12- Wurr, D.C.E. (1994). Effect of Seed Size and Spacing on the Yield and Grading of Two Main Crop Potato Varieties, J. Agric. Sci., Camb. 82: 37-52.

13- Wurr, D.C.E. & Allen, E.J. (1974). Some Effect of Planting Density and Variety on the Relationship Between Tuber Size and Tuber Dry-Matter Percentage in Potatoes, J. Agric. Sci., Camb. 82: 227-252.

14- Yuan, B.Z., Nishiyama, S., & Kang, Y. (2003). Effect of Different Irrigation Regimes on the Growth and Yield of Drip-Irrigated Potato, Agric Water Manage. 63: 153-167.

2

Investigation of Socio-Economic and Environmental Effects of Taleghan Dam Using Structural Equation Modeling

Vali Borimnejad [1] and Fatemeh Salimian[2]*

Abstract

Today water supply issue in large cities seems to be a big problem. This becomes more important with population growth and urbanization extension. To cope with this problem lots of irrigation and drainage projects have been implemented and operated that subsequently many socio- economic and cultural effects have been brought to the areas . The objective of the present study is to survey the socio - economic and environmental effects of Taleghan dam to compare the two periods before and after the dam construction. The needed data have been gathered by use of a questionnaire and by interviewing 400 samples in the irrigation and drainage network's region of Taleghan Dam, and encoded questions and outcome data have been analyzed in spss and lisrel software. The results showed an economic development in the region, but some factors like, lack of trust, lack of economic advantages and deficiency of facilities, decrease villagers' willingness to invest. 63.3% of region's environmental pollution has been increased mainly by the tourists. Not only is there no increase in Water sources but also they have been decreased. 98% of individuals declared that the people were not questioned in construction projects. About the coefficient of dam's effect on region's socio- economic and environmental development, the villagers have declared that the dam construction had no positive effect on region's development.

Keywords:
Dam construction, Structural equations, Taleghan Dam

[1] Associate Professor, Department of Agricultural Economics, Islamic Azad University, Karaj Branch, Karaj, Iran.
[2] MSc Student, Department of Agricultural Economics, Islamic Azad University, Karaj Branch, Karaj, Iran.
* Corresponding author's email: vali_borimnejad@Kiau.ac.ir

INTRODUCTION

The foot of the Zagros Mountains and the East Hills Farmers Mesopotamia were probably the first dam in the world. 6500 years ago the Sumerians, the plains between the Tigris and Euphrates brought under irrigation. The oldest dam, whose remains have been found, was built 3000 years BC. As with constructing the first pyramids, the Egyptian constructors built the "Alkafra" or "Pagams" dam at the bank of a seasonal river in Cairo. The Iranians learned how to use the surface and underground waters sooner than the others and they experienced the production increase with such irrigating methods.

The subterranean which is the most stable way in using the underground water, has been invented 3000 years BC in Iran. Collecting the rain water that simply apply the rain water of cold seasons to the hot seasons' cultivations, using very simple and stable methods, which is a sample of sustainable water supply, has been known since longtime ago in Iran. In all over the world more than 400000 km^2 have been used as a water reservoir. The widest land which has been used as a water reservoir is the 8500 km^2 delta behind the Akozomba dam, which contains 4 % of space of Ghana. The dams have 2 main functions, firstly, they save the water to compensate the volatility of rivers' flow and water and energy demand, and secondly, they increase the level of water in upstream so that in this way, they can deflect the water flow in a channel, or increase the hydraulic burden between the reservoir and the downstream. Building such this reservoir allows the dams to supply the required water in agriculture, industry and drinking water and also to control the floodwater, and it facilitates shipping in the river by regulating the water flow and by filling the river with water as soon as possible. As the other reasons of constructing a dam, the fishery and the entertaining purposes, like sailing, can be mentioned. The great dams mostly based on wide-spreading estimations, include about 1/3 of total crops harvest and 1/6 of total irrigated farms in whole world. The water reservoir of the dams for agricultural uses and generating energy, has been considered as a great wealth, that by using such this reservoir beside supplying the drinking and agricultural water, it makes it possible to generate hundreds megawatts of hydraulic energy. But despite lots of benefits and advantages in using the water which might be wasted, and its influence on human's life, ignoring the political issue in dam constructing may cause some disadvantages which are not easily compensable. Creating such this great wealth requires creating the lakes where so many human and village communities would be sunken in their banks. For instant, in Urmia city, constructing more than 40 dams in upstream has lead to the dehydration of Urmia lake and that expose the human and animal communities to a serious danger. The total amount of water in our planet, which is known as hydrosphere, is about 1360000 km^3, from which, 97.2 % is saline water and only 2.7 % is drinkable. And of this amount 2.14 % includes unavailable ices. The accessible amount is estimated about 12000 km^3 and of this amount a very low percentage can be used. And also the amount of underground waters is a little more than 8000 km^3 from which only 3/1000 is naturally renewed. The amount of the rainfall is 450000 km^3 annually that is justified with evaporation and of this amount only 110000 km^3 rains over the continents' lands. With the dry climates of countries like Iran, constructing the dams as an engineering way to control and regulate the flow to agricultural and drinking and other purposes, has become an interesting issue. And that's why the dam construction has got a historical background in Iran. The scientists have tried to save water and avoid water wasting by presenting some methods. Although the dam construction's history in the world, back to thousands years ago and the "Zolqarnein" dam in Iran is one of the oldest water reservoirs constructions in the world, but what is the impact of dam construction on the natural ecosystems in upstream and downstream and how this impact must be evaluated? The Taleghan project in order to achieve the national economic development goals and to optimally use the country's water resources' potential, and to use common usage of surface waters of Taleghan River in incorporation with Ziaran River, has being studied and designed since about 35 years ago. This project is considered as a multipurpose plan that has gained the below aims through constructing the Taleghan dam:

Control and regulation of river surface currents.

Water supply needed Qazvin

In order to achieve an optimum balance in the joint operation of surface water and groundwater Taleghan Qazvin

Provide drinking water for metropolitan Tehran

Use the hydroelectric potential of the dam and reservoir provides water transfer

The dam construction started on March 2001 and the impounding started on 16 January 2005. This dam transfer 298 million m^3 annually to the Qazvin field and supply 150 million m^3 of potable water of Taleghan (the Lar consulting engineers). The absence of participation of local communities and passive participation in water resources management, and also lack of local management themes and inefficient use of local communities in taking decisions, making decisions in the absence of beneficiaries and authorities and local decision makers in water and soil resources protection programs, has caused an inefficiency in state programs in protecting the natural resources. Such this negligence leads to a discouragement among villagers and farmers and the people who live in the bank of watersheds to the state projects, and lack of responsibility in preserving the natural resources and also a problem with local communities to accept natural resources projects who consider the government's strategies against their benefits. In order to study the environmental impact of dam constructing projects, the studies were fulfilled in 5 sections including: physical environment, biological environment, water resources, the contamination and dispersion of wildlife. At the time of dam construction, a village, some roads, several farms and fields and gardens were sunken whilst impounding the dam. In order to relate the upper villages to the central ones, several roads were built that consequently the farms and mountains were damaged or destroyed. Now the question is: was it worth to lose the resources by constructing the dam to achieve the goals of its operation? The impacts of dam construction on the region's economy were desirable or not? The value of lost resources was greater than the obtained facilities or not?

Hypotheses

There is a significant relationship between public participation and the rural development.

The local cooperation is effective in natural resources' protection.

The dam construction has lead to economic growth and consequently the people's satisfaction.

The dam construction has lead to prosperity in upstream fields.

There is a relationship between the needs of the project implementation progress.

Research purposes

Evaluating the economic impact of dam construction

Evaluating the social impact of dam construction

Evaluating the environmental impact of dam construction

Literature review

Falsafi Zade and Sabouhi Sabouni (2010), in their study determined the optimal exploitation of Kor river of Doroudzan dam. They applied a comprehensive economic- environmental model to study the irrigation and exploitation of surface waters in the plain around the Kor River, Doroudzan to Bakhtegan Lake. In this model economic, hydraulic and agricultural parts were considered. In economic part, the optimal allocation of water was investigated under 2 scenarios of presence and absence of regional water flow, using non-linear programming. The result of environmental simulated model in hydraulic part, were used as a preliminary data in economic part. In agricultural part, the real yield in wet, normal and dry years were determined by using the relationship between yield and the amount of irrigating water. The results showed that there is a difference between the quality of water, allocation due to wet, normal and dry years, with its optimal amount. A reason to such difference can be the lack of water market and price determining due to demand and supply. Using the model, the optimal cultivation model was determined to the region to preserve the surface water resources.

Borimnejad (2006) studied the effective factors on technical efficiency of wheat cultivators in Qom province, using a combined model of stochastic frontier method and path analysis.

This article provides a method to estimate the technical efficiency level of wheat firms using stochastic frontier method and testing the effective managerial factors to this efficiency. The needed data were the cross-sectional data of 2004 and for 149 firms in Qom province. The results showed that the variables like: the farmer's ability to read and write, being or not being a member of a cooperative or company, participating or not participating in informal educational courses, have a significant effect on explaining the technical efficiency levels of the firms.

Mohammadi Golrang *et al.,* (2006), studied the economic evaluation of watershed activities done in dams' area, in order to determine and identify the economic and social effect and watershed activities done in Torbat Heydarieh city (Khorasan province), and also to learn the attitude and viewpoints of direct beneficiaries of performed activities toward the quality of watershed activities returns.

Bani asadi and Mohseni (1995) in their study, investigated the influence of Ab-Barik project of Bam on economic and social life of residents of this region. The fulfillment of this project caused an employment for many residents of the region. Also, with performing this project, the immigration has been decreased appreciably.

Lee *et al.* (2009) fulfilled a research due to the integrated watershed economy and reservoir management. The dynamic optimizing frameworks for the complete management of reservoirs and appropriate suggested policies have been used. Soil conservation, free level of reservoir's sediments, downstream water allocation and the water quality are the goals under control. Applying the model to the "Aswan" dam showed that the international cooperation is needed to common water resources managements.

Hearnshaw *et al.* (2010) studied the quality of eco systemic services by constructing dam and water reservoir and its changes. Eco systemic services affect all around directly or indirectly. Most of water projects are being evaluated by cost-benefit analysis. Most of such benefits in cost-benefit analysis are being ignored because of the shortage in supplying centers, limited information, or lack of awareness about eco systemic services. Revising the water projects can explicitly and appreciably measure the evaluation of such projects. Such these projects can provide some natural benefits (increasing in water flow, decreasing in water contamination costs as a result of the increase in water flow). The water reservoir projects can bring positive or negative outcomes to the environment. An increase in water flow can lead to a growth in fishery, an enhancement in natural regularity, water dilution as a result of reduction in pollution, and make the aquatic plants ecosystem healthier. Irrigation causes an increase in production in the fields under the plan, since the land has got a good soil and other environmental factors out the river. So, according to quantifying the eco systemic services, they managed to find an appropriate and calm place for fixing the barometers of Opihi-Opuha River.

MATERIALS AND METHODS

The theories usually lead us to know how some variables are related to each other. So, developing an equation system is often favorable, like a model which shows the whole casual relationships between the variables. For example, a research wants to know how a family background, access to education, and other variables that form the socio-economic situation, might affect the lifestyle. This is one of the earliest situation development models (Sewell 1983). In present research we aim to study and estimate the casual relationships between variables in 3 groups of social, economical and environmental, as quantitative variables. As a matter of fact, each variable has been made of several elements that each element affects the others. Using path analysis, the casual relationships can be measured, and also the amount and the quality of each factor's influence can be determined. Path analysis is an advanced statistical method which examines the casual relationships between two or more variables. This system is based on a linear equation system that has been used for the first time by Sewell Wright in 1930 in phylogenetic studies. This method has been adopted by the sociologists in 1960 and was used in 1970 to study the increase in frequency in ecological literatures. In ecological studies, path analysis has been used mainly to realize the difference between the powers which are embedded directly or indirectly in variables. In path analysis, mediated pathway can be examined (y is the mediated variable in pathway of (X←Y←Z). The pathway in path models indicates the researcher's hypotheses which can never be statistically tested. Pathway analysis is a subset of structural equation modeling (SEM) which was defined by Ullman that allows us to examine the relationships between two or more independent variables, either continuous or discrete. To evaluate the variables, they must be exactly observable, measurable and they must clearly show the characteristics of variable. Latent variable is a variable which is not clear or obvious and it is determined by measuring other variables, that such these variables are created and known according to the factors like analytic factors of other observations. The structural equation modeling is a combination of several regressions and analytical factors, but path analysis contains only measured variables.

RESULTS AND DISCUSSION

In this step the independent variables of covariance's matrix, are being generated and estimated. Ullman and Holy have discussed the advantages and limitations of mentioned estimation and have come to this conclusion that ML and GLS are usually used for data which have got a normal distribution and their coefficients and errors are independent. ADF are most used for the data which have got an abnormal distribution. This method is used for the samples with more than 2500 observations. Ullman has declared that the best estimators for the data of abnormal distributions and error terms dependency is ML method.

Maximum Likelihood:

$$W=\Sigma^{-1}, pF_{ML}=\log|\Sigma|-\log|S|+tr(S\Sigma^{-1})-p \quad (1)$$

In this equation $=\Sigma^{-1}$, and p indicates the number of observable variables.

Generalized Least Squares (GLS):

$$W \; F_{GLS}=\frac{1}{2}tr[([S-\sum(\theta)W^{-1})^2] \quad (2)$$

That, *tr* shows the factor which is the sum of the elements in main diagonal matrix.

W^{-1} is the optimal weight which must be determined by researcher.

Asymmetric distribution Free (ADF):

$$FADF=[S-\sigma(\Theta)],W^{-1}[S-\sigma(\Theta)] \quad (3)$$

W in this function includes all of the elements which are derived from the highest statistical degree.

In order to study the internal stability of the questionnaire, Cronbach's Alpha coefficient was used.

$$\alpha=\frac{k}{k-1}(1-\frac{\sum s_i^2}{\sum s_{sum}^2}) \quad (4)$$

K is the number of questions

S^2_i is the variance of i th question

S^2 *sum* is the sum of the values of question's variance and covariance matrix

Cronbach's alpha coefficient varies between zero and one, that zero means lack of internal stability of questions and one means complete internal stability of a set of questions related to a fundamental concept. The validity of the questionnaire was examined using the theories. The results of validity testing showed a Cronbach's alpha coefficient of 0.906 which means a high validity of the questionnaire. Collecting information was fulfilled in supplementary phase through questionnaire and the questionnaires were sorted based on Likert scale.

Data distribution testing:

To perform the statistical methods and evaluating the appropriate statistics and a rational inference about research's hypotheses, the most important thing is to select a suitable method to the research. In that case being aware of data distribution is very important. In this case if the data are quantitative or ordinal we use Kolmogorov-Smirnov test and for two stages nominal data we use binominal test and for the data with more than two stages we use Chi-square test for compliance of distribution. Now according to the table, looking at the number corresponding to Sig., if the number is less than the target P, which is usually P=0.05, then we have an abnormal distribution, but if it is equal to P or more than that, then we have a normal distribution. In testing the ordinal data according to Kolmogorov and Shapiro test, we have got an abnormal distribution. It is the same for the nominal data and the distribution is abnormal. So, the method which was used in this study is maximum likelihood.

CONCLUSION

The data mentioned on the arrows connecting the latent variable to the observed one, are the factor loadings. The factor loadings between 0.3 and 0.6 explain a good estimation and the factor loading between 0.6 and 1.00 explain the best estimations. In fact, the more is the factor loading and the closer it is to 1, the better the observed variable will explain the independent variable. All of the estimated coefficients of economic variables are able to explain the latent variables by the observed ones. The negative sign of a coefficient shows a negative and inverse relationship, which we are going to analyze them. The variable of willingness to invest in the village has got a statistics about -5.10, and as it was mentioned above, the negative sign shows an inverse relationship between the latent variable and the observed variable. In this case, due to the questions answered by people, 0.75 % of people have willingness to invest in the village, but some factors like, lack of trust, lack of economic advantages, and deficiency of facilities, caused a less attempt of villagers in that case. The cultivated area and the crops and fruit yield were the other economic variables were considered in this study. Most of the lands in the region are being used by small holders so, they are being used to satisfy the individual's needs, and so, economic production with a high yield is very limited in the region. In this part, we are going to analyze the coefficients of environmental variables. 46.8% of villagers declared that the climate changed and the humidity increased in the region, however this variable is not known as an appropriate variable to describe the latent variable of environment. 62.8% and 63.3% of people agreed with this idea that the environmental damages and contaminations have been increased in the region, but such these damages are because of the tourists and not the dam construction, so such these damages can be considered as the outcome of dam construction. Another factor which was considered was the increase in water resources for irrigation, but since several years ago the water resources have been declined. 64% of villagers were against this concept that the agricultural water has been increased. However this variable was not known as an appropriate variable to explain the environmental latent variable. In the case of destroying the local plants, about 44% of population disagreed, 33% almost agreed, and 31% agreed with this hypothesis. According to the table the statistics of this coefficient is negative, as a reason, it can be said that the majority of population, were against this hypothesis. It is the same for the animal residential in the region. As it is shown in the table the estimated coefficient of the variables like, age, housing status, immigration to other cities, the quality of public participation in dam construction, the influence of dam on region's development, are the variables which can well explain the social latent variable which has got a positive and significant relationship with the latent variable. 50% of people are older than 55 years old and they are farmers and whose educational level is less than diploma. About 98% of people claimed that they are not investigated in constructing projects. The results of path analysis method, confirm such this claim. And the negative sign indicated the absence of investigation of local people. Regarding to the rural activities before and after dam construction it can be said that the rural main activities before dam construction were, farming 44%, husbandry 27%, after dam construction they faced a decrease about 20% and 5% for farming and husbandry respectively. In that case the villagers said that their activities have been stopped because of dam construction. As a reason it can be said that while constructing the dam, lots of farms and gardens were sunken and consequently their activities have been stopped. About the husbandry activities it can be said that regarding to the grasslands who supplied the livestock's feed, after dam construction they have been sunken, and it got more difficult than before to supply the livestock's feed and that caused stopping their activities. From 13.5% of livestock farmers, 20% have declared this subject. It must of course be mentioned that even before dam construction, because of the mountainous characteristics of the region, the feed supplying was difficult. But after dam construction this issue has been enhanced. The estimated coefficient of path analysis confirms that issue. About the variable of dam construction effect on economic, social, environmental development of the region, 34.4% of villagers declared that the dam construction had no positive effect on region's development. 14.3% of villagers have participated in dam construction and 85.8% of them have not. As it is shown in the table the estimated coefficient confirm well this subject. The estimated coefficients due to the reason "why have you participated in dam construction?" are zero and negative, that according to people's opinions it can be analyzed that most participation has been in land transfer and they are not satisfied with it, because most of the lands have been transferred compulsorily. The negative sign confirms such this dissatisfaction of land transferring and the value of zero indicates the very low percentage of people's participation in this subject. About the variable of "why haven't you participated in dam construction?" the estimated coefficient is about 0.01, the positive sign confirms the alignment between the observed variable and the latent variable. This estimated coefficient is small, due to people's feedback it might be analyzed in that way that about 50% of people said that they have not been asked to take part in dam construction. About the variable of villagers' participation in dam construction, 187 persons or in other words, 46.8% of villagers have not taken part in dam construction. The positive sign of this variable has no contradiction with the variable of "why haven't you participated in dam construction? This will confirm the previous issue that, neither the people aimed to take part in dam construction, nor did the authorities want them to do so. One of dam construction's outcomes in each region is job creation, but in constructing this dam no native labor was asked to take part and no employment has been created to the local people. In fact, all of these subjects are effective in people's dissatisfaction. In the topic of social variables the general attitude of villagers was asked and the estimated coefficient of path analysis is about -0.04. in this case about 45% of people declared that the environmental damages of dam construction were too high to be compensated and the negative sign shows such this dissatisfaction. In order to study the amount of development and to show the positive or negative changes of dam construction, the creating index method has been used. So, we considered an index to each question, thus, the indices above the value 3, are the appropriate indicators of development. The index of each variable is shown in the table.

RMSEA index

As it was mentioned before, so many indices can be used in model fitting. One of the most important indices is RMSEA index. Due to the importance of this index, right after evaluating the factor loading, this index it is shown below the graph. According to the results, the RMSEA index is about 0.086. Since the value is less than 0.1, so the model fitness is desirable.

Suggestions

According to the investigations, in general the outcomes of dam construction are: the region has developed economically and this development is obviously observable along the region. The region's environment has been damaged both because of dam construction and the tourists.

A considerable economic development is seen in the region but the issue is that despite such this economic development the people don't desire to invest because of lack of trust, deficiency of facilities and etc. So it is suggested to the region with such potential, be provided by necessary capacities to investment including: improving the facilities, increasing the villagers trust in the region, in order to not to lose the investment's opportunities.

Besides supplying the drinking water to Teheran and irrigation water to Qazvin fields, paying more attention to the problems of local people in supplying the agricultural and drinking water, is suggested.

According to the capacity of region it is suggested to plant the trees with high yield, and to construct the infrastructures facilities like the roads, and to decrease the transportation costs and to take the right policies which can lead to region's development and removing the problems like income deficiency, and increasing the production of fishery and gardening and farming.

Creating processing industries in the region is

suggested to increase the production and the value added in order to job creation.

Making a suitable context is suggested in order to achieve the sustainability and preserving the present and the future value of natural resources.

To supply the more income to the farmers, trying to identifying, introducing and promoting the suitable cultivars due to the regions situations and not to use the inappropriate and weak cultivars in the region, is suggested.

Creating capacities is suggested to attract more tourists and using the income caused by tourists in civil development and village buildings.

It is suggested to Increase the leisure facilities and to supply the target market of agricultural and husbandry's products which is effective in increasing the villagers' income and in decreasing the immigration to other cities in order to find jobs and income.

It is suggested to attract the villagers, and to create the facilities and to provide the credits in gardening and husbandry (industrial or semi-industrial), and to take the right policies to attract the young labors and to apply them in generating businesses in order to prevent their immigration to other cities.

REFERENCES

1- Applied Economics Association. (2009). AAEA & ACCI Joint Annual Meeting, Milwaukee, Wisconsin, July 26-29.

2- Azizi Khalkheili, T., & Zamani, GH. (2010). Factors affecting farmers' participation in irrigation management: Application of Path Analysis, Journal of Economics and Agricultural Development, Agricultural Science and Technology, 24: 83-90.

3- Bart, J., & Earnest, SL. (1998). Relative importance of male and territory quality in pairing succes of male rock ptarmigan. Behavioral Ecology and Sociobiology 45: 355-359.

4- Bodaghpour, S., & Jadidi, A. (2008). Environmental impacts resulting from the construction of Mijran dams, fourth national congress on civil engineering, Tehran University.

5- Borimnejad, V. (2006) .Factors affecting the technical efficiency of wheat farmers in Qom (a mixture of stochastic frontier model and path analysis), and development Agricultural Economics, 53: 23-39.

6- FalsafiZadeh, N., & Sabohi Saboni, M. (2010). Determine the optimal harvest of peripheral run off water from Kor river basin, dam Dorovdzan. Journal of Economics and Agricultural Development (Agricultural Science and Technology), 24: 415 – 424.

7- Haji Rahimi, M., & Torkamani, J. (2003). Investigation role of agriculture in economic growth model using path analysis, and Development Agricultural Economics, 41 - 42: 71-83.

8- Hearnshaw, E., Cullen, R., & Hughey, K. (2010). Ecosystem services review of water projects, Australian Agricultural and Resource Economics Society Annual Conference.

9- Holy, RH. Structural Equation Modeling. SAGE Publications, Inc. Thousand Oaks, CA. 1995.

10- Jaffari Eskandari M. (2011). Determine the causal relationship of science and technology development in the private sector industrial processes using path analysis, Journal - Science and Technology Policy Research, 4: 41-51.

11- Kalantari, KH., & Ebrahimi, M. (2005). Factors affecting the absorption of tea to Irrigation Associations case study: city clones, Journal of Agricultural Science, 36(5), 1273-1282.

12- Kelloway, EK. (1998). Using LISREL for structural equation modeling. Sage publications, inc. thousand oaks, CA.

13- Yoon, L., Yoon, T., & Shah, F. (1987). Economics of integrated watershed and reservoir management agricultural.

14- Loehlin, Jc. (2010). Latent variables of model. Lawrence erlbaum associates, inc. Hillsdale, NJ.

15- Ross, C., & Hearn shaw, E. The sustainability and Cost-Effectiveness of water storage projects on canterbury rivers: The Opihi river case, New Zealand Agricultural and Resource Economics Society Inc.

16- Tompkins, J. (2011). Evaluating the sustainability of impounded river systems and the cost-effectiveness of Dam projects: An ecosystem services approach, 55th Annual AARES national conference.

17- Ullman, JB. (1996). Structural equation modeling (In using multivariate statistics, third edition, B.G. tabachnik and L.S. fidel, Eds.)

18- Yavari, GH., Khalili, M., & Mirkiaiee, H. (2006). Social and economic effects of national projects Hablehrud PRA method village branch pilot city (village case study Behvard).

Appendices

Table 1: Indicators of environmental variables and sample size estimation

Variable	Estimated parameters	Results
Climate Change	4.06	Since the estimate of the index is greater than can be said about the climate change hypothesis is correct.
Environmental damage	4.30	Since the estimate of the index is greater than can be said about the hypothesis was correct environmental damage is imported.
Environmental pollution	4.2	Since the estimate of the index is greater than can be said of environmental pollution based on hypothesis were accurate.
Increasing water	1.6	Since the estimate of the index is smaller than can be said about the hypothesis of water resources have not increased due to the construction of dams and water resources have not increased.
Dispersal of animal species	3.2	Since the estimate of the index is greater than can be said about the dispersal of animal species hypothesis is correct.
Dispersal of bird species	2.7	Because the index is lower than the estimate of assumptions regarding the dispersal of species of birds can be said is not correct.
The loss of plant species	2.7	Since the estimate of the index is smaller than can be said concerning the loss of plant species hypothesis is not correct.
Socio - economic development - environmental	2.4	Since the estimate of the index is smaller than can be said about the hypothesis of socio - economic development - it is not proper environmental dam has not had a positive impact on regional development.

Table 2: Tests for determining the distribution of normal and abnormal kolmograph ordinal variables.

Shapiro-Wilk			Shapiro-Wilk			
p-value	df	Statistic	p-value	df	Statistic	
0.000	389	0.785	.000	389	0.265	Climate change
0.000	389	0.649	.000	389	0.358	Environmental damage
0.000	389	0.651	.000	389	0.364	Environmental pollution
0.000	389	0.655	.000	389	0.378	Water resources have increased
0.000	389	0.899	.000	389	0.173	Animal species scattered
0.000	389	0.893	.000	389	0.151	Species of birds scattered
0.000	389	0.884	.000	389	0.156	Native plant species have disappeared

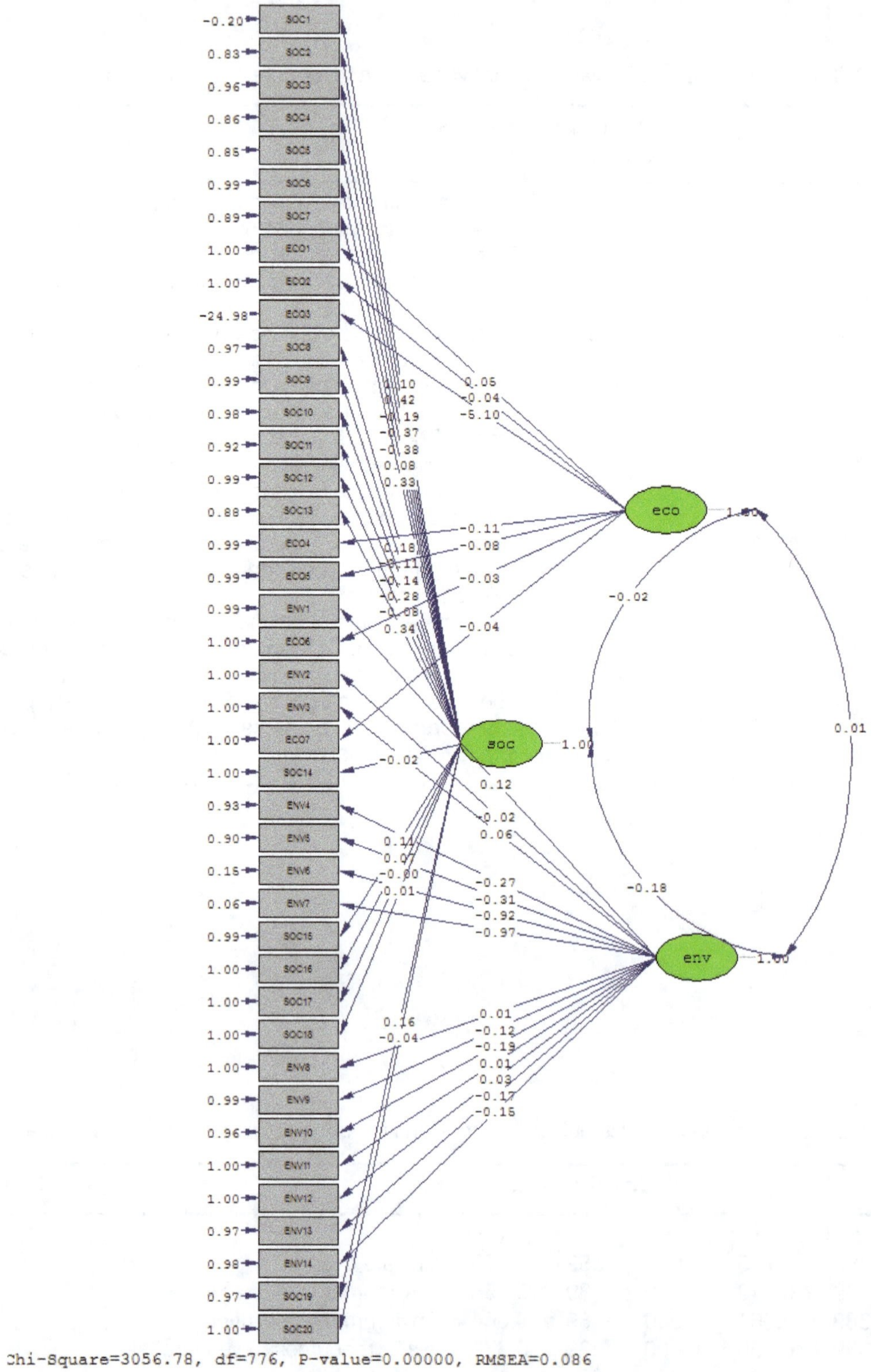
SOC1
SOC2
SOC3
SOC4
SOC5
SOC6
SOC7
ECO1
ECO2
ECO3
SOC8
SOC9
SOC10
SOC11
SOC12
SOC13
ECO4
ECO5
ENV1
ECO6
ENV2
ENV3
ECO7
SOC14
ENV4
ENV5
ENV6
ENV7
SOC15
SOC16
SOC17
SOC18
ENV8
ENV9
ENV10
ENV11
ENV12
ENV13
ENV14
SOC19
SOC20
eco
soc
env
Chi-Square=3056.78, df=776, P-value=0.00000, RMSEA=0.086

3

Comparative – Analytical Study of Economic Productivity of Water between Smallholding and Rural Production Cooperative Utilization System

Mohammad Reza Bakhshi [1*], Behnaz Nejati [2] and Mofid Shateri [3]

Abstract

In recent years, farming section of South Khorasan Province faces different limitations such as increasing shortage of water resources and continuous reduction of weather showers. Considering consecutive droughts and water crisis in agricultural plain of the province, it is necessary to use water resources optimally and increase productivity of water shortage input. Comparison of water productivity among the available utilization systems in agricultural section of the province can cause recognition of suitable and efficient utilization system for optimal use of water shortage input and increase water productivity in production of crops. In the present research, different indices of water productivity for production of crops in two small holding utilization system and Rural Production Cooperative in Khosef County have been calculated and compared. The required data have been collected with a sample of 247 farmers and with two-staged cluster sampling and with questionnaire. To calculate and compare water productivity, Benefit Per Drop, Crop Per Drop and Net Benefit Per Drop indices have been used. Results showed that the said indices were different for similar products in two utilization systems and in most crops, the said indices in rural production cooperative system were higher than small holding system. Therefore, gathering of smallholder's farmers as rural production cooperative can lead to more desirable utilization of water resources and reduction of drought effects and water crisis.

Keywords:
Farming, Utilization system, Cooperative, Water, Economic productivity

[1] *Associate professor of Agronomy Department, Faculty of Agriculture, University of Birjand.*
[2] *Former Graduate Student Department of Geography, Faculty of Literature and Humanities, University of Birjand.*
[3] *Associate professor of Department of Geography, Faculty of Literature and Humanities, University of Birjand.*
* *Corresponding author's email: mbakhshi@birjand.ac. and irmbakhshi@birjand.ac.ir*

INTRODUCTION

Today, drought and water resources shortage are regarded as important issue and one of the constraints of farming section. Due to recent droughts, South Khorasan Province is ranked 24 in terms of rainfall and on top of the drought pyramid of the country (South Khorasan Province Agricultural Jihad Organization, 2009). Agricultural section plays important role in economy of the province and particularly in rural regions of the province and life of people is dependent on it. Now, when drought and water shortage occur, water input shortage increases and water has special and important place among the resources and production inputs and perhaps, water can be regarded as the most important production source in farming. Therefore, improvement of water productivity in production of crops is mentioned as the main agricultural issue in the future. Under such conditions, it is necessary to conduct extensive and effective researches for evaluation and study of the condition of using water and optimality of this input and based on suitable strategies to increase productivity of supply and consumption of this rare and valuable input (Ghasemian and Eshraghi, 2012).

Goal of increasing water productivity in farming is to maximize profit of water resources consumption in farming section. Therefore, limited water resources of the country should be allocated to the crops which have higher return for each cubic meter of water to obtain higher productivity in farming section. Although this issue cannot mean negligence of other long-term goals such as food security and employment, it is necessary to consider financial and economic efficiency to increase productivity beside these goals (Zeid Ali *et al*., 2008). Increase of productivity increases production in surface area on the one hand and reduces production costs and cost price of the product on the other hand (Ghasemian and Eshraghi, 2012). Utilization systems always play important role in application of production factors and increase of input productivity such as water shortage inputs. Variety of utilization systems in farming section caused different water productivity in production of crops. Therefore, comparison of water productivity in different utilization systems can cause recognition of suitable and efficient utilization system in water consumption and increase its productivity in production of crops.

Based on importance of water input in production of crops, water productivity measurement has been studied very much inside and outside the country. Poor Sani *et al.* (2008) in a research entitled "analyzing indices of agricultural water productivity in a rural production cooperative" studied indices of BPD and NBPD in a rural production cooperative in Kerman Province. The obtained results indicate promotion of these indices by performing infrastructural operations and they can be used in major planning of this section, selection of cultivation pattern, irrigation method and analysis of economic value of water for production. Rohani (2005) in a research compared water productivity in farmlands of the member and non-member farmers of Hamedan Province rural production cooperatives. The obtained results show that mean water productivity for the crops is almost equal in member utilizations of rural production cooperatives and non-member utilizations and membership in production cooperative has no effect on water productivity. Zwart and Bastiaanssen (2004) in a research studied physical productivity of water input in production of wheat, rice, cotton and corn crops in different countries. Based on results of this research, mean physical productivity of water for the mentioned crops are 1.09, 1.09, 0.65, 0.23, and 1.80 kg/m^3. Henry *et al.* (2006) estimated physical productivity of water to be between 0.4 and 0.70 kg/m^3 for corn in Tanzania by performing farm experiments. Eshraghi and Ghasemian (2012) in a study calculated and evaluated water economic productivity in production of important crops in counties of Golestan province such as cotton, rapeseed, soy and rice with water productivity indices. Results of the research showed that the highest economic productivity of water has been 134960 Rials/m^3 of water for cotton in Gonbad Kavoos, 5620 Rials/m^3 of water for rapeseed in Agh Ghola, 3940 Rials/m^3 of water for summer soy, 21080 Rials/m^3 of water for high quality long grain rice, 9880 Rials/m^3 of water for fruitful rice in Ali Abad and 9600 Rials/m^3 of water for medium grain rice in Kalaleh. Cai *et al.* (2003) in a research measured physical and economic efficiency of water consumption in agricultural areas around a river in Chile and show that both types of the said efficiency in farmlands with pressurized irrigation systems have significant difference from other farmlands. Liu *et al.* (2008) have calculated and reported water physical productivity in corn for 124 countries. Findings of this research show that the highest physical productivity of water relating to America and China is 1.5 kg/m^3 of water and the lowest value relating to African countries is below 1 kg/m^3 of water. Sanei and Hassan Poor (2009) analyzed water productivity in irrigation and drainage system in Khoozestan with CPD and BPD indices and suggested that corn should be excluded from the available cultivation pattern and barley crop should be included to improve water consumption productivity. Soltani *et al.* (2007) studied agricultural water productivity in Marvdasht –Karbal. Results showed that cultivations with high water consumption and low economic return such as sugar beet should be excluded from cultivation pattern of the region and some cultivations such as forage maize or one-year period of native cultivations such as wheat and barley and vegetables such as tomato which decrease water consumption and also create high economic benefits for the farmers and agricultural users should be used. Results of study by Singh *et al.* (2006) on physical productivity of water for wheat, rice and cotton in the country show that there is significant difference between physical productivity of the above products so that physical productivity of wheat is four times as much as the cotton. Based on results of this study, physical productivity of the said crops is equal to 1.04, 0.84 and 0.21 kg/m^3 of water. Zeid Ali *et al.* (2008) evaluated and compared water productivity in three sections of Moghan agro-industry, Pars agro-industry and private section lands. Results of this research showed that average indices of CPD, BPD, and NBPD in private section lands are better than irrigation and drainage system of Moghan and Pars agro-industry.

Studies show that water productivity in different time periods and regions are different from each other and are affected by production method and production region and generally, type of utilization system. On this basis, the present research has compared agricultural water productivity in common crops of two smallholding and Rural Production Cooperative Utilization systemin Khosef County located in South Khorasan Province. The main hypothesis of this research is that different indices of water productivity are different in production of crops between utilization systems and water consumption can be reduced and aquifer of the studied zone has been empowered through policymaking for expansion of utilization system by identifying utilization system with higher water input productivity for production of common crops of the region.

MATERIALS AND METHODS

Agricultural water productivity measurement and analysis have special position in Iran due to quantitative and qualitative limitations of this valuable substance. Productivity is related to the extent and manner of using inputs or production factors in a specific production process, definite period and specified geographical zone to achieve the determined goals (Ehsani and Khaledi, 2003). Productivity generally means ratio of outputs to inputs. In other words, productivity means average production for each unit of total inputs. So, if average production increases for each unit of inputs, productivity will increase and vice versa. To calculate productivity considering type of the input which is applied in production process, types of productivity indices can be defined. Generally, productivity indices are divided into two partial productivity indices and whole productivity indices of production factors (Vali Zadeh, 2005). Productivity index indicates ratio of production volume or value to one or more factor value which has been used for producing it. In other words, any relationship between output and data as ratio is productivity index. Indices have different types and application of each one of them depends on goal of the research. In this research, considering the mentioned goals and hypotheses, CPD (Crop Per Drop), BPD (Benefit Per Drop) and NBPD (Net Benefit Per Drop) indices have been used:

$$CPD = Y(kg)/W(m^3) \quad (1)$$

$$BPD = P_{y*}Y(kg)/W(m^3) \quad (2)$$

$$BPd = p_{y*}(kg) - TC/W(m^3) \quad (3)$$

Where, Y is quantity of crop, W is water consumption, Py is price of each crop unit and TC is total cost of production. Relation (1) shows that consumption of each m^3 of water obtains some kilograms of crop. This index has evaluated physical productivity of water from agricultural viewpoint and higher index for a crop indicates higher physical productivity in production of that crop. BPD and NBPD indices which are applied in economic analyses have paid attention to economic and monetary aspect of water in addition to physical aspects and measure economic productivity of water. The said indices show how much Rial value has been created for each m^3 of water. According to Relation (3), any product which can have higher net earning with lower consumption of water will have higher economic productivity (Ehsani and Khaledi, 2003, Liu *et al.*, 2008; Henry *et al.*, 2006). These indices can be calculated for comparison of a type of definite crop in different regions or for a special region (farmland) over time. In other words, these indices can be applied for an external or interregional comparison (between farmlands) and also for an internal comparison (time trend) (Ehsani and Khaledi, 2003). In this research, all three induces have been calculated and analyzed from both agricultural and economic viewpoints to compare water productivity among the crops.

Statistical population of the present research includes farmers of Khosef county located in South Khorasan Province. Khosef plain has been regarded as one of the important agricultural poles of South Khorasan Province and farming as one of the important economic sections plays considerable role in employment and earning living in Khosef county. The most important crops of this county include wheat, barley, alfalfa and beet (South Khorasan Province Agricultural Jihad Organization, 2009). The studied scope has smallholding, rural production cooperative and agro-industry utilization systems. Considering that the prevailing systems in this county are smallholding and rural production cooperative utilization systems, this research has compared agricultural water productivity of common crops of two utilization systems in the studied zone.

The data used in the present research has been obtained with questionnaire and interview with farmers. In this research, crops production cost questionnaire which has been prepared by Ministry of Agriculture and is used every year in crops statistics has been used with some alterations. The questionnaire has been prepared by the agricultural experts and have suitable validity and reliability considering its use in the consecutive years. Statistical sample size was determined 245 samples with Cochran's formula:

$$n = \frac{Nt^2S^2}{Nd^2 + t^2s^2} = \frac{5600 \times (1/96)^2 \times (0/817)^2}{5600 \times (0/1)^2 + (1/96)^2 \times (0/817)^2} = 245$$

In the above formula, n is sample size, s standard deviation of the desired trait in the population, N is population size and d is desirable probable accuracy. To determine standard deviation for the desired trait in the studied population and also determine desirable potential accuracy with pretest method, 30 samples of the studied statistical population were selected randomly and the questionnaire was completed. To determine the number of sample in two utilization systems, Proportional allocation method was used, on which basis, the number of sample is equal to 207 and 38 in smallholding and production cooperative utilization systems. Sampling method was based on two-staged cluster method and in the first stage, some villages were selected in the studied region (first cluster) and then the desired sample was randomly selected among farmers of the said villages (second cluster).

RESULTS AND DISCUSSION

Results of calculating productivity indices of CPD, BPD, and NBPD for crops of two studied utilization systems including wheat, barley, alfalfa, cotton, fodder beet, millet, and sorghum are presented in Tables 1 and 2. Based on information of Table 1, the highest and lowest water consumption per hectare among the crops of smallholding utilization system related to alfalfa and barley, which was equal to 12042 and 6566 m^3/hectare. Based on CPD index, beet crop has the highest production rate for each m^3 of water. On the contrary, cotton has had the lowest production for consumed water, which has significant difference from each other and other crops.

Maximum and minimum value of BPD index which shows gross profit for the consumed water per hectare related to barley and cotton, respectively. Based on results of calculating

Table 1: CPD, BPD, and NBPD Indicators for products of smallholding systems

item	crops					
	wheat	barley	cotton	alfalfa	fodder beet	millet
Water consumption (Cubic meters per hectare)	7025	6566	10700	12042	11851	8080
Yield (Kg per hectare)	2770	2790	1530	5357	17464	2322
Gross income per hectare (IRR)	7966850	6685440	10074070	11789440	11945650	12378570
Cost per hectare (IRR)	6492320	5829960	4693620	4900050	10531020	8878700
Net income per hectare (IRR)	1474530	855470	5380460	6889380	1414640	3499870
CPD (Kg per m³) *	0.57 bc	0.86 bb	0.15 d	0.51 cc	1.54 a	0.31 ddc
BPD (IRR per m³)	1660 aabb	2460 aa	1020 b	1140 bb	1030 bb	1630 abb
NBPD (IRR per m³)	227 abb	573 aa	556 aa	564 aa	128 b	145 bb

*Means with the same letter(s) in each row have not significantly different based on Duncan's test ($p<0.05$)

NBPD index, barley and beet with 573 and 128 Rials/ m³ of water had the maximum and minimum net benefit per drop.

Considering the calculated value of water productivity indices, smallholding-farming priorities can be determined. In the agricultural year studied in smallholding utilization system, the highest cultivated area has been allocated to crops such as wheat, barley, cotton, alfalfa, beet and millet. Considering value of the said indices, it can be said that in case goal of farmers is to maximize quantity of the produced crop for each m³ of water, cultivation priority is placed on beet, barley, wheat, alfalfa, millet and cotton. To obtain maximum profit, smallholding cultivation priority includes barley, alfalfa, cotton, wheat, millet, and beet, respectively. Although beet is in the first priority of cultivation based on CPD index, it is in the last priority of cultivation based on NBPD productivity index. This

Table 2: CPD, BPD, and NBPD Indicators for products of rural production cooperatives

item	crops				
	wheat	barley	cotton	alfalfa	Sorghum
Water consumption (Cubic meters per hectare)	8206	6567	12451	12061	13748
Yield (Kg per hectare)	3120	4080	2388	5680	7866
Gross income per hectare (IRR)	9126190	6049480	20865940	13832500	20020000
Cost per hectare (IRR)	6347060	5520980	7242300	5778270	11906500
Net income per hectare (IRR)	2779120	528500	13623640	8054220	8113500
CPD (Kg per m³)*	0.422 c	0.673 aa	0.212 d	0.482 bcc	0.604 abb
BPD (IRR per m³)	1260 abb	1770 aabb	1840 aa	1180 b	1530 aabb
NBPD (IRR per m³)	404 b	900 abb	1200 aa	697 bb	612 bb

*Means with the same letter(s) in each row have not significantly different based on Duncan's test ($p<0.05$)

Table 3: Comparison of CPD indicator among joint products of smallholding systems and rural production cooperatives

Crop	Farming system		t-value	p-value
	Smallholding System	Rural production cooperative		
wheat	0.574	0.422	0.626	0.534
barley	0.864	0.673	1.233	0.221
alfalfa	0.512	0.482	0.565	0.575
cotton	0.154	0.212	-2.98*	0.004

*p<0.05

result is in line with studies by Soltani *et al.* (2007) and Sanee and Hassan Poor (2009). It is not surprising that the information used in each index is different from another index. However, considering conditions of the region which have experienced consecutive droughts in the past years and face crisis of water shortage and drop of aquifer level, it seems that the mentioned cultivation pattern is more suitable based on NBPD index which considers costs and incomes resulting from consumption of each m3 of water in offering of the crops of the region.

Results of calculating CPD, BPD and NBPD about major crops in rural production cooperative system are given in Table 2. Based on information of the above Table, the maximum consumable water per hectare related to cotton with 12451 m^3 and the minimum one relates to barley with 6567 m^3. The maximum value of CPD index relates to barley with 0.637 kg for each m^3 of water and the minimum value of index related to cotton with 0.212 kg for each m^3 of water. As information of the above Table shows, value of CPD index has significant difference among crops of this utilization system. The calculated value of the said index for wheat is close to result of studies by Soltani *et al.,* (2007) and Sanee and Hassan Poor (2009).

Results of calculating BPD productivity index show that the maximum and minimum gross profit for each m^3 of water in this system is equal to 1840 Rials and 1180 Rials, which relate to cotton and alfalfa, respectively. Value of this index for cotton has significant difference from other crops and this difference is not significant for other crops. The maximum NBPD (Net Benefit per Drop) in this system relates to cotton and is equal to 1200 Rials per m^3 of water. In addition, the maximum value of NBPD index relates to wheat and is 404 Rials per m^3 of water.

In the present cultivation pattern in production cooperative system in the studied year, the largest cultivated area relates to wheat, sorghum, barley, alfalfa and cotton, respectively. Information included in Table 2 shows that the present cultivation pattern of rural production cooperative system is not consistent with the proposed priorities obtained from the indices considering value of different indices of water productivity. In Tables 3 to 5, water productivity indices have been statistically compared with each other among the common crops between two utilization systems. As the results of table 3 shows, value of CPD index has no statistically significant difference between two utilization systems among crops of wheat, barley, and alfalfa and only value of this index has statistically significant difference in level of 5% for cotton.

Table 4: Comparison of BPD indicator among joint products of smallholding systems and rural production cooperatives

Crop	Utilization system		t-value	p-value
	Smallholding System	Rural production cooperative		
wheat	166.448	126.022	0.593	0.555
barley	246.413	177.106	1.618**	0.109
alfalfa	114.843	118.333	-0.259	0.797
cotton	101.590	184.068	-5.286*	0.000

*p<0.05, **p<0.10

Table 5: Comparison of NBPD indicator among joint products of smallholding systems and rural production cooperatives

Crop	Utilization system		t-value	p-value
	Smallholding System	Rural production cooperative		
wheat	22.752	40.403	-1.128	0.264
barley	57.368	90.010	-1.397	0.169
alfalfa	56.476	69.776	-0.936	0.355
cotton	55.686	120.408	-4.601*	0.000

*p<0.05

Tables 4 and 5 show BPD and NBPD indices in two smallholding and rural production cooperative utilization systems in the studied zone. The said indices are different in crops of two utilization systemsto be studied. However, difference of the said indices is statistically significant for cotton. It means that although gathering of farmers in rural production cooperative could increase water economic productivity, the said effect was lower for productivity of wheat, barley and alfalfa and its special effect is found in cotton.

It is clear that the obtained results will be more accurate and documented by increasing statistics and information and conducting similar studies in some consecutive periods. However, results of the present research can clarify conditions in utilization systemsofKhosefplain in South Khorasan Province, therefore, as Ghasemian and Eshraghi (2012) mention, finding the best opportunity to create higher value added can be a step for more effective growth of farming and economy of the region and the country considering shortage and high value of water input in the country.

CONCLUSION

To compare water productivity in production of crops in peasant and production cooperative utilization systems, CPD, BPD, and NBPD indices were calculated. Results showed that value of the said productivity indices in rural production system was higher than that of small holding system. On the other hand, cultivation priorities in two said utilization systems are not consistent with prioritization of productivity indices. Considering the said fact, combination of cultivation can be changed to improve condition of farmers in two utilization systems on the one hand and save water on the other hand. In this case, production efficiency of crops can be improved in the region by deleting cultivations with high water consumption and low yield and replacing them with cultivations with lower water consumption, higher yield and higher economic profit. Considering the problems which drought has created for the farmers in recent years, it is recommended to avoid more water-more yield option in irrigation management and planning of the studied region and consider less water –more yield option and perform cultivation pattern based on maximum economic productivity of plants' water consumption i.e. NBPD to increase effectiveness of using water resources. On this basis, gathering of smallholder's farmers in production cooperatives is recommended as one of the optimal utilization strategies for water resources and reduction of drought effects and water crisis.

REFERENCES

1-Cai X., Rosegrant M., & W. Ringler C. (2003). Physical and economic efficiency of water use in the river basin: Implications for efficient water management. Water Resour Research. 39(1): 1013-1025.

2-Ehsani, M., & Khaledi, H. (2003). Agricultural water productivity. Iran Irrigation and Drainage National Committee. Tehran.

3-Ghasemian, S., & Eshraghi F.(2012). Studying economic productivity of water consumption in Golestan Province. Agricultural Water Research Journal. 26(3), 317-322.

4-Henry, E.I., Henry, F.M., Andrew, K.P.R., & A.S.Baanda. (2006). Crop water productivity of an irrigated maize crop in Mkoji sub-catchment of the Great Ruaha River Basin, Tanzania. Agricultural Water Management, 85(2), p. 141-150.

5-Hussain, R.Z., & Young R.A. (1985). Estimates of the Economic Value Productivity of Irrigation Water

in Pakistan from Farm Surveys. Journal of The American Water Resources Association. 21(6), 1021–1027.
6-Khaki, G.H. (2003). Productivity Management (its analysis in organization). Islamic Azad University Scientific Publication Center, Tehran.
7-Kijne, J.W., Tuong, T.P., Bennett, J. Bouman, B., & Oweis, T. (2002). Ensuring food security via improvement in crop water productivity. Available online at: http://citeseerx.ist.psu.edu/viewdoc/. Accessed 25 January 2012.
8-Liu, J., Zehnder, A. J. B., & Yang, H. (2008). Drops for crops: modelling crop water productivity on a global scale. Global NEST Journal, 10(3), p. 295-300.
9-Poor Sani, A., Akbari Toodehi, D., & Arya Vand, H. (2008). Analyzing Agricultural water productivity indices in a rural production cooperative. The second conference on irrigation and drainage systems management. Shahid Chamran University of Ahwaz.
10-Randhir, S., & murthy, K. (1999). Productivity variation and use in farm of madratkam takfed area of changal patuu district tamilnadu.Indian Joumal of Agriculture Economics. 45(1), 56-60.
11-Rohani, S. (2005). Comparing water productivity in farmlands of member and non-member farmers of production cooperatives (case study: Hamedan Province). Journal of Agricultural Economy and Development. Pp: 53-84.
12- Sanee Dehkordi, KH., & Hassan Poor, A. (2009). Studying condition of water productivity based on CPD and BPD index in irrigation and drainage systems in Khoozestan and presenting suitable strategy for increasing them. 12th conference of National Iranian Irrigation and Drainage Committee.
13 -Singh, R., Van Dam, J.C., & Feddes, R.A. (2006). Water productivity analysis of irrigated crops in Sirsa District. Indian Agricutural Water Management, 82: 253-278.
14- Soltani, GH.R., Akbari., S.M.R., & Mohammadi, H. (2007). Studying agricultural water productivity in regions with drought (case study: Marvdasht-Kerman). Collection of papers of the six[th] conference of Iranian Agricultural Economy. Mashhad. Iran.
15- South Khorasan Province Agricultural Jihad Organization.(2009). Statistical Yearbook. Statistics and Information Unit.
16 -Vali Zadeh, P. (2005). Studying productivity in economy of Iran. Department of Economic Investigation and Policies. Central Bank of Islamic Republic of Iran. Publication No. 24.
17- Zeid Ali, S., Khaledi, H., & Kholghi, M. (2008). Studying condition of water productivity in Moghan irrigation and drainage system. Collection of papers of the 11th conference of National Iranian Irrigation and Drainage Committee.
18- Zwart, S.J., & Bastiaanssen, W. G.M. (2004). Review of measured crop water productivity values for irrigated wheat, rice, cotton and maize. Agricultural Water Management, 69(2), p. 115-133

4

Economic Pricing of Water in Pistachio Production of Sirjan

Nasrin Ohadi*[*1] and Jaleh Kurki Nejad[2]

Abstract

Given the strategic remarkable rank of pistachio in non-oil exports, inputs' management in its production is so important. As the scarcest input in agricultural sector, water is considered to be among the most important inputs of pistachio production. Water inadequate supply and climate conditions increase water demand in pistachio growing areas. It is necessary to determine the real value or price of water for establishing a balance between its demand and supply. Therefore, this study has aimed at estimation of water economic value. The method used in this research is the production function approach. Requested data sets were obtained from the questionnaires was filled out for crop years of 2013-2014. The results show that the average economic value of irrigation water is 50360 RLS but the average price paid by farmers is 1771 RLS per cubic meter of water. Thus, there is a wide gap between water value and the price paid by farmers with which appropriate pricing of water based on its economic value can be eliminated.

Keywords:
Economic value of water, Pistachio, Production function approach, Sirjan

[1] *Master of Agricultural Economics, University of Sistan and Baluchestan, Iran.*
[2] *PhD student of Agricultural economics, Faculty Member, University of Payam- Noor Sirjan, Iran.*
* *Corresponding author's email: Nasrin.ohadi@yahoo.com*

INTRODUCTION

The water is considered as a valuable scarce input because of its capability of creating goods and services flows over the time. Water usage is limited by physical, economic and spatial terms and has a very significant impact on yield and farmers' income under controlled conditions. Water as the most precious production inputs, almost assign high percentage of subsides itself in most countries. Remained water from irrigation projects usually has costs such as operating and capital expenditures which paid by consumers. On one hand, low pricing of water leads to its inefficient use. On the other hand, it causes income transfer from dry farmed lands to irrigated ones and thus, increases inequality of income distribution. In addition, there are no costs or fines for water contaminating through wasting it, fertilizers or chemicals. As a result, any fines won't be considered for environment damages leading to resources abuse. Chosen method for cost based on pricing of water is, requiring water value knowledge and information and varies for plant type, area and water quality. This method seems to be more complicated than the cost pricing and leads to a more equitable distribution of costs and efficient usage of scarce resources such as water (Kashakoglu and Erol Cakmak, 1997). Iran is the world's largest producer and exporter of pistachio; 40 percent of production, 63 percent of the cultivated areas is of Iran (Food and Agricultural Organization, 2009). According to the latest statistics of the ministry of agriculture, pistachio cultivated areas of Iran are about 420 thousands hectares. Kerman province contains 70 percent of Iranian pistachio cultivated areas based on these statistics (Abdollahi Ezatabadi, 2008). Having 60 thousands hectares of pistachio cultivated areas, Sirjan is one of the main pistachio producers of Kerman (Jahad-Keshavarzi of Sirjan region, 2013). Pistachio cultivation is done in desert and arid areas where rainfall is very low and the weather has a borderline state (very cold winters and very hot dry summers). High salinity and inadequate contents of agricultural water resources in many areas have been major constraints of pistachio production in recent years (Sedaghat, 2006). Kerman province experiences hot and dry climate and has the problem of water scarcity. Due to unequal rainfall distribution, neighborhood of the Lout Desert and Kavir plains, high evaporation and exceeded usage of resources; water scarce issue in desert regions is severely obvious. Declining ground-water is over 50 cm in Kerman the year. Because of illegal utilization, aquifer levels have been declined in some regions especially pistachio growing areas accompanying with dangerous problem of salinity. Thus, it's necessary to consider optimum using of this vital resource (Mehrabi Boshrabadi, 1995). Drop of aquifer level of ground-water in world is yearly between 750 to 800 billion cubic meters and drop of aquifer level in Iran is equivalent to 7 billion cubic meters. That is 1 percent of drop aquifer level of world's ground-water. According to report of Iranian water resource management corporation in water year 2001-2002 (Departmen of Energy, 2005). The economic valuation of water input, therefore, is essential for its efficient usage. Water- extracted values are categorized in two groups of use values and non- use values:

1- Use values include: this classifications is based on :A- consumptive and non-consumptive values. B- Position type. C - Economic role.

2- Non -use values include: A Existence value. B - Option value. C – Bequest value.

Economic valuation is a complex topic of economics. In absence of markets or their sufficient performance, one can utilize non-market methods to determine the value. The methods used for determining the economic value of water can be divided into two categories of inductive and deductive ones. All methods are practically practicable but some of them are more common than others because of limitations. Observation of agricultural land market transactions and wastewater are examples of simple methods. In complex practical methods, one can mention mathematical programming procedure and production function estimation approach (Department of Energy, 2011). The method used in this study is production function estimation. Since the parametric method allows statistical testing of econometric models' estimated parameters, it is used with production function technique in this study. Hence, the water value obtained can be more reliable. There is no need to specify the water restriction threshold and supply type in parametric methods. Among these, production function approach was chosen because of the impossibility of applying the profit and cost functions (Dashti *et al.*, 2010).

Hayati *et al.* (2009) with using translog production function estimated that the actual price of irrigation water in wheat and barley for North Khorasan, Razavi Khorasan and South Khorasan. The results showed that irrigation water price in North Khorasan, Razavi Khorasan and South Khorasan for wheat were 122.14, 2882.24 and 456.30 RLS per cubic meter and for barley were 703.01, 1343.67 and 112.67 RLS per cubic meter, respectively. That is higher than the prices which was paid by farmers.

Dashti *et al.* (2010) estimated the economic value of water for Damghan. In this study they used method of production function. The results of analysis showed that the economic value of irrigation water for wheat was 403.2 RLS, the estimated amount is higher than current value of water in region for study.

Shamsoldini *et al.* (2010) determined the economic value of water through analysis of the production function among sugar beet farmers of Marvdasht city. Based on the prices of 2006, the price was assessed 211.6 RLS per cubic meter of water in sugar beet production.

Khaje Roshanaei *et al.* (2010) applied production function method in wheat production of Mashhad to determine the economic value of agricultural water. In order to assess production function coefficients, the classic and generalized maximum entropy models were used. According to the results, the classic method and Translog functional form led to the best outcomes. The economic value of water for wheat was calculated 1870 RLS.

Ehsani *et al.* (2011) with using the production function approach and the Dual cost function, determined the water economic value from the viewpoint of wheat suppliants in areas irrigated via Qazvin plain irrigation network in crop year of 2007-2008 .They set the economic value of water 586 and 609 RLS per cubic meter.

Dehghanpour and Sheykhzeinodin (2013) determined the economic value of agricultural water in Ardakan- Yazd plain of Yazd province. In this study the economic value of agricultural water was calculated by using production function. Also, water economic value and water unit price per cubic meter were calculated 997.5 and 530,8 RLS, respectively. The difference between economic value and unit price of water can be one of the reasons of excessive and inefficiency water use in wheat production.

Karthikeyan (2010) determined the economic and social value of irrigation water. The results showed that the comparison of the economic value of estimated water with using different methods, strongly suggests that the present water use pattern will not lead to sustainable use in the tank command areas.

In a study aimed at determining the optimal cropping pattern and water shadow price calculating under hazardous conditions in Baft region, Zare Mehrjerdi (2011) used a combination of mathematical programming methods of under risk target Motad and the residual value procedure. In this study, water shadow price was determined 944 RLS per cubic meter.

Given the importance of ground-water in the city of Ravar, Sherzei and Amir Teimouri (2012) determined the economic value of water using input productivity value calculation method. The economic value of Ravar ground-water was obtained 19870 RLS per cubic meter in their study.

Garcia and Reynaud (2004) studied water pricing method using econometric patterns and simulation of social variables. They concluded that there is a significant difference between the observed market prices and the marginal production value of water.

In their study "increasing of irrigation through water demand management by implementing different policy methods of pricing", Yousef *et al.* (2008) resulted that various pricing methods leads to encourage the farmers in selecting and cultivating of water scarcity resistant crops but pricing policy is not a reliable tool for improving irrigation efficiency.

Mesa-Jurado *et al.* (2010) estimated marginal value of water for irrigated olive grow with the production function method. Net marginal value of water obtained were € 0.60 m^{-3} for the allocation of 100 m^3 ha^{-1} and € 0.53 m^{-3} for the water right allowance of 1500 m3 ha^{-1}.

All of the above studies emphasized on determining economic value of water in their studied region and results show that the price paid by farmers is lower than estimated economic value of irrigation water.

The aim of present study is estimation of economic value of water in Sirjanian pistachio producers.

MATERIALS AND METHODS

Determination of the water economic value has several procedures dividing into two overall categories of parametric and nonparametric methods. Given water usage in agriculture sector as an input, production function and mathematical programming methods are the most famous ones in estimating its economic value. Since the parametric method allows statistical testing of econometric models' estimated parameters, it is used with production function technique in this study. Hence, the water value obtained can be more reliable. There is no need to specify the water restriction threshold and supply type in parametric methods. Among these, production function approach was chosen because of the impossibility of applying the profit and cost functions (Dashti *et al.*, 2010).

Production function method

Production function is a completely physical concept showing simply the relationship between production outputs and inputs. The most general form of the production function is (Chambers, 1988):

$$Q=f(x.z) \tag{1}$$

Where Q is the production rate; f denotes the functional relationship, x shows variable inputs vector and z represents the vector of fixed or quasi- fixed inputs.

i-th input is calculated through differentiating of expression (1) with respect to the desired input of production function:

$$MP_{xi} = \frac{\partial f(x)}{\partial x_i} \tag{2}$$

Average production rate representing production per input consumption unit is:

$$APP_i = \frac{Y}{X_i} \tag{3}$$

The economic value of water, therefore, is obtained from the following equation (Mousa Nejad and Najarzadeh, 1997):

$$VMP_w = MPP_w \times Py = P_w \tag{4}$$

Where MPP_w is marginal production of water input, P_Y denotes the product price, VMP_w represents the value of marginal water input production and P_w is the economic value of water. The marginal production rate can be calculated using the following equation (Ehsani *et al.*, 2010):

$$MP_W = E_W \times AP_W = \frac{\delta \ln(y)}{\delta \ln(w)} \times AP_W \tag{5}$$

Given dependence of water economic value to the production function, Cobb-Douglas, Transcendental and Translog production functions were tested to select the best form for estimating the water economic value in this study. Three forms of these functions are shown in expressions (6), (7), and (8) respectively:

$$\ln(y) = \alpha_0 + \alpha_i \sum_{i=1}^{9} \ln(x_i) \tag{6}$$

$$\ln(y) = \alpha_0 + \alpha_i \ln(x_i) + \beta_i \sum_{i=1}^{9} x_i \tag{7}$$

$$\ln(y) = \alpha_0 + \sum_{i=1}^{9} \alpha_i \ln(x)_i + 1/2 \sum_{i=i}^{9} \beta_i (\ln(x_i))^2 + \sum_{i=1}^{9} \sum_{i=1}^{9} \gamma_{ij} \ln(x_i) \ln(x_j) \tag{8}$$

Where Y is the production (kg), X_1 is N fertilizer consumption (kg), X_2 represents PO_4 fertilizer consumption (kg), X_3 denotes K fertilizer consumption (kg), X_4 is animal manure consumption (ton), X_5 is pesticide consumption (lit), X_6 indicates rented labor (Individual – day), X_7 is family labor (Individual – day), X_8 represents machines using (hours) and X_9 is the water consumption (m^3).

Population of the study is Sirjan city pistachio Growers. In this study, the use of simple random sampling 67 pistachio growers were chosen and necessary information obtained from a questionnaire.

RESULTS AND DISCUSSION

In order to choose the best form of production function, Cobb-Douglas, Transcendental and Translog functions were estimated in this study. Since among non-flexible production functions such as Cobb-Douglas and Transendental have no limitation on production function and depict more suitable real behavior of economic factors (Salami and Mohammadi Nejad, 2002). In this study translog production function has been used. Since among flexible production functions translog form although has less limitation production function in compare to other forms of functions, this shape of function is proper for presenting technology of production (Kumbhakar, 2004).

Table 1: The results of production functions

Function	Cobb-Douglas	Transendental	Translog
Coefficients' numbers	9	18	53
Significant Coefficients number	2	5	31
R^2	0.20	0.44	0.95
F	1.6*	2.12**	4.88***
D.W	1.9	2.6	2.6
J-B	16.4 (0.00)	0.03 (0.98)	2.9 (0.22)
W	13.6 (0.000)	9.00 (0.000)	1.7 (0.14)

***p<0.01 ** p<0.05, *p<0.1

Table 2: The results of Translog production function.

Parameter	Coefficients	Standard Deviation	t-statistics
$Ln(x_1)$	11.01	2.59	4.24***
$Ln(x_2)$	-15.48	3.84	-4.02***
$Ln(x_3)$	9.16	3.74	2.44**
$Ln(x_4)$	-7.14	3.08	-2.31**
$Ln(x_9)$	3.07	1.16	2.64***
$(Ln(x_1))^2$	-1.49	0.32	-4.56***
$(Ln(x_6))^2$	1.34	0.67	1.97**
$(Ln(x_8))^2$	-1.84	0.92	-1.98**
$Ln(x_1)\,Ln(x_2)$	1.07	0.32	4.62***
$Ln(x_1)\,Ln(x_3)$	0.62	0.32	1.93**
$Ln(x_1)\,Ln(x_5)$	2.94	0.82	3.57***
$Ln(x_1)\,Ln(x_6)$	-2.86	0.71	-4.02***
$Ln(x_1)\,Ln(x_7)$	-1.31	0.41	-3.14***
$Ln(x_1)\,Ln(x_9)$	-1.71	0.49	-3.44***
$Ln(x_2)\,Ln(x_4)$	-1.02	0.46	-2.21**
$Ln(x_2)\,Ln(x_5)$	-2.27	0.84	-2.67***
$Ln(x_2)\,Ln(x_6)$	3.76	0.72	5.17***
$Ln(x_2)\,Ln(x_7)$	1.09	0.5	2.17**
$Ln(x_2)\,Ln(x_9)$	2.15	0.69	3.09***
$Ln(x_3)\,Ln(x_5)$	1.12	0.58	1.92**
$Ln(x_3)\,Ln(x_8)$	-1.15	0.54	-2.11**
$Ln(x_3)\,Ln(x_9)$	-2.55	0.89	-2.86***
$Ln(x_4)\,Ln(x_5)$	-1.92	0.74	-2.59***
$Ln(x_4)\,Ln(x_7)$	3.20	0.79	4.03***
$Ln(x_4)\,Ln(x_8)$	-3.09	1.01	-3.05***
$Ln(x_4)\,Ln(x_9)$	3.7	1.04	3.55***
$Ln(x_5)\,Ln(x_6)$	-1.14	0.62	-1.83**
$Ln(x_5)\,Ln(x_7)$	-1.44	0.57	-2.52***
$Ln(x_5)\,Ln(x_8)$	4.2	1.19	3.53***
$Ln(x_6)\,Ln(x_9)$	-1.57	0.87	-2.006**

***$p<0.01$ ** $p<0.05$, *$p<0.1$

The results are shown in table 1.

Translog flexible function was chosen as the best function. High value of R^2 statistic indicates appropriate fitness of the model. It shows that 95% of production changes are explained by considered variables and their interactions. According to F statistics, whole of the model is significant in probability level of 99%. This model includes the most significant coefficients shown in table 2.

Partial elasticity of water was calculated using model derived coefficients as 3.04 % indicating pistachio growers' presence in the first and non-economic production area. Due to insufficient production of the water input and adequate land as the agricultural constant input in the first area, this is an acceptable result for our agricultural sector. Thus increasing of changing input leads to production increment (Abounouri, 2002). Providing water input increment of 10 %, production rate will increase by 30%. Increment of water input doesn't mean necessarily its consumption per area unit. Instead, this indicates efficient water to increase available water for pistachio trees because traditional irrigation methods cause water loss in the studied region. Using modern irrigation methods, therefore, water loss can be prevented and water availability for trees will be increased. Once water partial elasticity was calculated and placed in equation (5), marginal production value of water was obtained. Accordingly, minimum, average and maximum economic value of water is 22500, 50360 and 68000 RLS respectively but the average price paid by farmers is 1771 RLS per cubic meter of water. Due to inappropriate pricing of water in agriculture sector, there is a wide gap between water actual price and the price paid by pistachio growers. Since rising price of the product leads to mounting of water marginal production value based on the equation (4), significant increment of pistachio product price in cropping year of 2012-2013 can be considered as another reason of this broad gap.

CONCLUSION

Economic value per cubic meter of water to pistachio production is higher than its average exchange value in the region leading to non-optimal using of water. Since the studied area is arid and because of recent droughts and aquifers levels' declining, inputs' insufficient using causes serious complications for the pistachio as the most important agricultural export product of Iran. Therefore, the following suggestions are represented based on research results:

1-Appropriate water pricing based on its economic value, to eliminate the gap between the actual price and the paid one. Implementation of this policy should be done in long run.

2-Using of tax implementation policies based on different income categories in order to reduce the water consumption by the government.

3-Increasing of water prices may encourage farmers to use modern irrigation methods but due to low income, some farmers are not able to use these methods. The government can provide long-term low-interest loans and grants to assist them in applying these methods. Thus, water consumption is directed to optimal amount leading to cost saving for farmers.

4-Organizing training courses and informing the farmers about modern ways of irrigation.

REFERENCES

1-Abounouri, A.(2002). Microeconomics (Vol. 1), Tehran: Islamic Azad University, Central Tehran Branch, p. 437.

2-Abdollahi Ezatabadi, M. (2008). Role of nonintegrated policy making in pistachio unstable cultivation with emphasis on water resources. *Agricultural Economics and Development*, 16(63), 117-137.

3-Chambers, R.G. (1988). Applied production analysis: A dual approach, Cambridge University Press.

4-Dashti, G.h., Aminian, F., Hossein zad, J., & Hayati, B. A. (2010). Estimation of water economic value in wheat production (Case study: Damghan underground resources). *Journal of Sustainable Agriculture Knowledge,* 2 (20), 121-131.

5-Dehghanpour, H., & Sheykhzeinodin, A. (2013). Determining the economic valuation of agricultural water in Ardakan-Yazd plain of Yazd province. *Journal of Agricultural Economics and Developmenl,* 21(82), 45-68.

6- Ministry of Energy. (2005). Special Issue of water. http://lib.moe.gov.ir/

7- Ministry of Energy. (2011). A Guidance to determine the economic value of water for agricultural purposes. http://lib.moe.gov.ir/

8-Ehsani, M., Hayati, B., & Adeli, M. (2010). Estimation of economic value of water in maize production (case study of the central sector of Alborz city, Qazvin province). *Agricultural Economics and Development,* 18 (72), 76-93.

9-Ehsani, M., Dashti, G.H., Hayati, B., & Ghahreman zade, M. (2011). Estimation of water economic value of Qazvin plain irrigation network: using of Dual approach. *Journal of Agricultural Economics and Development* (Food Science and industry), 25 (2), 237-245.

10-Food and Agricultural Organization. (2009). Statistical Yearbook.

11-Garcia, S., & Reynaud. A. (2004). Estimating the benefits of efficient water pricing in France. *Resource and Energy Economics,* 26,1-25

12-Hayati, B., Shahbazi, H., Kavoosi Kalashmi, M., & Khodaverdizadeh, M. (2009). Estimation of irrigation water actual price in wheat and barley: production function approach (case study: North Khorasan, Razavi Khorasan and South Khorasan provinces). *Journal of Agricaltural Knowledge and Sustainable Production*, 19(1), 143-155.

13-Jahad-Keshavarzi of Sirjan Region. (2013). Statistical Yearbook.

14-Karthikeyan, C. (2010). Economic and Social value of irrigation water: Implications for sustainability. Fourteenth International Water Technology Conference, IWTC 14 2010, Cairo, Egypt, 823-835.

15- Kumbhakar., S.C. (2004). Productivity and technical change measurement and testing. *Empirical Economics*, 29,185-191.

16-Kashakoglu, H., & Erol Cakmak., H. (1997). Economic value and pricing of water in agriculture, *CIHEAM-Options Mediterraneennes*, 137-145.

17-Khaje Roshanaei, N., Daneshvar kakhaki, M., & Mohtashami, G.H. (2010). Determination of the water economic value in production function approach using classic and entropy models (Case study: Wheat in Mashhad). *Journal of Agricultural Economics and Development (Agriculture Sciences and Industry)*, 24 (1), 113-119.

18-Mehrabi Boshrabadi, H. (1995). *Evaluation of Productivity of Pistachios Production Factors in Rafsanjan City.* Unpublished Thesis in Agricultural Economics Engineering. Faculty of Agriculture. Tarbiat Modarres University.

19-Mesa-Jurado, M. A., Berbel, J., & Orgaz, F. (2010). Estimating marginal value of water for irrigated olive grow with the production function method. *Spanish Journal of Agricultural Research*, 8(S2), S197-S206.

20-Mousanejad, M., & Najarzadeh, R. (1997). Economics of agricultural production. Prepared by D. L. Aldbertin. Institute of Economic Research, Tarbiat Modarres University.

21-Salami, H., & Mohammadinejad. (2002). determining the value of water using flexible production function (Case study:Central Plains of Saveh). *Journal of Agriculture Sciences and Industry*, 2(16), 85-96.

22-Shamsodini, A., Mohammadi, H., & Rezaei, M. (2010). Determining the economic value of water in sugar beet cropping. *Journal of Sugar Beet*, 103 (1), 26-93.

23-Sedaghat, R. (2006). *An Economic Analysis of Pistachio Production, Processing and Trade in Iran*, unpublished dissertation, Department of Agricultural Economics, University of Agricultural Sciences, Bangalore, India.

24-Sherzei, G.H., & Amirteimouri, S. (2012). Determination of the ground-water value (Case study: Ravar city (Kerman province). *Journal of Economic Research*, 98, 113-128.

25-Yousef, H., Francois, M., & Venot, J. (2008). Irrigation in the Jordan Valley: Are water pricing policies overly optimistic? *Agricultural Water Management,* 95, 427-438.

26-Zare Mehrjerdi, M. (2011). Determining the optimal cropping pattern and water valuation using a combination of under risk mathematical programming and residual value methods (Case study: Orzouieh Region, Baft. *Journal of Water Research in Agriculture,* 25 (2), 121-132.

5

Factor Affecting the Sustainable Management of Agricultural Water

Masoud Samian[1], Karim Naderl Mahdei [2], Heshmatollah Saadi [3], Hamid Balali [2] and Reza Movahedi [2]*

Abstract

The main purpose of the study was to investigate the factors affecting the sustainable management of agricultural water in Hamedan. The study population included all wheat farmers possessing irrigated farms in Hamedan city (N=1800). Of these farmers a sample of 317 people has been selected by using randomized multi-stage sampling method. The data were collected through a questionnaire's tool with help of the interview technique. Accuracy of the questions in the questionnaire was face validated by a panel of specialists. To test the reliability of the questionnaires, the questionnaires were first given to 30 farmers and Cronbach's Alpha was calculated (Alpha=0.92) then the questionnaire was finalized. Data analyzing methods such as Multiple Regression and the coefficient of variation (CV= standard deviation /mean) were used in this study. To determine the level of sustainability of the farms Bossel method proposed for classification and grading the fields was used. The results showed that variables agronomic factors, policy factors and institutional factors were able to explain 34 percent of the dependent variable's changes (sustainable management of agricultural water). According to the results, 95.3 percent of the farmers were categorized into unsustainable group, 4.1 percent into semi-sustainable and only 0.6 percent in sustainable group.

Keywords:
Sustainable management, Agricultural water, Multiple regression,Farmers irrigated wheat, Hamedan city

[1] *Ph.D. Student, Agricultural Extension and Education Department, College of Agriculture, Bu-Ali Sina University, Hamedan, Iran.*

[2] *Assistant Professor Agricultural Extension and Education Department, College of Agriculture, Bu-Ali Sina University, Hamedan, Iran.*

[3] *Associate Professor, Agricultural Extension and Education Department, College of Agriculture, Bu-Ali Sina University, Hamedan, Iran.*

* *Corresponding author's email: knadery@yahoo.com*

INTRODUCTION

Several studies indicate that the amount of water allocated for agriculture will be further restricted in the entire world due to continuous population growth, development of Urbanization and Industrial Development till 2025 (Ehsani and Khaledi, 2003) which results in the lesser agricultural production, especially in dry areas. The problem in many arid and semi-arid countries has led to the importation of food (Hamdy *et al.*, 2003). Iran is among the countries where facing a water crisis based on the classification that is done by International Water Management Institute (Ehsani and Khaledi, 2003). In addition, it is predicted that by 2030, Iran will be among those countries that renewable water resources per capita will be less than 1500 cubic meters per year that it will lead to a critical water situation (Rijsberman *et al.,* 2006). What is emphasized is that the negative social effects of resource scarcity, including migration, conflicts and competitiveness, reduce production and income and thus reduction of the life quality, will disappear through increasing the community's capacity to deal with these shortages (Bocchi *et al.*, 2006).

With the knowledge that in a national scale, the most important use of water is for agriculture, so, better management of water in agriculture will have the greatest effect on the availability of water resources (Frozani andKarami, 2012). Iran water use efficiency is about 30% at the national level. Thus, water scarcity and misuse of currently available water resources are the major threats to the sustainable development of the agricultural sector (Hamdy *et al.*, 2003). Today, water management challenges and shortcomings have increased due to excessive pumping of groundwater which cause a rise in water salinity. Notion of water consumption in most countries shows the increasing pressure on water resources and the need for new approaches to managing the consumption has increased drastically (Sullivan, 2002). The challenge of water scarcity and droughts in recent years has doubled the focus to the issue of water management in agriculture sector; which, consequently, is introducing a wide range of related approaches to water management at the farm level (Frozani and Karami, 2012).

During the past few decades, increasing degradation of natural and environmental resources reveals that, the development did not have a rhythmic motion and it causes instability and worsens environmental hazards in the environment. Water as part of the environment – is the environmental, basis of life and the fundamental component of each development model, and it has a central and important role in sustainable development. Unfortunately, in the last few years, as a result of human activities, despite its limited resources, this valuable resource has suffered from irreparable damage and its quality and quantity in many communities declined sharply (Poorasghar, 2001). In this regard, experts say, growing demand for water and the severe limitations of this vital element in the global scale will be considered as the main challenge to the world in the next few years (Sayer and O'Riordan, 2000). Supply management and development of water resources, as a major dynamic and effective element in sustainable development, draws its attention to the environmental, political, legal and organizational issues (Dungumaro, 2003; Serageldin, 1995).

Agriculture has an important role in planning and policy as the largest consumer of water (Borimnejad and Yazdani, 2004). Because more than 90 percent of total water consumption in the country is devoted to agriculture and less than 7 percent is devoted to urban and industrial uses (Godarzi *et al.*, 2009). Globally, agriculture accounted for the highest consumption of fresh water so that in 2002 about 70 percent of the water used in the world has been devoted to agriculture (Anonymous, 2003). Therefore, proper management of water in agriculture can influence the development of sustainable agriculture in Iran (Godarzi *et al.,* 2009). Due to limited renewable water resources and allocating the bulk of its production in agricultural products, all of the growth in world population and food shortages, the importance and sensitivity of water resources management has increased (Howarth *et al.*, 2005) and pronounced the need for attention to mechanisms and programs to increase productivity and optimal use of water resources in the agricultural sector (Hartley, 2006).

Distribution of precipitation in the world is very heterogeneous and its distribution is such that has placed Iran among the countries with severe water restrictions. The average annual precipitation is about 250 mm in Iran which this rate is about a quarter of the global average (Hasheminia, 2004). Based on Meteorological Organization studies, Iran is among the countries that are already in water stress and this stress intensity remain is increased each year (Ansari *et al.*, 2010). This issue requires much attention to the sustainable management and optimal use of water resources. Sustainable management of water resources is a systematic process for sustainable development allocation and is used for the monitoring of water resources for social, and environment economy purposes (Taylor *et al.*, 2008).

Review of literature:

Without a doubt, understanding the factors affecting agricultural water management is a leading provider of water management in agriculture. Azizi (2001) classifies these components in several categories as management factors (age, history of agriculture, education, frequency of attendance extension ...), physical (cropping pattern, number of components, climate, irrigation ...), economic (insurance, difficult access to inputs), community (neighbors consumer behavior, time dependent, non-agricultural income), institutional (ownership of water resources, lack of enforcement of laws and regulations, rental of water ...).

Zahtabian (2005), considers lack of irrigation management due to the low efficiency of knowledge and believes that the role of education in promoting and improving irrigation management irrigation efficiency is significant. Ehsani and Khaledi (2003) considered role of agricultural extension and education in increasing water productivity. Khadem Adam (2003) mentioned factors such as: irrigation at the wrong time and flooding of agricultural land and the influence of water on earth as the most important problems related to water loss. Borimnejad and Yazdani (2004) analyzed sustainable water management of calibration irrigation efficiency and water stability index by multi-objective fractional programming, was.

In this study, it was found that when the performance values of the ratio of gross efficiency of water is 35%, the stability index equals to 15; while in irrigation efficiency of 75%, this index increases to 31.2. Zarei Dastgerdi *et al.* (2006), in a research, to improve agricultural water management, considered the instruction, research and awareness, understanding agricultural water management skills and new technologies important. Saadi (2008) in a study entitled need for training farmers in water management to cope with drought basis for coping with drought, acknowledged that farmers themselves should be familiar with appropriate methods of irrigation and figuring out ways to use less water. If in a long-term program, irrigation to be diverted and drip irrigation to optimum use of water, we won't face is the water limitation in the coming years. Farmers should be ready to cope with drought and the necessary training in coping with drought.

Ansari *et al,* (2010) study the water management in parallel with sustainable development anf they concluded at the end of their study that significant reductions in precipitation, degradation of water resources, lack of systematic and efficient management of water are inefficient factors to sustainable development in the world. Ommani (2010) investigated the factors affecting the sustainable management of water resources and concluded that crop payments and the extension of educational activities, economic characteristics, farmers' knowledge, social activities and government support are the most important factors affecting the sustainable management of crop water requirements. Frozani and Karami (2012) as in the study of water management: wheat producers across city near Shiraz, Fars Province reached to the conclusion that n general, farmers have limited knowledge and poor management in relation to agricultural water in their study area. Despite the significant positive relationship knowledge between farmers and water management and optimal management at the farm level, there is the total amount of available water and agricultural knowledge of water management there is a significant negative relationship between management and their optimization.

According to research of Regner *et al,* (2006) failure to provide necessary training for farmers on irrigation management is an important problem in the field of successful water management. Riesgo and Gomez (2006) in a study analyzed the effect of different policy scenarios of agricultural policy and water pricing policies in the agricultural sector in Spain using several criteria of mathematical programming models. The results showed that cover the full cost pricing of water than normal conditions in the price of water are zero can be used in about 50 percent and reduce water demand.

Gruber *et al.* (2009), in their study, considering water demand agricultural season, reached to the conclusion that improving knowledge about better managing administrative structures, technical knowledge and access to credit for farmers should be used in national strategies for improving agricultural water.

Osooli *et al.* (2011), in a study entitled, factors affecting the sustainable management of agricultural water in dry areas, reached to the conclusion that the most important factors affecting the sustainable management of agricultural water are educational, agronomic, technical, social and economic factors. Ommani (2011), in a research entitled factors affecting the sustainable management of water resources in agriculture, concluded that factors related to farm size, factors of agricultural knowledge, social factors and mechanization are the factors affecting the sustainable management of water resources in agriculture.

MATERIALS AND METHODS

This research is a survey. In this method, by selecting and studying samples chosen from the abundance, and distribution, relationships between variables are reviewed and evaluated. The population of the survey are irrigated wheat farmers in Hamadan city engaged in farming which are 1,800 among which a total of 317 cases have been selected by sampling (multistage sampling).

For data collection, a questionnaire containing 21 open questions and 60 closed questions was used. The accuracy or validity of indicators and items in the questionnaire has been confirmed by specialists. To evaluate the reliability of the research instrument the questionnaire was given to a number of farmers in Hamadan city. Obtained results for Cronbach's alpha was 92% which is rather an acceptable figure. Due to the nature of the research, the data are described and analyzed in this study. In data description, the descriptive statistics such as frequency, percentage, median, mode, standard deviation, variance, mean, and others were used. In the data analysis, multiple-regression was used. The dependent variable in this study named composite index of sustainable management of agricultural water of irrigated wheat farmers is presented below:

$$S = \sum_{i=1}^{9} Xi - \sum_{j=1}^{5} Yj$$

In this formula, S is stability, X1 to X9 are respectively agricultural water productivity, crop rotation, use of legumes, the use of animal manure, the use of green manure, conservation tillage, transmission channels of agricultural water, deep water wells, irrigation and Y1 to Y5 are intake of pesticide spraying, the amount of herbicide, pesticide, phosphate fertilizer consumption, the rate of chemical fertilizer nitrogen and the amount of potash chemical fertilizer (Ommani, 2010). Independent variables in this study are agronomic factors, policy and institutional factors, education and extension factors, social factors and economic factors.

RESULTS

Individual characteristics of the study population

Based on the findings mean age was 41.50 years (SD = 13 years). The youngest of samples was18 and the oldest of them was 79 years old. Among the total respondents, the most common examples are illiterate (36.9%). The depth of the water wells in the study is as the following: thus, 9 farms with less than 70 m depth wells, 238 fields with wells with depths of 70 to 100 meters, and the number farms with more than 100 deep wells are 70; and about 60 percent of farms are with electric pump and about 40 percent of them are with diesel pump; and about 40 percent of farms use polyethylene pipes, about 33 percent of them use cement channel and about 26 percent of them use soil creeks to convey water from source to field.

Pesticides, herbicides and pesticides are discussed as the two main factors in the sustainable

Table 1. Description of Farming on sustainable water management in the target Farms, Farmers' perceptions

Priority	Variable	M	SD	CV
1	Prevent the accumulation of waste in the water channel	4.22	0.689	0.163
2	Fertilization performed using water	4.14	0.810	0.195
3	Fight timely and proper weed	4.09	0.809	0.197
4	Removing the winding water channels on the farm	4.00	0.882	0.220
5	Proper distribution of water to the land surface	4.10	1.01	0.246
6	Water storage tanks, water storage and seasonal rainfall	3.87	1.00	0.258
7	Reuse of waste water and excess	3.59	1.05	0.292
8	Sponsorship agricultural rotation	3.64	1.10	0.302
9	Early planting plants in order to escape from the stress and tension	3.65	1.12	0.306
10	The use of drought resistant wheat cultivars	1.95	0.627	0.321
11	Irrigation in the morning or evening	3.52	1.19	0.338
12	Use of pressurized irrigation	3.47	1.46	0.420
13	Integrating land to prevent water loss	3.94	1.81	0.459

Note: very low=1, low= 2, moderate= 3, much=4, very much= 5; M=mean, SD=standard deviation.

management of water in agriculture distribution of the samples show that the cache-offs in terms of pesticides. Most of the study population between 1.5 to 2 liters per hectare consumption of pesticide has and distribution of samples based on the amount of chemical herbicides indicates that the population studied more than 2 liters per hectare consumed poison.

The use of chemical fertilizers has considerable influence on the sustainable management of agricultural water. To evaluate the amount of chemical fertilizers in the fields of study we have classified phosphate fertilizer (Black) and N (white) in three categories: describing the following consumption of 200 kg per hectare, 200 to 400 kg ha and more than 400 kg per hectare, potash fertilizer also classified as the following three categories: 150 kg ha 150 to 200 kg ha. We have more than 200 kg ha category: distribution of samples based on the amount of chemical fertilizer phosphate (black) shows that most of the study population have consumed between 200 to 400 kilograms per hectare fertilizer. Frequency distribution of samples based on the amount of chemical fertilizer N (white) indicates that most of the study population have consumed between 200 to 400 kilograms per hectare fertilizer and the distribution of samples based on the amount of chemical fertilizer potassium indicates that the population have studied more than 150 kilograms per hectare fertilizer consumption.

Variables affecting the sustainable management of water in terms of priority:

A. Agronomic Factors

The most important agronomic factors are those that its variation coefficients are less. The most important elements farmers observed in their fields were: prevent the accumulation of waste in the Water Channel with the coefficient of variation of 0.163, fertilization using water with the coefficient of variation of 0.195 and timely and accurate struggling with weeds with the coefficient of variation of 0.197 (Table 1).

B. Policy and Institutional Factors

The most important policy and institutional factors are those that its variation coefficients are less. The most important policy and institutional factors from the perspective of farmers' are: reduction rules to obtain bank credit with the coefficient of variation of 0.175, longer term of returning the financing system with the coefficient of variation of 0.179 and lowering interest rates on loans related to agricultural water management high productivity with the coefficient of variation of 0.180 (Table 2).

C. Education and Extension Factors

The most important education and extension factors are those with less variation coefficients. The most important factor from the viewpoint of education and extension farmers are: visits to successful farms with the coefficient of 0.149, teaching sessions for optimal management of

Table 2. Description of Policy making and Institutional Factors, the Farmers' Perceptions

Priority	Variable	M	SD	CV
1	Reduction rules to obtain bank credit	4.21	0.740	0.175
2	Longer repayment system facilities	4.15	0.746	0.179
3	Lowering interest rates on loans related to water resources management	4.13	0.746	0.180
4	Investments or for modification and repair facilities Rivers	4.13	0.769	0.186
5	Increased use of irrigation facilities with high productivity	4.07	0.766	0.188
6	Funds allocated by the government for long-term reconstruction and rehabilitation of water sources	4.16	0.791	0.190
7	Public investment to improve the condition of a water conveyance canals	4.09	0.816	0.196
8	Treatment and use of wastewater in agricultural production	4.16	0.824	0.198
9	The diesel-electric wells	4.15	0.824	0.198
10	Coordination between government agencies concerned with water	3.86	0.956	0.247
11	Support private sector investment in agricultural water management sector	3.88	1.00	0.257
12	Operations aquifer	3.79	1.07	0.282
13	Regarding the privacy and legal distance between wells	3.59	1.14	0.317
14	Blocking unauthorized wells	3.63	1.18	0.325
15	Installation of water meters on wells	3.62	1.21	0.334
16	Obtaining fine harvest unauthorized users	3.68	1.24	0.336

Note: very low=1, low= 2, moderate= 3, much=4, very much= 5; M=mean, SD=standard deviation

agricultural water with the coefficient of variation of 0.187 and teaching farmers the necessity for efficient use of water with the coefficient of variation of 0.198 (Table 3).

D. Economic Factors

The most important economic factors are those with less variation coefficients. The most important economic factors in farmers' views are: reduction of costs by using modern irrigation methods with a coefficient of variation of 0.184, financial ability of the farmer and providing sufficient initial capital for new irrigation methods with a coefficient of variation of 0.210 and quotas to limit production with high water demand with CV of 0.210 (Table 4).

E. Social Factors

The most important social factors are those with least variation coefficients. The most important factors in the farmers' perspective are: public participation in decision-making in agricultural water management projects with a coefficient of variation of 0.182, people's participation in agricultural water management projects with a coefficient of variation of 0.185 and the formation or group of people to oversee the efficient use of water resources with a coefficient of variation of 0.188 (Table 3).

Analysis of the stability fields of study

To analyze the level of stability of the fields of study, standard deviations from the mean is

Table 3. Description of Education and Extension Factors, the Farmers' Perceptions

Priority	Variable	M	SD	CV
1	Views from the field and successful demonstration	4.23	0.633	0.149
2	Teaching sessions for optimal management of agricultural water	4.13	0.776	0.187
3	Training farmers to the need for efficient use of water	4.11	0.817	0.198
4	Regular visits of new irrigation systems and rural education	4.02	0.857	0.213
5	Distribution of educational brochures and magazines	3.90	0.955	0.244
6	Expertise experts agricultural extension	3.90	0.957	0.245
7	Educational film on water management	4.07	1.01	0.248
8	Expertise in irrigation sector experts	3.83	0.966	0.252
9	Increase participation in agriculture, agricultural water management	3.81	1.04	0.272
10	Farmers referred to counseling centers for agricultural water management problem	3.76	1.04	0.276
11	Farmers encouraged to attend training classes	3.78	1.06	0.280

Note: very low=1, low= 2, moderate= 3, much=4, very much= 5; M=mean, SD=standard deviation

Table 4. Description of Economics Factors on basis of the Farmers' Perceptions

Priori	Variable	M	SD	CV
1	Reduce costs by using modern irrigation methods	4.14	0.763	0.184
2	Level of funding sufficient financial ability of farmers to use modem irrigation methods	4.04	0.849	0.210
3	Quotas to limit production with high water requirements	3.96	0.835	0.210
4	Rationing of energy supply	3.94	0.873	0.221
5	Production quotas on the amount of available water	3.81	1.00	0.263
6	Increased price of chemical inputs as inputs to supplement water	3.64	1.06	0.291
7	Local water supply rationing	3.70	1.11	0.300
8	Increasing the price of water rights (water pricing)	3.58	1.08	0.301
9	Energy prices in groundwater as a supplemental water input	3.57	1.10	0.308
10	Reduce subsidies for agricultural inputs	3.55	1.14	0.321

Note: very low=1, low= 2, moderate= 3, much=4, very much= 5; M=mean, SD=standard deviation

used (Taghipoor *et al.*, 2013). In this method, the converting methods of the obtained scores based on the proposed Bossel table are estimated in 3 categories:

A < Mean – St.d: A unsustainable

Mean- St.d < B < Mean: B semi-sustainable

C > Mean + St.d: D sustainable

To analyze the stability fields of study Bossel proposed method for classification and grading of the fields are used. Bossel proposed method is shown in table 6.

Based on the Bossel proposal, stratification for the stability of unsustainable field values is between 0 and 0.4, unsustainable farmland values are between 0.4 and 0.6 and sustainable farmland values are between 0.6 and 1. Given the stability of the proposed Bossel and the mean of the combined formulation of sustainable management of agricultural sustainable management of water, land leveling is studied (Table 7).

Most fields of the study were in a unsustainable agricultural water management means that roughly about 95 percent of the farms in terms of sustainable management of agricultural water are unsustainable which this is a serious problem in the agricultural sector if that trend continues this practice in the near future there would be problem regarding agricultural water supply.

Table 5. Description of Education and Extension Factors In the Farmers' Perceptions

Priori	Variable	M	SD	SD/M
1	Public participation in decision-making in agricultural water management projects	4.21	0.769	0.182
2	People's participation in agricultural water management projects	4.12	0.764	0.185
3	Focus formation or group of people to oversee the efficient use of water resources	4.19	0.790	0.188
4	The use of local traditions and local laws among farmers for water management	3.93	0.820	0.208
5	Identification and use of indigenous knowledge for agricultural water man-	3.94	0.830	0.210
6	agement practices			
7	Coordination between government agencies and community organizations	3.94	0.849	0.215
8	Public participation in the preservation of agricultural water management projects	4.06	0.898	0.221
9	Local norms regarding the proper utilization of water resources	3.86	0.940	0.243
10	Cooperative formation of water users to improve water management in rural areas	3.72	1.02	0.274
	Together with the growing possibility of increased cooperation and participation of non-governmental organizations	3.68	1.09	0.296

Note: very low=1, low= 2, moderate= 3, much=4, very much= 5; M=mean, SD=standard deviation

Table 6. Sustainability Level

Level	Values
unsustainable	0 - 0.4
semi-sustainable	0.4 – 0.6
sustainable	0.6 - 1

Reference: (Bossel, 1999)

Table 7. Sustainability Level of the target Farms

sustainable	Equivalent	Number	Percent
unsustainable	0 - 0.4	302	95.3
semi-sustainable	0.4 – 0.6	13	4.1
sustainable	0.6 - 1	2	0.6

Table 8. ANOVA and Brief Model of Correlation and Determination Coefficient

correlation	R^2	F	Sig
0.542	0.340	17.581	0.000**

Regression analysis of factors affecting the sustainable management of agricultural water

To analyze the factors affecting the sustainable management of agricultural water, multiple regression analysis was used. In this method, first, all independent variables that have a significant relationship with the dependent variable enter into the equation simultaneously.

Regarding factors affecting the sustainable management of agricultural water multiple correlation coefficients (R) equals to 0.542 and coefficient of determination (R^2) equals to 0.340 (table 8). In other words, 34% of the variability is explained by the independent variables and other factors are related to changes that have not been studied in this research.

Factors that influence the sustainable management of agricultural water in this study were including crop factor policy and institutional factors that have significant relationship with the dependent variable which is the sustainable management of agricultural water; that the amount of crop factor equals to 0.462 means that a unit changing in the standard deviation causes the standard deviation of the dependent variable to be changed 0.462. And policy-taking and institutional level is 0.241 and the results of the obtained coefficients are given in table 9.

But according the results of the regression, significant relationship with educational and extension of social, economic was not observed.

According to the results in table 9, the linear regression equation is as follows:

$Y = 2.078 + 0.119X_1 - 0.061 X_2$

in which:

Y = Sustainable agricultural water management

X_1 = farming factors

X_2 = Policy and institutional factors

Therefore, the agronomic factors have a larger proportion in comparison with other variables in predicted the dependent variable. So, one unit change in standard deviation cause the standard deviation of the dependent variable to be 0.462. While the standard deviation of a unit change in policy and institutional strength to standard deviation of the dependent variable in size equals to 0.241.

DISCUSSION

The correlation between the sustainable management of agricultural water and each of the 5 factors (independent variables of research) the results showed that there is a positive and significant relationship between agronomic factors

Table 9.The Calculated Coefficient Related to Influencing Factors on Sustainable Agricultural Water Management

Variable	(B)	Beta	t	Sig.
Constant factor	2.078	-	2.175	0.000**
Agronomic Factors	0.119	0.462	4.521	0.000**
Policy and Institutional Factors	-0.063	-0.241	-2.610	0.005**
Education and Extension Factors	-0.021	-0.105	-0.876	0.298
Economic Factors	0.036	0.124	1.156	0.243
Social Factors	-0.031	-0.095	-1.361	0.124

and policy factors institutional framework and sustainable management of surface water; and there is no correlation significant between education and extension agents, social and economic factors and the level of sustainable management of water in agriculture. In other words, there is a 99 percent probability of a significant relationship between the dependent variable the sustainable management of water in agriculture and agronomic factor; the more improved the agronomic factor, the more sustainable management of water in agriculture will be. In addition, policy and institutional factors are significantly associated with the sustainable management of agricultural water in the present study (95%).

The results of the grading fields of study in terms of sustainable management of agricultural water showed that the majority of farms in terms of sustainability in water management are unsustainable that is a problem in the agricultural sector. As was previously stated in conjunction with the stability fields of sustainable management of agricultural water Ansari *et al.* (2010) showed there is a low level of water sustainability in the fields of and in a research done by Chatorudi (2001) the same problem is seen in less developed countries.

The problem reminds us that water consumption in agriculture in the studied population is highly unsustainable and if this trend continues it causes destruction and loss of the agricultural sector over time; and this requires deep and careful study to solve this big problem in the agricultural sector. Perhaps a quick solution to this problem is continuous government support for the creation and operation of pressurized irrigation methods and determination of governments to avoid drilling new deep wells and determination of harvest levels for farmers This can temporarily relieves these great problems in the agricultural sector and it requires a long-term planning in order to solve this problem. Other solutions to this problem culture-building in rural communities that help of local leaders, clergy and mass media use are highly appreciated.

CONCLUSION

The results of the regression analysis showed that an agronomic factor has a larger proportion in comparison with other variables in predicting dependent variable. So, one unit change in standard deviation cause the standard deviation of the dependent variable to be 0.462. While the standard deviation of a unit change in policy and institutional strength to standard deviation of the dependent variable in size equals to 0.241. The results of this study on agronomic factors in predicting the dependent variable is consistent with the results of Azizi (2001), Noruzi and Chizari (2007), Ommani (2011) and Gruber *et al.* (2009). The role of the policy and institutional and agronomic factors were positive in the works done by Khalilian Mehrjardi (2006), Ommani (2010). Shortt *et al.* (2006), also stated that it is better to rely on social factors instead of applying the laws and policies.

The results of the regression again indicated the important things we can do on the farm regarding the sustainable management of agricultural water. One of the simplest but most important, as pointed out by farmers, for example, prevent the accumulation of waste in the water channels. It should be noted that other factors must be met, for example, a family of legumes in crop rotation can significantly contribute to the sustainable management of water to maintain soil moisture. Unfortunately, in the fields we study, this was not done due to the habit of sole harvesting. However, the role of policy and institutional factors the results indicate negative effect of the policy and perhaps one of the reasons is that in the agricultural sector the things cannot be relied upon by force and, as stated by Shortt *et al.* (2006), social factors has better effects instead of applying the rules to achieve sustainability. Another guidelines in this section is to assign water management institutions and community organizations to the local people; and the government observe the affairs and shouldn't involve directly in water matters. However, this study showed that other factors have no significant role in predicting the dependent variable and other studies are needed. In general, it can be said that many factors involved in the management of agricultural water sustainable in which some of these factors were expressed in this study.

REFERENCES

1- Ansari, H., Salehnia, N., & Boostan, A. (2010). Evaluation of water management in sustainable development. Proceedings of the scientific conference on water challenges in Qom. Qom University (In Persian).

2- Azizi, J. (2001). Sustainability of agricultural water. *Journal of Agricultural Economics and Development,* 9(36), 137-113 (In Persian).

3- Borimnejad, V., & Yazdani, S. (2004). Sustainability analysis in water resources management in agriculture tiny fraction of the program: the case of the province. *Research and Development in Agriculture and Horticulture*, 63, 16-2 (In Persian).

4- Bocchi, S., Disperati, S. P., & Rossi, S. (2006). Environmental security: A geographic information system analysis approach-The case of Kenya. *Environmental Management*, 37(2), 186–199.

5- Bossel, H. (1999). I*ndicators for sustainable development: theory, method, applications* (pp. 1-124). Winnipeg: International Institute for Sustainable Development.

6- Chaturvedi, M.C. (2001), Sustainable development of India's waters: some policy issues, *Water Policy,* 3, 297–320.

7- Dungumaro, E. W., & Madulu, N. F. (2003). Public participation in integrated water resources management: the case of Tanzania. *Physics and Chemistry of the Earth, Parts A/B/C*, 28(20), 1009-1014.

8- Ehsani, M. & Khaledi, H. (2003). Understanding and improving water productivity in agriculture, food security and water, Eleventh Congress of the Iranian National Committee on Irrigation and Drainage, 660-658 (In Persian).

9- Frozani, M. & Karami, A. (2012). Knowledge of water management: wheat producers across city near Shiraz, *Fars province Agricultural Education Management Journal, Summer Special Issue*, 34-43 (In Persian).

10- Gruber, I., Kloos, J., & Schopp, M. (2009). Seasonal water demand in Benin's agriculture. *Journal of Environmental Management,* 90, 196-205.

11- Godarzi, S., Shabanalifami, H., Movahed Mohamadi, H., & Jalalzadeh, M. (2009). Investigating factors influencing the perception of personal and professional Karaj city farmers towards agricultural water management problems, *Journal of Economics and Agricultural Development (Agricultural Sciences and Technology)*, 23(2), 62-55 (In Persian).

12- Hamdy A., Ragab, R., & Scarascia-Mugnozza, E. (2003). Coping with water scarcity: water saving and increasing water productivity. *Irrigation Drainage,* 52, 3-20.

13- Hartley, T. W. (2006). Public perception and participation in water reuse. *Desalination*, 187(1), 115-126.

14- Hasheminia, S. (2004). water management in Agriculture. *Ferdowsi Mashhad University Publications, Mashhad.*

15- Howarth, S.E., Parajuli, U.N., Baral, J.R., Nott, G.A., Adhikari, B.R., Gautam, D.R., & Menuka, K. C. (2005). Promoting good governance of water users' associations in Nepal. Department of Irrigation of His Majesty's Government of Nepal.

16- Khadem Adam, N. (2003). Policies in the agricultural economy of Iran. Publication of Information, Tehran (In Persian).

17-Khalilian, S. & Zare Mehrjardi, M. (2006). Valuation of groundwater in agricultural operation (Case Study of Wheat Farmers in Kerman city). *Journal of Agricultural Economics and Development,* (51), 1-22.

17- Koh, M.H. (2002). Status of Agricultural Water in Korea: water use and water quality. National Institute of Agricultural Science and Technology (RDA), Korea.

18- Noruzi. O., & Chizari, M. (2007). Study of social and cultural structures of the city Skinheads wheat attitude towards the development of irrigation. *Journal of Agricultural Extension and Education,* 2, 71-59.

19- Ommani, A. (2010). Investigating factors affecting the sustainable management of agricultural water resources in the basin north of Madras, Khuzestan province. *Watershed Research (Research and Development),* 88, 27-34 (In Persian).

20- Ommani, A, & Chizari, M. (2011). Identifying appropriate models to predict crop adoption of sustainable management of water resources in the city Wheat farmers in Ahvaz, *Agricultural Economy and Development,* 19 (73), 77-100 (In Persian).

21- Ommani, A.R. (2011). Affecting Factors on Adoption of Sustainable Water Resources Management in Agriculture. *Research Journal of Environmental Science*, 5(1), 36-48.

22- Osooli, N., Hosseini, M., Hosseini, J. F., & Haghiabi, A. H. (2011). A Survey: Factors Affecting the Sustainable Water Recourses Management (SWRM) in Agriculture under Drought Conditions in Lorestan, Iran. *World Applied Sciences Journal,* 12(4), 476-484.

23- Poorasghar, F. (2001). The use of economic instruments for environmental protection, environmental, *Journal of the Environmental Protection Agency*, 35, 85-122.

24- Regner Jochen, H., Salman, A.Z., Wolff, H.P., & Al-Karablieh, E. (2006). Approaches and impacts of Participatory Irrigation Management (PIM) in complex, centralized irrigation systems-experiences and results from the Jordan Valley. Conference on International Agricultural Research for Development, University of Bonn, October 11-13, 2006.

25- Riesgo, L., Gomez-Limon, J.A. (2006). Multi-Criteria policy scenario analysis for public regulation of irrigated agricultur. *Journal of Agricultural Systems,* 91, 1-28.

26- Rijsberman, F., Manning, N., & Silva, S. D. (2006). Increasing green and blue water productivity to balance water for food and environment. 4th World Water Forum Water, Food, and Environment Theme Baseline document, Mexico, 139-143.

27- Saadi, H. (2008). Agricultural education necessary for coping with drought, Retrieved from http:// www. aiiri.gov.ir.

28- Sayer, M., & O'Riordan, T. (2000). Climate change, water management and agriculture. Center for social and economic research on the global environment, University of East Angelia and University Collage, London.

29- Serageldin, I. (1995). Toward Sustainable Management of Water Resources (Directions in Development). The International Bank for Reconstruction and Development/THE WORLD Bank 1818 H Street. N.W. Washington. D.C. 20433.

30- Shortt, R., Caldwell W. J., Ball J., and Agnew P. (2006). A participatory approach to water management: Irrigation advisory committees in Southern Ontario. *Canadian Water Resources Journal,* 31(1), 13-24

31- Sullivan, C. (2002) Calculating a water poverty index. *World Development*, 30 (7), 1195-1211.

32- Taghipoor, M. Abbasi, A. Chizari, M. & Heydari, A. (2013). Gachsaran views of city farmers to form cooperatives, water users associations, Cooperative and Agriculture, 6, 1-22.

33- Taylor, P., Gabbrielli, E., & Holmberg, J. (2008). Economics in Sustainable Water Management, Retrieved from Cap-Net web site:www.cap-net.org.

34- Zahtabian, G.H.(2005). Causes of low efficiency in Varamin, presented to the Seventh Biennial of Iranian National Committee on Irrigation and Drainage, Retrievev from http://www.irncid.org/seminars/7.htm (In Persian).

35- Zarei Dastgerdi, Z., Mokhtari, A., & Shabanalifami, H. (2006). Participatory management of irrigation networks, a new approach in the transfer of agricultural water management. Proceedings of the First National Conference on Irrigation and Drainage Networks, Shahid Chamran University, 1753 - 1743 (In Persian).

36- Zareie, A., Irvani, H. Shabanalifami, H., & Mokhtari, A. (2007). Analysis of improving farmers' participation in the management of water supply network Jarghouyeh Hearst. *Journal of Agricultural Extension and Education.* 3(2), 45-55 (In Persian).

6

Effective Factors on Rural People's Non-Participation of Mahabad's Dam Catchment in Watershed Management Projects

*Soleiman Rasouliazar * and Saeid Fealy*

Abstract

The purpose of this descriptive-correlation study was to investigate effective factors on rural people's non-participation of Mahabad's dam catchment in Watershed Management Projects. The research instrument was structural questionnaire with close-ended questions, which its validity confirmed by panel of academic staff and reliability of questionnaire was confirmed. The target population of this study consisted of all householders who lived in Mahabad's dam catchment (N=2458) out of them, according to Cochran's formula 175 people were selected by using cluster sampling in a simple randomization method (n=175). The descriptive results showed lack of financial ability for participating in watershed management, lack of awareness about watershed management efficiency, the long-term rate of return on the investment attracted in watershed management were the main variables related to rural people's non participation in watershed management. By applying Factor Analysis Explanatory Technique, effective factors on rural people's non-participation in watershed management were reduced to five factors namely weakness of agricultural extension services, getting watershed management out of governmental control, no achieving success to implement another rural projects by government, and no considering local individuals or organization by government. These five factors expressed 84% of the total variance of the non-participation people on Mahabad's dam catchment in watershed management projects. Therefore points to these factors could solve the barriers of non-participation people on Mahabad's dam catchment in watershed management projects.

Keywords:
Non-Participation, Catchment, Watershed Management, Mahabad Township, Iran

Department of Agricultural Management, Mahabad Branch, Islamic Azad University, Mahabad, Iran.
** Corresponding author's email: rasouli88s@yahoo.com*

INTRODUCTION

Increasing demand for food in the world was the result of increasing population and changing consumption patterns. Therefore this problem caused to more attention on water resources and soils in agriculture for obtain sufficient food for population. On the other hand the lack of proper and efficient management of water and soil resources in the fields of agriculture, caused to increase the process of water resources and soil degradation. Besides these issues, the problems associated with slope, made up of rocks fragile areas, the flood and sediment dams, reservoirs and irrigation networks, caused to that authors and researchers were interesting to find ways the prevent of erosion in water and soil resources. This is beginning to take shape as a new science as watershed management. Watershed management is called to mechanical, biological and management practices in the watershed area in order to promote social and economic status of the residents and based on the sustainable use of resources takes place (Darghouth *et al.*, 2008).

Undoubtedly, watershed is concerned as one of the major activities has fundamental role to the management of land, water and vegetation resources, optimum utilization of resources and the preservation of fundamental capital (Sadeghi *et al.*, 2004). Also, watershed management can be properly managed life conditions (Ghanbari and Ghodosi, 2008). The watershed can be attributed to the utilization of these resources (water and soil) in order to increase agricultural production, and watershed management is essential for sustainable livelihood of rural people for increase their income (Yoganand and Gebremedhin, 2006), increase soil moisture levels (Shah, 2001), increased crop diversity (Renfro, 2004), erosion control (Kerr *et al.*, 2002), employment for rural people (Reddy *et al.*, 2004), increases the groundwater levels (Wani *et al.*, 2005), and reduce the migration of rural people to urban areas (Yoganand and Gebremedhin, 2006).

Watershed management plan were depending on the success of public participation in decision-making, implementation and maintenance of a project. Evaluation of projects shows that when people in different stages was participated, and conducted to the different needs of rural people, were succeed and continuity. People's participation in various projects such watershed management projects greatly has increases the successful projects rate (Movsaey, 2009). Yaganand and Gebremedhin (2006) in the same study showed that 60 percent of the watershed projects participants are convinced that stakeholder participation was essential to the success of watershed management project. Durham and Brown (1999) have reported the success of watershed management projects don't dependents to the government support or a particular structure, but that popular participation was very essential.

Participatory watershed management can be defined as a process that aims to create a sustainable system that is self-support is necessary (Wani *et al.*, 2005). Despite the importance of watershed management, and participation of local people in watershed planning, this question arises that why the villagers and residents of the watershed regions don't have participate in watershed management projects?

Movsaey (2009) investigated the causes of this subject in the watershed management plans of Fars province, and stated that economic factors) Non-economic watershed management plans, lack of funds needed to implement the projects, long-term returns and financial inability) were the main obstacles to the participation of local people in the watershed projects. Also Movsaey (2009) by use of multiple regression analysis showed that variables such as lack of confidence among beneficiaries of the project, lack of funds needed to implement the project, non-accordance projects with the needs of the rural poor and lack of proper training methods related to watershed management plans have main role in to the explaining participation stakeholders in watershed management plans. Hematzadeh and Khaliqy (2006) showed that 87.70 percent of stakeholders who don't participated in watershed projects have expressed: 39 percent had no knowledge about plans, 35 percent lack of capital and more than a quarter of them points to lack of awareness of the benefits of the project as their reasons for non-participation in watershed management plans. Bagaey *et al.* (2006) also points to the reasons for non-participation in watershed management projects in the rural: The difficulty in obtaining loans, the content isn't consistent with the needs of rural projects, financial inability, lack of respect and lack of trained manpower and skilled experts. Hosseinpour (1993) showed that revealed of attention to educational activities in the region, rural illiteracy and low literacy, lack of coordination among agencies providing services, the presence of troublesome rules and regulations on the use of credits and investments were the most important soil conservation and watershed management issues for villagers in the Hezar River in Iran. By these studies concluded that the economic, social, cultural, education-advocacy, policy making and formulation of poorly designed (in terms of purpose, content, implementation, etc.) were affective components on rural people non-participation in watershed management plans.

The present study has been done in the field of rural non-participation in the watershed management for investigate the factors affecting the non-participation of rural people in Mahabad dam watershed management projects. Mahabad dam in West Azerbaijan province has an important role on agricultural activities. This dam is known as one of the major arteries in the basin of West Azerbaijan province in Iran.

This area is composed of two sub home as name Kavtar and Bytas. Also Mahabad dam is covered 82 villages, 15,374 people (2,458 households) and 79,300 hectares. Annual precipitation of Mahabad dam laying around 1.8 million cubic meters of container per year and loss ratio of it is about 0.61 percent and at the critical situation in the country is located in the fifth dams (Talibpourasl and Khezri, 2010). Continue this process led to the irreparable damage and will cause damage to the local economy of rural people. Therefore in recent years some of the watershed projects like biological projects (transplant, grass lading, keeping pasture) and mechanical projects (construction sediment retention mortar, gabion dam) has been carried out in rural areas.

The Main purpose of the paper was to study effective factors on non-participation rural people of Mahabad's Dam catchment in watershed management projects in Iran. And also:

Identify Priority variables on non-participation of stakeholders in the watershed projects.

Describe the demographic characteristics of respondents.

Identification the effective factors on rural people's non-participation of Mahabad's dam catchment in watershed management projects.

MATERIALS AND METHODS

The methodology used in this research involved a combination of descriptive and quantitative research and included the use of correlation and descriptive analysis as data processing methods. The target population of this study consisted of all householders who lived in Mahabad's dam catchment (N=2458) out of which, according to Cochran's formula 175 people were selected using cluster sampling in a simple randomization method (n=175). Cronbach's Alpha coefficient was 0.84 which demonstrated that the questionnaire was highly reliable.

The questionnaire was an instrument to collect data. The instrument was divided to two sections. The first section focused on evaluating the affective factors on non-participation of respondents' in Dam catchment in watershed management projects (26 items). The second section was designed to gather data about of respondents' characteristics such as sex, age, educational level and etc. In both of these section five-point Likert-type scale was used to quantify responses which ranged: 1=very low, 2=low, 3=medium, 4=high, and 5=very high. The face validity was established by a panel of experts consisting of faculty members in natural resources and extension-education specialist in university and agricultural officers of Mahabad Township. The data were coded and analyzed by using the Statistical Package for the Social Science (SPSS 21) for windows. Descriptive statistics (frequencies, means, standard deviation, range, minimum, and maximum) were used to describe analyzed data. Also, explanatory factor analysis was employed to identify effective factors on rural people's non-participation of Mahabad's dam catchment in watershed management projects.

RESULTS AND DISCUSSION

Describe the demographic characteristics of respondents

The average age of respondents was 36 years, that the majority of them (n=85 or 48.60 %) ranged from 41-53 years old. Respondents' agricultural experience was 24 years, that the majority of them (n=65 or 37.10 %) ranged from 18-32 agricultural experience years. Also, finding showed that the average cultivated land by respondents was 7 ha. In addition, the average of respondents' education was 5 years; also 31.4 percent of them were illiterate (Table 1).

Priority of variables associated with non-participation of villagers in the watershed plan

Table (2) shows the mean, standard deviation, coefficient of variation and prioritizing the variables related to non-participation of stakeholders in the watershed plans. Results shows that economic variables such as "lack of financial ability to participate in the project watershed" (CV=0.220) was the main important economic variable in non-participation on the watershed plan and "uncertainty and lack of knowledge about the economic performance of the watershed" (CV=0.230) and "long-term outcome of watershed management plans" (CV=0.245) located in the 2th and 3th important priorities economic variables in non-participation on the watershed plan.

Also, results showed that "none according of watershed management projects with the requirements of farmers" (CV=0.475) and "ignore the social and cultural characteristics of rural areas in watershed projects" (CV=0.490) and "different stakeholder opinions together" (CV=0.526) located in the least important priorities economic variables in non-participation on the watershed plan (Table 2).

Results Identification of factors influencing villagers' Non-Participation

Explanatory Factor Analysis is a generic name for some multivariate statistical procedure whose main purpose is to summarize the data. This method of dealing with large number of variables within the correlation and finally, they are limited in the general category and explaining. Thus, in the study exploratory factor analysis approach was used to summarize the data. The variables were analyzed associated with non-

Table 1: Describe the demographic characteristics of respondents

Variables	Level of variable	Frequency	Percent	Mean	Standard Division	Min.	Max
	28-40	52	29.7	45.82	8.22	28	66
Age (year)	41-53	85	48.6				
	54-66	38	21.7				
	3-17	62	35.4				
Agricultural	18-32	65	37.1	23.65	12.40	3	46
experience	33-66	48	27.4				
(year)							
Land under	2-13	102	58.3	7.12	4.31	2	27
cultivated (ha)	14-25	63	36				
	26-36	10	5.7				
		55	31.4				
	Illiterate	36	20.6				
	Elementary	37	21.1	-	-	-	-
Education	Guidance school	30	17.1				
level	High school	10	5.7				
	Diploma	7	4				
	University degree						

Table 2: Priority variables on non-participation of stakeholders in the watershed projects

	Mean	SD.	CV	Rank
Lack of financial ability to participate in the project watershed	4.03	0.89	0.220	1
uncertainty and lack of knowledge about the economic performance of the Watershed	4.04	0.93	0.230	2
Long-term outcome of watershed management plans	4.04	0.98	0.245	3
Lack of funding for watershed projects	3.93	1.01	0.256	4
Lack of clearly participation of farmers in the watershed plan	3.92	1.04	0.265	5
Illiteracy and illiteracy farmers	3.70	1.00	0.270	6
The government ignored the ownership status of the beneficiaries of watershed management plans	3.75	1.06	0.282	7
Not specify the trustee to keep watershed plan implementation process	3.82	1.12	0.293	8
Neglecting the role of local leaders in informing watershed plan	3.86	1.14	0.295	9
The weak relationship among stakeholders, Experts and advocates of natural resources	3.29	1.00	0.303	10
Lack of educational methods for information resource management plans	3.01	1.03	0.342	11
Inadequate monitoring of the proper implementation of watershed management plans	3.56	1.24	0.348	12
Lack of incentives for property	3.07	1.09	0.355	13
Lack of cooperation between the governmental agencies in the watershed plan	3.36	1.23	0.366	14
Neglecting the role of rural councils in informing watershed plan by government	3.03	1.11	0.366	15
Lack of the trust from stakeholders to watershed management plans	3.44	1.32	0.383	16
Difficult borrowing conditions for participation in watershed projects	3.96	1.53	0.386	17
Failure of other projects at the village level administrative	3.15	1.22	0.387	18
Lack of investment by stakeholder in some operation that their future is unclear	3.37	1.31	0.388	19
Ignorance of the contents of watershed management plans	3.27	1.38	0.422	20
Disagreement elders and adults with watershed management plans	2.94	1.26	0.428	21
Lack of perspective and local knowledge of farmers in the watershed plan	2.88	1.33	0.461	22
Lack of social security in rural areas	2.58	1.19	0.461	23
Non according of watershed management plans with the requirements of farmers	2.90	1.38	0.475	24
Ignore the social and cultural characteristics of rural areas in watershed projects	2.53	1.24	0.490	25
Different stakeholder opinions together	2.79	1.47	0.526	26

participation in the rural watershed projects. In this study, the KMO value was 0.754 obtained, in that order, and higher than the 0.75 it indicates that the data (correlation existing among data) for analysis as well. Bartlett statistic is significant at the 99% level (χ=3542.65 and p=0.000) which shows that the correlation matrix based on the factor analysis are not equal to zero.

The number of extracted factors with specific amounts of each of them, percentage of variance and cumulative percentage variance of each factor are presented in Table 3.

The first factor referred to "economic inhibiters factors" with a principal component of (8.86), which is higher than other factors, explains 45.61% of the total variance (Table 4).

This factor was also identifying in other researches that affective on non-participation of stakeholders in projects (Baghaey *et al.,* 2006; Hematzadeh and Khaleghy, 2006; Mosaey, 2009). The second factor was named weakness of extension-educational activities. This factor according to the specific amount 3.93 could explain 15.15% of total variance (Table 4). This finding was according to Hematzadeh and Khaleghy, 2006; Mosaey, 2009; Baghaey *et al.,* 2006 findings.

Table 3: Extracted factors with Eigen values, percentage of variance and cumulative variance

Factors	Eigen values amount	Percentage of variance	Cumulative variance
1	8.860	45.617	45.617
2	3.939	15.151	60.767
3	2.453	9.436	70.204
4	1.945	7.482	77.686
5	1.614	6.209	83.894

Table 4: Factors and variables with loadings factor on non-participation people on watershed projects

Factor	Variables	Factor loading
Economic barriers	Lack of financial ability to participate in the project watershed	0.78
	Lack of funding for watershed projects	0.71
	Difficult borrowing conditions for participation in watershed projects	0.66
	Long-term outcome of watershed management plans	0.59
	Neglecting the role of rural councils in informing watershed plan by government	0.57
	uncertainty and lack of knowledge about the economic performance of the Watershed	0.59
Weakness of extension-educational activities methods	Lack of educational methods for information resource management plans	0.77
	The weak relationship among stakeholders, Experts and agents of natural resources	0.72
	Ignorance of the contents of watershed management plans	0.70
	Illiteracy and illiteracy farmers	0.68
	Lack of clearly participation of farmers in the watershed plan	0.58
Weakness of government on monitoring the projects	Inadequate monitoring of the proper implementation of watershed management plans	0.77
	Not specify the trustee to keep watershed plan implementation process	0.68
Lack of successful of government in implementation other rural plans	Failure of other projects at the village level administrative	0.79
	Lack of the trust from stakeholders to watershed management plans	0.65
	The government ignored the ownership status of the stakeholders of watershed management plans	0.60
Lees attention of government in use of local individual and local organizations	Neglecting the role of local leaders in informing watershed plan	0.59
	Disagreement elders and adults with watershed management plans	0.55

The third factor was named weakness of government monitoring on project. These factors according to the specific amount 2.45 could explain 9.45 % of total variance. The fourth factor was named Lack of successful of government in implementation other rural plans. This factor according to the specific amount 1.94 could explain 7.48% of total variance (Table 4).

The five factors were named "Lees government attention in use of local individual and local organizations in projects". This factor according to the specific amount 1.61 could explain 6.20% of total variance. Therefore people's local knowledge in the protection of natural resources and watershed management projects were playing an important role in the development and preservation of natural resources (Table 4).

Also, these results showed that the majority of the variables intercommunicated with the external factors in non-participation in watershed management projects. For example the difficulty of obtaining a loan.

CONCLUSION

Studies showed that community participation in watershed management plans.has an importance role. The main goal of this study aimed to identify affecting factors on the non-participation of stakeholders in Mahabad's Dam Catchment in Watershed Management Projects. Descriptive results showed that economic variables include the "lack of financial ability to participate in the project watershed" and "uncertainty and lack of knowledge about the economic performance of the watershed" and "long-term outcome of watershed management plans" located in the important priorities of inhibited economic variables in non-participation people on the watershed projects in Mahabad Township.

In these section authors suggested that government must be supported (by delivery loans) and encouragement rural people for doing watershed projects. Also extension agents must be conducted educational class for rural people for increasing their knowledge about benefits of watershed projects. Also extension agents should be considering extension visits on other areas that similar these watershed projects has been implemented.

The results of explanatory factor analysis showed that affective factors in non-participation of stakeholders in Mahabad dam watershed projects were classified in five factors. These factors were includes "inhabit economic factors, weakness of extension-educational activities methods, weakness of government on monitoring the projects, lack of successful of government in implementation other rural plans. These factors could explain 84 percent of total variances in affective inhabit factors on non-participation of stakeholders in Mahabad dam watershed projects. Also, finding shows that "lack of financial ability to participate in the project watershed, lack of educational methods for information resource management plans, inadequate monitoring of the proper implementation of watershed management plans, failure of other projects at the village level administrative and neglecting the role of local leaders in informing watershed plan were the main important variables in these factors.

Authors suggested delivery government funds like loans and financial assistance grants, and facilitate bureaucracy to getting loans could solve some economic barriers on non-participation rural people in watershed management projects.

Also one of the important factors effects on non-participation of rural people was educational barriers. Therefore authors suggested that extension agents and broadcasting center Mahabad (Radio and TV programs) could have significance role on increases of rural people about beneficiary of watershed programs. In these programs natural resources experts could delivery educational superstations for stakeholders.

Also natural resource organization and extension agents must invite local leaders and Islamic council in their policymaking and implementation of watershed projects. By invite rural people in projects, the knowledge and awareness of stakeholders about the dangers of soil erosion were increased.

But besides these guidelines government must be points to economic strategies such as facilitate in obtain loan, increase the duration of repayment, low income allocated to loans and attention to pay awards to people who activated in implementation watershed projects. Authors of this article will hope that the results of research could be used in the formulation and implementation of appropriate strategies for land utilization of water and soil by planers.

REFERENCES

1- Bagaey, M., Chizari, M., Fealy, S., & Merzaey, A. (2006). Investigation inhabits factors on non-participation of rural people in watershed projects (case study: Zarcheshmeh Honchan). Seminar of participatory development planning of soil and water. Semnan.Iran.

2- Cochran, W. G. (1977). Sampling techniques (3rd Ed.). New York: Wiley.

3- Duram, L. A. , & Brown, K. C. (1999). Assessing public participation in U.S watershed. *Planning Initiatives Society and Natural Resource*, 12, 455-460.

4- Darghouth, S., Christopher, W., Gambarelli, G., Styger, E. & Roux, G. (2008). Watershed Management Approaches, Policies, and Operations: Lessons for Scaling Up. Water resource board discussion page series. Paper Number 11.

5- Ghanbari, Y., & Ghodsi, J. (2008). Investigation the socio-economic effects of watershed activities on rural stakeholders in Tangkhoshk Samirom Township. *Journal Research of Esfahan University*, 29(1), 195-204.

6- Hematzadeh, Y., & Khaliqy, N. (2006). Investigation affective factors on non-participation of stakeholders in watershed projects (case study: Kachekn Golestan Province). *Journal of Agricultural Science and Natural Resources*, 13(4), 88-101.

7- Hossenpour, A. (1993). *Investigation necessary of participation and determining the training needs of rural rancher's farmers for contribute to soil conservation and watershed management in Hezar. In Amol Township*. Unpublished Thesis in Agricultural Extension and education, Faculty of agriculture, Tarbiyat Modares University.

8- Kerr, J., Pangare, G., & Pangare, V. L. (2002). Watershed development projects in India: An evaluation. Research Report 127, International Food Policy Research Institute.

9- Mosaey, A. (2009). Investigation affective factors on non-participation of stakeholders in watershed projects (case study: Fars Province). *Journal of Agricultural Extension and Economic*, 2(2), 65-85.

10- Reddy, V. R., Reddy, M. G., Galab, S., Soussan,

J. & Baginski, O. S. (2004). Participatory watershed development in India: Can it sustain rural livelihoods. *Development and Change*, 35 (2), 297-326.

11- Renfro, R. Z. (2004). The value of participation in development- relevance to soil and water conservation. Retrieved from http:// www. adbi.org/ files / the_value_of_participation_in_development_keynote_final.pdf

12- Sadeghy, S. H., Sharify, F., Frotan, A., & Rezaey, M. (2004). Quanitative appraisal of watershed projects. (Case study: catchment watershed of Keshar). *Journal of Research and Constriction in Natural Resources*, 65, 96-102.

13- Shah, A. (2001). Who benefits from participatory watershed development? Lessons from Gujarat, India. London: International institute for Economic Development (IIED), Gatekeeper series no.97.

14- Talibpourasal, D., & Khezri, S. (2010). Investigation the land application and Sediment in the southern of Mahabad River. *Journal of Rangeland and Watershed,* 63 (3), 314-358.

15- Wani, S. P., Singh, H. P., Sreedevi, T. K., Pathak, P., Rego, T. J., Shiferaw, B., & Iyer, S. R. (2005). Farmer-Participatory Integrated watershed Management: Adarsha Watershed, Kothapally India.

16-Yganand, B., & Gebremedhin, T. (2006). Participatory watershed management for sustainable rural livelihoods in India. Selected working paper prepared for presentation at the Southern Agricultural Economics Association Annual Meetings Orlando.

7

Analysis of Factors Affecting Adoption and Application of Sprinkler Irrigation by Farmers in Famenin County, Iran

Hassan Afrakhteh [1]***, Maryam Armand*** [2*] ***and Fatemeh Askari Bozayeh*** [2]

Abstract

Due to its location in the dry belt, the vast country of Iran has always been faced with the problem of water shortages. In such a climate one should take measures to increase the productivity and improve the efficiency of irrigation through expansion of water-conserving technology in the agricultural sector, as this is the largest and most important consumer of water. Therefore this study aimed at investigating factors affecting the adoption and application of sprinkler irrigation technology by farmers in the county of Famenin, Iran. Survey research was the dominant approach. The statistical population consisted of farmers in the Famenin County, and, using Cochran's formula, 280 of them were selected through multi-stage stratified random-sampling equally among adopters and non-adopters. In order to assess the validity of the questionnaire the opinions of experts were used and to measure the reliability of the questionnaire Cronbach's alpha statistic was used (82%). The results showed the adoption of sprinkler irrigation systems to be influenced by environmental factors such as the area under cultivation, access to water, water quality, and non-environmental factors such as the workforce number in the family, employment diversity, and participation in extension education and courses on agricultural water management. Ultimately the logit model is estimated in this paper. Adopters have also expressed reasons such as failure to adequately train farmers for maintenance after system installation, scientific and practical justifications, lack of availability of efficient repairs, inappropriate design and implementation by companies, low-quality components and fittings, clogging of sprinklers due to salty water, non-compliance with environmental conditions, difficulty using machines, the large area occupied by these systems, transportation, parts shortage, lack of security and the possibility of the theft of parts and fittings, and communal ownership of water resources as their dissatisfaction factors.

Keywords:
Adoption, Sprinkler Irrigation, Beneficiaries, Famenin County

[1] *Professor in the Department of Rural Planning, Kharazmi University, Tehran, Iran.*
[2] *PhD Student of Rural Planning, Kharazmi University, Tehran, Iran.*
* *Corresponding author's email: maryam.armand@yahoo.com*

INTRODUCTION

The agricultural sector is the largest consumer of water worldwide, constituting approximately 70% of total water exploitation from surface and groundwater resources (Frenken and Kiersch, 2011). Thus, for more than a decade, the main objective of farm management especially in countries with limited land and water resources is centred on increasing food production using less irrigation (FAO, 2002). In Iran, due to being in the arid to semi-arid areas of the Earth belt, water scarcity has always been one of the main problems towards development and economic development. The average annual precipitation in Iran is about 225 mm, which is much lower than the global average rainfall (860 mm) (Farzande Vahy, 2002). This volume of precipitation disperses in an unsuitable way, and rainfall often occurs during the non-agricultural seasons (Hasheminia, 2004). On the other hand, 59% of land needs to be irrigated. This figure, compared with the worldwide average, i.e., 17%, is very high (Information Center of World Bank, 2006); also in addition, it is estimated that fertile, suitable land for agriculture about 35 to 50 million hectares covers roughly 20 to 30% of the country's total area (Ghasemzadeh, 2001). In these conditions, traditional methods of irrigation with a low efficiency of only 32% on farms has interrupted the balance between groundwater aquifer exploitation (55 billion m3) and nutrition (46 billion m3), and the underground water balance sheet is negative (UL Hassan *et al.,* 2007). In addition, the rate of water loss in transmission channels exceeds 85% (Keshavarz and Sadeghzadeh, 2000). Therefore, investment in pressurized irrigation systems, designed to achieve up to 70% efficiency in sprinkler irrigation and 95% efficiency of trickle irrigation, encourages and trains farmers and provides the necessary inputs and credits to help the country to produce more food through increasing the irrigated cultivation area and water conservation (Abdolmaleki and Chizari, 2009).

In recent years, due to low rainfall, Hamedan Province has experienced groundwater and surface water resource shortages. Aquifers a thick layer of rock and soil at a depth of 10 m that can store water in the plains of the Province have been severely deficient in groundwater resources. Famenin, Kabudarahang and Qahavand plains in the northern parts of the Hamedan Province are among the plains that are facing sharply declining groundwater levels. The Famenin plain, which starts from south of Razan County and extends to Qahavand to embrace Famenin County has 1341 deep wells, 904 semi-deep wells (excluded from unauthorized wells), 106 aqueducts and 95 springs that provide the required water for manufacturing and service activities. However, following uncontrolled exploitation and power plant activity with 200 wells and high discharge, the Regional Water Organization of Hamedan declared this plain as a 'Banned plain' in 1993 (Water Organization of Hamedan Province, 2011). However, unfortunately despite this valuable declaration, both authorized and unauthorized digging activities have largely continued. In this regard, one of the most important and effective ways to tackle the water crisis and water management in the county is to increase irrigation efficiency through the application of pressurized irrigation systems.

Wang (2008), in a study aimed at the economic analysis of areas accepting water conservation measures in northern China, concluded that when the availability of water is low, farmers tend to be more likely to use water-saving techniques, and access to water has a major impact on the probability and intensity of this adoption. In addition, government actions such as agricultural extension-education, demonstration farms, and the provision of loans and subsidies, leads to a higher rate of adoption. In addition, farmers with more arable land are pioneers in this field. Michailidis *et al.* (2011) examined the past, present and anticipated behaviours of farmers in relation to sprinkler irrigation technologies. The results show that yield levels and product quality has a decisive impact on the adoption of irrigation types, and the paper ultimately recommended that if decision-makers give more attention to the attitudes and perception of farmers and improvement of irrigation management rather than the type of technology can achieve more favourable results towards improving water productivity. Nguyen (2008) concludes that factors such as income, farm size, the education level of farmers and extension services influence the satisfaction of farmers in relation to the use of pressurized irrigation technology. Adeoti (2009) reported that access to employment and higher attendance in extension-education courses are among the factors that increase the probability of adoption, and, in areas with a higher number of adopters, there exist signs of poverty reduction. Kohansal *et al.* (2009) examined the factors affecting the adoption of sprinkler irrigation in Khorasan Razavi, concluding that variables such as farm size, education level, agriculture being the main form of employment, land slope, soil heterogeneity and access to credits and facilities have a significant and positive impact on the adoption of sprinkler irrigation. Yigezu *et al.* (2013), in a study considering changes in irrigation technology and its impact on the effectiveness of resource use in Syria, stated that farmers' education and their familiarization with modern irrigation methods increases their efficiency compared to traditional methods, and protects water resources through the wider adoption of these systems. Dinar and Yaron (1990), in a study investigating the impact of the quality and scarcity of inputs on the adoption of modern irrigation technology, demonstrated that this type of technology is being adopted faster in regions with poor quality soil and high evaporation, and areas with sensitive root plants and limited water resources. The consequences stated include the impact of management, private capital, and farm size on the adoption rate; finally the authors acknowledge the price of water to be an effective tool towards increasing the adoption rate. According to Jahannama (2001) personal and social characteristics such as age, work experience, education level, knowledge and financial options are effective in the adoption of pressurized irrigation systems. He also enumerates rigid loan terms and conditions, poor performance of design and implementation firms, and low quality of receiving equipment as reasons for the rejection of these systems. He notes also that the major reason for the dissatisfaction of adopters is the mismatch of irrigation systems with farm and climate conditions. Pezeshkirad *et al.* (2011), examined the factors affecting the discontinuation of the use of sprinkler irrigation systems by Mahabad County farmers, and using discriminate analysis concluded that five variables including rivers as water resources, the rate of economic implications of systems, level of educational and extension support, maintenance services and age, placed 99% of farmers into two groups: either those continuing, or those discontinuing, sprinkler irrigation systems. Nowruzi and Chizari (2006), using stepwise multivariate regression, showed variables including irrigated wheat yield, technical expertise in agronomic water management, use of communication channels and the age of farmers to determine approximately 70.7% of changes in the attitudes of farmers to the development of sprinkler irrigation. Shrestha and Gopalakrishnan (1993) found increasing yield and income, water and labour-saving to be the most important factors contributing to the rapid adoption of trickle irrigation in Hawaii State in the US.

In fact, several factors affect behavioural change and the acceptance of new ideas by people. According to Lionberger and Gwin (1995), these agents vary from one person to another and from community to community. Among these factors, they state the role of the situations that the people in question are experiencing whether real or imagined) types of help that they can receive from professionals and external agencies, resources at their disposal, instructional strategies that have been exposed to the treatment and behaviour of the social system, and values that refer people to change. Swanson (1991) argues that the factors associated with the adoption of new innovations and technologies include personality characteristics as well as the characteristics of the environment in which people are working.

According to the above-mentioned details, this paper is going to demonstrate which factors are effective for the adoption and application of pressurized irrigation systems by farmers in Famenin County.

The following points show the objectives of the research:

Identify the personal and professional characteristics of adopters and non-adopters of the sprinkler irrigation system.

Identify the effects of non-environmental factors age, education, technical knowledge, the labour force number within a family, agriculture as the main job, participation in the education and extension courses and environmental factors soil type, land slope, area under cultivation, water access, and water quality on the adoption of sprinkler irrigation systems and estimation of logit model.

Identify factors contributing to dissatisfaction of adopters.

MATERIALS AND METHODS

This study from the point of view of nature ,goal and evaluation method is a quantitative-qualitative ,applied and descriptive-correlative study respectively. The statistical population consisted of 280 farmers in the Famenin County, were selected using the Cochran formula and through multi-stage, stratified sampling from both groups of adopters and non-adopters equally with each group comprising 140 farmers. The required data were collected using questionnaires through structured interviews in summer 2014.

The questionnaire was designed in two parts: the first part included the personal and professional characteristics of farmers and agricultural factors associated with the acceptance and use of sprinkler irrigation systems. The second part of the questionnaire included some open-ended questions to determine the causes of dissatisfaction with the operation of sprinkler irrigation systems: the respondents in this section were those only farmers who accepted sprinkler irrigation systems.

In order to measure the validity of the questionnaire the comments and recommendations of university professors and experts in the Famenin Soil – Water Deputy of Agricultural Jihad Ministry were sought, and modifications were made to the questionnaire accordingly. To determine the reliability of the survey instrument, the questionnaires were distributed to and completed by 30 farmers in both groups in the sample villages. Then, using SPSS software, a Cronbach's alpha coefficient of 82% was obtained, which represents the acceptable reliability of the research tool. Data analysis includes descriptive and inferential statistics. Descriptive statistics included frequency percentage, cumulative percentage, standard deviation, and mean average. In this study, in order to assess the technical knowledge of farmers in agricultural water management practices, 18 items were used, and correct answers and incorrect answers were assigned the numbers 1 and 0, respectively. Therefore, the total scores of farmers' technical knowledge were between 0 and 18. Then, to describe the variables qualitatively the standard error of the mean was calculated (Sedighi and Mohammadzadeh, 2003). Scores obtained in this way were divided into four levels:

A = Poor: A < Mean – SD

B = Average: Mean – SD < B < Mean

C = Good: Mean < C < Mean + SD

D = Very good: Mean + SD < D

Inferential statistics including the Chi–squared test, Cramer's and Phi correlation coefficients were used to investigate the relationship between variables and logistic regression analysis was performed. The condition for multiple regression usage is that the dependent variable should be quantitative. The dependent variable was nominal, and so was dichotomized with a value of 1 if the farmer was an adopter of the sprinkler irrigation system and 0 if he was not an adopter. In this case logistic regression should be used instead of multiple regressions. In the logistic regression the concept of odds ratio the ratio of presence or absence a property $(P_i /1-P_i)$ was used: the logarithm of the Odds ratio is calculated according to equation 1, and is known as the logit model (Momeni, 2007).

$$\text{Ln } (P_i /1-P_i) = \beta_0+\beta_i x_i \quad (1)$$

To investigate the factors with the greatest impact on adoption of sprinkler irrigation system, backward conditional logistic regression was used. In the qualitative part of the research, in order to determine the causes of dissatisfaction of adopters, first open-ended questions were asked, the results analysed, and then tabulated. The units of analyses include words, phrases or sentences hidden within the text, with the researcher intending to fit them into a value (Holsti, 1994).

Research Study Area

The studied area was one of the nine counties of Hamedan Province, which is located in the northeast of the province. This county is limited to Razan in the North, Hamedan in the South, Markazi Province in the East, and Kabudarahang in the West (Figure 1). Famenin, with an area of 1360 km^2, comprises 6-7% of the total area of Hamedan Province (Famenin's Census Bureau, 2011), and according to Census if Statistical Centre on Iran in 2011, the population of the county is estimated at 42,485 (Statistical Centre of Iran, 2011), of which 6,000 are engaged in agricultural activities. Of a total 88,200 ha of arable land in Famenin, with 73% dry farming and 23% irrigated farming, only 6,500 ha i.e., 21% of the land in this county is covered by a pressurized irrigation system, which includes 268 adopters (Agricultural Jihad of Hamedan Province, 2012).

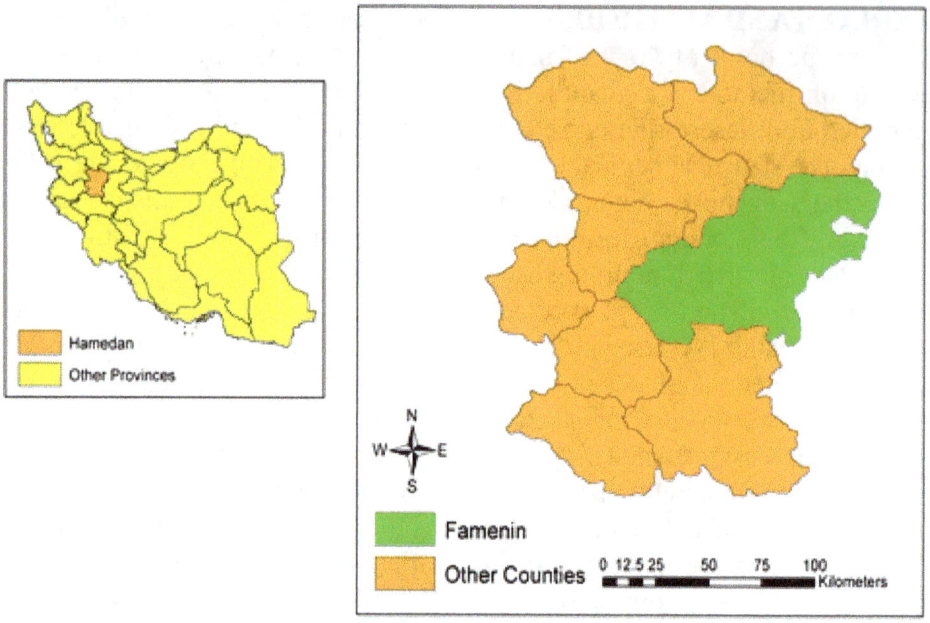

Figure 1: Site of study

RESULTS

Personal and professional characteristics of the respondents

Farmers' age is variable between 25 (minimum age) and 80 years (maximum age). Therefore, the study population in terms of age covers a wide range of young to old people. The average age is 50.54 (SD=11.34). Findings indicated a low level of education among farmers, so that considering a variety of education ratings (Illiterate, primary, secondary school, Diploma and Higher education); approximately 2.17% achieved a maximum of elementary level. More than half of the farmers (76%) stated agriculture to be their main business and main source of income without any side or second jobs. The average area under cultivation is 11.58 ha and 87% of the sites have no product diversity. The major crops are alfalfa, barley, wheat, and beets, respectively. In order to assess the technical knowledge of farmers regarding agricultural water management practices, in this study 18 items (Table 1) were used, with a value of 1 for correct answers and a value of 0 for incorrect or no answers.

As Table 2 shows, the technical knowledge of

Table 1: Items related to the evaluation of technical knowledge of agricultural water management.

Rank	Items	M	SD
1	What should be done to reduce water waste in traditional brooks and streams?	0.93	0.246
2	What positive impact does land smoothing have in a region with less water?	0.89	0.309
3	What is the relationship between ploughing the field and weeds with soil moisture?	0.85	0.356
4	What is the effect of the density of plants and sown seeds in low-water seasons on agricultural products?	0.75	0.434
5	In low-water seasons, what kind of seeds should be used depending on the type of the agricultural product?	0.72	0.447
6	In low-water seasons, how should fertilizers be used depending on the type of crop?	0.66	0.473
7	Which irrigation method is widely used on dry and low-water land?	0.61	0.488
8	What is the advantage of early-maturing cultivars in low-water conditions?	0.57	0.496
9	When is the most critical time of water requirement depending on the crop type?	0.51	0.501
10	During low-water seasons, what kinds of fertilizers should be used to increase plant resistance?	0.44	0.497
11	What is the difference between irrigation in the early- and late-maturing cultivars?	0.32	0.468
12	What is the relationship between burning of agricultural residues after harvesting and soil moisture?	0.29	0.457
13	What is the effect of grazing the previous season's plant residues in low-water seasons on soil moisture?	0.23	0.423
14	In drought conditions, what are the benefits of stubble in farms after harvesting?	0.15	0.361
15	What is the role of the supplemental irrigation of dry farming for plant growth?	0.07	0.252
16	What kind of plough should be used in low-water seasons?	0.04	0.203
17	In low-water seasons, which kind of spraying should be used against weeds?	0.04	0.203
18	In weed control, to reduce soil stirring and the evaporation of moisture, what kind of cultivators should be used?	0.03	0.177

Table 2: Frequency distribution of farmers according to their technical knowledge of water management.

Technical knowledge level	Frequency	%	Cumulative percentage
Poor	54	19.3	19.3
Average	122	43.6	62.9
Good	65	23.2	86.1
High	39	13.9	100
Total	280	100	-

Minimum: 3, Maximum: 16, Mean: 7.13, SD: 2.91

more than half of the farmers is good and moderate and only 19.3% of farmers have poor technical knowledge.

Relationship between variables

Table 3 shows the results concerning the relationship between the dependent variable adoption of the sprinkler irrigation system and independent variables. According to Table 3, two variables age and education level among environmental factors have no effect on the adoption of sprinkler irrigation system by farmers, and three variables of agriculture being the main job, education participation and extension courses and technical knowledge of adopters on water management have significant and positive effects on the adoption of the sprinkler irrigation system. The family labour force number has a negative and significant impact at the 0.01 (1%) significance level, and among the environmental factors two variables soil type and land slope have been ineffective. The variable area under cultivation, at the 0.01 (1%) significance level, has a significant and positive effect. Two variables access to water and irrigation water quality have a significant negative effect on adoption.

logit model

Firstly, of a total of 11 variables, three of these the number of family members in the labour force, area under cultivation, and technical knowledge which are defined based on the interval scale, and four variables agriculture as the main job, participation in extension-education courses, access to water and water quality based on Table 4 are defined as dummy, and had a significant relationship with the dependent variable adoption of sprinkler irrigation system were entered into the logistic regression equation.

Based on the results (Table 5), and according to significance level by the Wald test, three-step regression analysis removed the independent variables with less impact at each stage, so that in the second step the variable of access to water, and in the third step, the variable of participation in extension-education courses, were removed from the regression equation. The indicator of the Chi-squared test in the third step

Table 3: The relationship between the research-dependent variable with independent variables.

	Independent variable	Dependent variable	χ^2	p value	V	p value	Phi	p value
Non environmental factors	Age	Adoption	2.133	0.545	-	-	-	-
	Education	Adoption	7.503	0.112	-	-	-	-
	Agriculture as the main job	Adoption	27.983**	0.000	-	-	0.316	0.000
	The number of labour force	Adoption	15.735**	0.001	-0.237	0.001	-	-
	Technical knowledge	Adoption	25.699**	0.000	0.303	0.000	-	-
	Participating in education & extension courses	Adoption	5.758*	0.012	-	-	0.143	0.016
Environmental factors	Soil type	Adoption	0.862	0.353	-	-	-	-
	Land slope	Adoption	1.400	0.237	-	-	-	-
	area under cultivation	Adoption	53.308**	0.000	0.428	0.000	-	-
	Access to water	Adoption	6.303*	0.012	-	-	-0.150	0.012
	Water quality	Adoption	48.557**	0.000	-	-	-0.416	0.000

Minimum: 3, Maximum: 16, Mean: 7.13, SD: 2.91

Table 4: Dummy variables affecting the adoption of sprinkler irrigation.

Variable	Value	
	1	0
Agriculture as the main job	Main job	Secondary job
Participation in education courses	Participating	Not participating
Access to water	High	Low
Water quality	Saline	Non-saline

Table 5: Regression

SteplevelFrequency	Nagelkerke R Square	Chi square	p value
1	0.618	156.994	0.000
2	0.625	155.932	0.000
3	0.632	154.967	0.000

is equal to 154.967, which is significant at the 0.01(1%) significance level and is indicative of the overall significance of the model. Nagelkerke R Square, which is equal to R^2 in linear regression, is 0.632 in the third step, indicating that 63.2% of the variability of the dependent variable (adoption) is presented by variables of labour force numbers, area under cultivation, technical knowledge, agriculture as the main job and water quality.

Variable coefficients, the Wald test and significance levels are shown in Table 6. Based on the B-values in the table and constant value in the last step of the regression analysis equal to 2.998, the regression equation is obtained as follows:

$ln\ (p/1-p)=2.998-0.927x_1+0.870x_2+2.077x_3+1.578x_4-2.473x_5$

Dissatisfaction factors among adopters

One way to encourage the increased participation of farmers is to attain a high level of satisfaction in current sprinkler irrigation users. Of the total 140 surveyed, 75% of farmers (105 in total) used a classical fixed underground system

Table 6: Independent variables with the most significant effects on the dependent variable (second step).

Independent variable	B	SD	Wald	df	p value
Labour force numbers (X_1)	-0.927	0.234	15.638	1	0.000
Technical knowledge (X_2)	0.870	0.202	18.450	1	0.000
Agriculture as the main job (X_3)	2.077	0.402	26.741	1	0.000
Area under cultivation (X_4)	1.578	0.337	21.949	1	0.000
Water quality (X_5)	-2.473	0.445	30.921	1	0.000

Table 7: Frequency distribution of reasons for adopter's dissatisfaction.

Responses	Priorities	Frequency	%
Failure to adequately train farmers as to maintenance after system installation and their scientific and practical justification	1	128	91.43
Non – availability of efficient repairs	2	69	49.28
Inappropriate design and implementation by companies	3	42	30
low – quality of components and fittings	4	34	24.28
Clogging of sprinklers due to salty water	5	32	22.86
Non-compliance with environmental conditions	6	24	17.14
Difficulty using machines& large area occupied by the systems	7	20	14.28
System transportation	8	15	10.71
Parts shortage	9	12	8.57
Lack of security and the possibility of the theft of parts and fittings	10	8	5.71
Communal ownership of water resources	11	6	4.28

(solid set) and the remaining 25% used either centre-pivot systems (20 farmers) or wheel move systems (15 farmers). The highest rate of satisfaction was for the classical fixed underground system due to its security and impossibility of the theft of its parts. A total of 85% of water sources are obtained from wells, followed by springs, aqueducts and rivers.

In order to investigate dissatisfaction factors among adopters, content analysis of open-ended questions is performed. According to Table 7 these include: failure to adequately train farmers as to maintenance after system installation and their scientific and practical justification, non-availability of efficient repairs, inappropriate design and implementation by companies for example system mismatch with the farms' size low quality of components and fittings, clogging of sprinklers due to salty water, non-compliance with environmental conditions for example installation of Wheel move systems in windy areas difficulty using machines and large areas occupied by the systems, system transportation, parts shortages, lack of security and the possibility of the theft of parts and fittings, and finally communal ownership of water resources had the greatest impact on the dissatisfaction of farmers.

DISCUSSION AND CONCLUSIONS

According to the results obtained among non-environmental factors, two variables of age and education had no effect on the adoption of the sprinkler irrigation system by farmers. The variable of the number of family members in the labour force had a significant and negative impact on adoption at the 0.01 (1%) significance level, which is due to the use of family labour force for irrigation in particular. The obtained results are in consistent with Bakhshudeh (2008) and Kohansal (2009). Agriculture being the main job of a participant had a significant and positive impact on adoption, so that this leads to their effort to improve their farming level which is in consistent with Shahpasand (1998) and Bakhshudeh (2008) studies as well. Participation in education and extension courses had a positive and significant impact on adoption at the 0.05(5%) significance level, which is consistent with studies performed by Adeoti (2009), Nguyen (2008), Pezeshkirad *et al.* (2011), and Yigezu *et al.* (2013). Among other variables that had a positive and significant influence on the adoption of the sprinkler irrigation system is technical knowledge of farmers regarding agricultural water management, and the results obtained in this study confirm those by Amirkhani *et al.* (2010) and Nowruzi and Chizari (2006). Among the environmental factors, soil type had no effect on adoption, despite, according to soil tests, there being cases where the high percentage of clay (over 48%) ruled out the possibility of using sprinkler irrigation. The variable of land slope appears to have no significant impact on adoption, because Famenin is located in the plains. The variable of the area under cultivation had a significant and positive impact on adoption at the 0.01(1%) significance level, mostly due to more efficient sprinkler irrigation systems in medium and large farms and the use of equipment and methods that lead to savings in the labour force and increased efficiency. These findings confirm research by Albrecht and Ladewing (1999) and Dinar and Yaron (1990). Two variables of access to water and irrigation water quality had a significant and negative impact on adoption. It can be said that farmers who are faced with high water restrictions are more likely than other farmers to accept this method. On water quality we can say that one of the serious problems in the surveyed region towards adoption of the sprinkler irrigation system is the salinity of water, which, because of excessive exploitation of groundwater and digging of unauthorized wells, aggravates the seriousness of the situation. The results are consistent with Karami and Rezaei Moghdam (2002), and Wang (2008).

The results of the logistic regression show that 63.2% of the variability of the dependent variable (adoption) is explained and determined by the variable of labour force number, area under cultivation, technical knowledge, agriculture as the main job, and water quality.

By considering the dissatisfaction of the adopter using content analyses of open-ended questions, the results obtained show that failure to adequately train farmers in maintenance after system installation, their scientific and practical justifications, non-availability of efficient repairs, inappropriate design and implementation by companies for example system mismatch with farm size low quality of components and fittings, clogging of sprinklers due to salty water, non-compliance with environmental conditions for example installation of wheel move systems in windy areas difficulty using machines and large area occupied by the systems, parts shortages, transportation, lack of security and the possibility of stealing parts and fittings, and communal ownership of water resources had the greatest impact on the dissatisfaction of farmers. The results support the findings obtained by Amiri Ardakani and Zamani (2003), Azami *et al.* (2011), Dechmi *et al.* (2003), Karami and Rezaei Moghadam (2002), Jahannama (2001), Jalali and Karami (2006) and Lahannama (2002).

In this regard, in order to achieve the sustainability, development and expansion of the sprinkler irrigation system, the expectations and needs of its adopters should be considered.

In the following some recommendations are offered:

In order to solve farmers' problems, it is suggested that in addition to design and installation companies, other organizations should be established to provide services and training courses to farmers and their scientific justification after installation, and, for this, funding should be allocated.

Considering government grants in recent years to implement sprinkler irrigation systems and a greater tendency of large farm owners to do so, it is suggested that studies on the implementation of the system's compatibility with small farms be researched.

Given that inappropriate design and implementation is one of the dissatisfaction factors of adopters, the companies that install and run these systems should give more attention to engineering principles and make an arrangement to compensate farmers' losses in the case of inappropriate and/or poor design and implementation.

In order to control the salinity of water and excessive groundwater exploitation by farmers, it is necessary that the Department of Water Affairs installs volumetric flow meters as soon as possible.

Considering groundwater dropping in the surveyed area and high water consumption crops like alfalfa, barley and wheat it is recommended to change farming patterns and move toward crops with lower water consumption, such as canola.

REFERENCES

1- Abdolmaleki, M., & Chizari, M. (2009). The influences of social – economical characteristics on the attitudes and information of farmers to adopt and apply the pressurized irrigation system in the Lorestan Province, *Journal of Modern Agricultural Science,* 5 (15), 77-87. (In Persian).

2- Adeoti, A. I. (2009). Factors influencing irrigation technology adoption and its impact on household poverty in Ghana, *Journal of Agriculture and Rural Development in the Tropics and Subtropics*, 109(1), 51-63.

3- Agricultural Jihad of Hamedan Province. (2012). Annual Performance Report. Deputy of Water and Soil. (In Persian).

4- Albrecht, D., & Ladewing, H. (1999). Adoption of irrigation technology, *Journal of Extension*, 34 (2), 26-41.

5- Amiri Ardakani, M., & Zamani, Z. (2003). Difficulties and obstacles to utilize pressurized irrigation systems in Kohgiluyeh and Boyer-Ahmad Province, *Journal of Soil and Water Science*, 17(2), 230-245. (In Persian).

6- Amirkhani, S., Chizari, M., & Hosseini, M. (2010). Education & extension factors affecting on the transfer and increase technical knowledge of wheat growers in the County of Varamin, *Research Journal of Agricultural Education Management*, 15, 57-68 (In Persian).

7- Azami, A., Zarafshni, K., Dehghani, H., Sanich, V., & Gorji, A. (2011). Analysis of farmers' satisfaction with pressurized irrigation system in Kermanshah Province, *The Journal of Soil and Water*, 4, 845-853 (In Persian).

8- Bakhshudeh, M. (2008). Factors affecting on the development and adoption of sprinkler irrigation system (case study in Isfahan Province), Iranian *Journal of Agricultural Economics and Development /Research,* 39(1), 21-30 (In Persian).

9- Dechmi, F., Playan, E., Cavero, J., Faci, J. M., & Martinez-Cob, A. (2003). Wind effects on solid set sprinkler irrigation depth and yield of maze (Zea mays), *Irrigation Science*, 22(26), 142-148.

10- Dinar, A., & Yaron, D. (1990). Influence of quality and scarcity of inputs on the adoption of modern irrigation technologies, *Western Journal of Agricultural Economics,* 15(2), 224 –233.

11- Duc, N. M. (2008). Farmers satisfaction with aquaculture–A logestic model in vietnam, *Ecological Economics*, 68(1), 525–531.

12- Famenin's Census Bureau. (2011). General Information County, Retrieved from: http:// www. sabteahval-.ir/fa/showpages.aspx? mid = MjA4&spi =NTM= (In Persian).

13- FAO. (2002). International workshop on irrigation management, FAO-ICID. Proceedings Retrieved from: http: // www. fao. Org/ ag/ agl/aglw/ ias/> 24th July, Monterall.

14- Farzande Vahy, J. (2002). *Factors influencing on the development of pressurized irrigation system in Kermanshah Province*, Unpublished thesis, Department of agricultural education & extension, Tarbiat Modares University. (In Persian).

15- Frenken, K., & Kiersch, B. (2011). Monitoring agricultural water use at country level: Experiences of a pilot project in Benin and Ethiopia, FAO Land and Water Discussion Paper.

16- Ghasemzadeh, F. (2001). Sustainable development of agriculture by creation modern irrigation methods, *Abrar Eghtesadi*, 24: 12. (In Persian).

17- Hasheminia, S. M. (2004). *Water Management in Agriculture,* Mashhad: Ferdowsi University Press. (In Persian).

18- Holsti, L. R. (1994). *Content analysis of Humanities and social sciences*, translated by Salarzadeh Amini, N., Tehran: Allameh Tabatabai University Press.

19- Information Centre of World Bank (2006). Investments in Agricultural Water, Retrieved from: http://siteresources.worldbank.org/EXTAGISOU/Resources/Module9_Web.pdf

20- Jahannama, F. (2001). Social and economic factors affecting the adoption of sprinkler irrigation system, *Journal of Agriculture Economics and Development*, 36: 237-258. (In Persian).

21- Jalali, M., & Karami, A. (2006). Discontinuity of sprinkler irrigation technology: a case study of farmers in the Billo Village, Marivan County, *Journal of Village & Development,* 35: 131-154. (In Persian).

22- Karami, A., & Rezaei Moghadam, K. (2002). The use of sprinkler irrigation: issues and problems, *Journal of Agricultural Economics and Development,* 10 (37), 221-245. (In Persian).

23- Keshavarz, A., & Sadeghzadeh, k. (2000). *Recommendations to improving water use efficiency in the farm lands of the Iran*, Tehran, Technical Publications of Extension Deputy of TAT Organization, 1-15. (In Persian).

24- Kohansal, M. R., Ghorbani, M., & Rafiee, H. (2009). Environmental and non – environmental factors affecting on the adoption of sprinkler irrigation system, (case study: Khorasan Razavi Province), *Journal of Agricultural Economics and Development,* 17 (65), 97-112. (In Persian).

25- Lahannama, F. (2002). Socio-economic factors related to acceptance of pressurized irrigation systems (A case study in Tehran), *Journal of Agricultural Economics and Development*, 4(36), 237-260.

26- Lionberger, H. F., & Gwin, P. H. (1995). *Technology transfer, from researchers to users* (M. Chizari, Trans), Tehran: Tarbiat Modares University Press. (In Persian).

27- Michailidis, A., Koutsouris, A. & Nastis, S. (2011). Adoption of sustainable irrigation practices in water scarce areas, *Bulgarian Journal of Agricultural Science,* 17(5), 579-591.

28- Momeni, M. (2007). *Analysis of statistical data using SPSS.* Tehran, Ketab – e – Now Publication.

29- Nowruzi, A. V., & Chizari, M. (2006). Cultural and social structures affecting on the attitudes of wheat growers in Nahavand County concerning development of sprinkler irrigation, *Journal of Iran Agricultural Education & Extension*, 2, 59-71. (In Persian).

30- Pezeshkirad, Gh., Feli, S., Rasuli Azar, S., & Yaghubi, A. (2011). Factors affecting on non – adoption of sprinkler irrigation system by farmers in Mahabad County, in the West Azarbaijan Province, *Journal of Agricultural Economic and Development,* 19(73), 1-20.

31- Sedighi, H., & Mohammadzadeh, G. (2003). Extension professional staff, attitude toward participatory approach of extension activities and rural development, *Journal of International Agricultural and Extension Education*, 9 (2), 7- 17.

32- Shahpasand, M. (1998). *Factors influencing on the adoption of the pressurized irrigation system in Tehran Province,* Unpublished Thesis, Department of agricultural education & extension, faculty of agriculture, Tehran University. (In Persian).

33- Shrestha, R., & Gopalakrishnan, Ch. (1993). Adoption and diffusion of drip irrigation technology: An econometric analysis, *Economic Development and Cultural Change*, 41(2), 407 – 418.

34- Statistical Centre of Iran. (2011). General census of population and housing. Retrieved from: http:// www.amar.org.ir/Default.aspx?tabid=1160. (In Persian).

35- Swanson, B. E. (1991).*Agricultural extension (A reference manual)* (E. Shahbazi & A. Hajaran, Trans.), Tehran: Agricultural Extension Organization Press. (In Persian).

36- UL Hassan, M., Qureshi, A. S., & Heydari, N. (2007). A proposed framework for irrigation management transfer in Iran: lessons from Asia and Iran, Colombo, Sri Lanka. *International Water manage-*

ment Institue (IWMI Working Paper 118).

37- Wang, X. (2008). *Economic analysis of adoption of water-saving land improvements in Northern China,* Unpublished dissertation, Clemson University. http://tigerprints.clemson.edu/all_dissertations/ 292.

38- Water Organization of Hamedan Province. (2011). Feasibility study report of water, soil and agriculture workforce, Famenin County, Water resources office. (In Persian).

39- Yigezu, Y. A., Ahmed, M. A., Shideed, K., Aw-Hassan, A., El-Shater, T., & Al-Atwan., S. (2013). Implications of a shift in irrigation technology on resource use efficiency: A Syrian case, *Agricultural Systems*, 118: 14-22.

8

Irrigation Water Pricing in Iran: The Gap between Theory and Practice

Morteza Tahamipour [1], Mohammad Kavoosi Kalashami [2] and Amirhossein Chizari [3]*

Abstract

Pricing policies play an important role in water demand management and its optimal allocation. Determining proper water price leads to optimal allocation of water especially in agricultural consumptions. Applying new subsidy targeting law in Iran which insists on pricing water based on its supply cost, will effects considerably on water resource management in agriculture sector. So, in this study, different Irrigation water pricing methods is investigated and proper irrigation water price is determined using survey data for 2010-2011 farming year in Golestan Province of Iran. At the first step using econometric approach, economic value of irrigation water in different agriculture crops is determined that shows demand side price for water. Then, supply cost of surface and ground water is calculated using accounting approach which shows supply side price for irrigation water. Finally, economic value and supply cost of irrigation water compared and different water pricing methods is evaluated. Results indicated that, weighted average of economic value and supply costs of irrigation water in Golestan province were 1795 and 1399 IRR per cubic meter, respectively. So, improvement of water demand and supply management could be achieved using price policies.

Keywords:
Water, Economic Value, Pricing, Supply cost, Golestan Province

[1] *Assistant Professor, Department of Economics, University of Shahid Beheshti.*
[2] *Assistant Professor, Department of Agricultural Economics, Faculty of Agricultural Sciences, University of Guilan.*
[3] *Assistant Professor of Agricultural Economics, Department of Economics and Agricultural Development, University of Tehran.*
* *Corresponding author's email: m_tahami@sbu.ac.ir*

INTRODUCTION

Iran is one of the countries in the world that suffer from water shortages. Water scarcity is one of the key problems affecting northern Iran. Price of water plays an important role in world's water resource management. One of the main goals of water pricing discipline is incentives creation and responsibilities impose for consumption pattern improvement, increasing water consumption productivity, covering part or financial fund need in water systems investments and operation and maintenance costs, reducing environmental degradation and preserving future generations' rights. Connecting pricing discipline to costs of water service providence is the first step in water pricing system improvement. For determining tariff discipline, three principals include costs covering, economic efficiency and equity are very important.

Table 1 shows the status of groundwater and surface water resources between different uses in Golestan province. As can be seen in table 1, seventy-two percent of surface water and 82 percent of groundwater resources consumed in agricultural uses. Therefore, irrigation water has the largest proportion of water resources and managing water in this kind of consumption is very important.

In this study, irrigation water pricing in Golestan province in Iran has considered. The most important agronomy crops in this province reported in table 2.

Table 2 shows that, wheat, soybean and tomato covered about 63 percent of irrigated acreage in Golestan, and so, managing water demand for these crops can be useful for water sources of this province. In this regard, estimating cost of building dams and water supply facilities costs and economic valuation of irrigation water as an input could be useful criteria for choosing water selling tariffs to water supply costs in future years.

Table 1: Water allocation system up to 2008 (million cubic meters).

Sector		Water	Share
Surface	Agriculture	578.2	0.72
	Industry	30.8	0.04
	Domestic	34.8	0.04
	Fishery	80.7	0.10
	Tourist	0.0	0.00
	Packing	0.1	0.00
	Other	79.8	0.10
	Sum	804	1
Groundwater	Agriculture	790.7	0.82
	Industry	24.9	0.03
	Domestic	144.3	0.15
	Fishery	0.0	0.0
	Tourist	0.0	0.0
	Packing	0.0	0.0
	Other	0.0	0.0
	Sum	960	1

Source: Water Regional Organization of Golestan Province (2010).

National development documents in Iran insist on mentioned goal. Water equity allocation law insist that water tariff in municipal, agriculture, industry and other use should be determined according to quantity and quality of consumption. Also, in the case of adjusted water systems all variable costs include management, maintenance and depreciations should be including in water tariff considering social-economic conditions of each regions. Also, long-run development strategy of Iran's water resource and Fourth Iran's development Plan, economic and social program insist on economic valuation of water and calculating supply costs.

Previous studies of optimum water pricing in agriculture such as Guohau (1986), Hussain and Young (1985), He and Tyner (2004), Renzeti and Dupont (1999) and Seagraves and Easter (1983) focused on supply cost or operating and maintenance cost as bases for determining optimum price of ir-

Table 2: Acreage and production of important crops in Golestan (farming year 2009-2010)

Crop	Acreage (hectare)			Production (tone)		
	Irrigated	Dry land	Sum	Irrigated	Dry land	Sum
Wheat	164170	221217	385387	501522	564213	1065735
Soybean	52010	5722	57732	111748	8295	120043
Tomato	6703	609	7312	221392	10505	231897
Sum	355616	339461	695077	2155614	1001744	3157358
Share %	63	67	65	39	58	45

Source: Ministry of Jihad-e-Agriculture of Iran.

rigation water in, while price of water in demand side also is important and optimum price of water must show ability of farmers to payment. Singh (2007) reported that there is a big gap between supply cost as water price and economic value of water. Also, Huang *et al.* (2006) showed that farmers were quite responsive if the correct price signal was used and policy makers must increase water price to the level of VMP of water. Also, Guerrero *et al.* (2010), Hussain *et al.* (2009) and Sadeghi *et al.* (2010) have measured economic value of irrigation water in Texas, Iran and Pakistan respectively. So, in this study, different Irrigation water pricing methods is investigated and proper irrigation water price is determined.

MATERIALS AND METHODS

Economic value of irrigation water represent the price of demand side and show the maximum willing to pay of consumers for each unit of irrigation water. In general, methods of estimating economic value of irrigation water as an input, classified in parametric and nonparametric methods. Linear programming method in nonparametric approach and production function method in parametric approach named as the most famous methods in each classification (Deacaluwe *et al.*, 2004 and Young, 2005). Present study use production function method for determining economic value of irrigation water in investigated region. A production function like Y shows the technical relationship among inputs ($x_1, x_2 \ldots x_{n-1}$, water) and output. If water considered as an input in the production of an agricultural product, then the value of marginal product of irrigation water could be interpreted as its economic value (Debertin, 1997).

$$Y = f(x_1, x_2, \ldots, x_{n-1}, water)$$

$$VMP_{water} = p_y \times MP_{water} = p_y \times (\partial Y / \partial_{water}) \quad (1)$$

Where, Y is output quantity, water is irrigation water input quantity, p_Y is output price, MP_{water} is the marginal product of irrigation water and VMP_{water} is the value of marginal product of irrigation water or its economic value. For showing neoclassic conditions a production function should follows some characteristics like homogeneity, continuity, concavity and twice differentiating (Chambers, 1988). All functional forms which follow the mentioned characteristics could be used for estimating production function and the determination of regression coefficients. Functional forms divided into two groups include flexible and inflexible. Quadratic, Translog and Leontief are examples of flexible functional forms. These functional forms preferred to inflexible ones because of their proper characteristics (Chambers, 1988). Flexible functional forms have the same characteristics in many aspects and acceptable in theoretical point of view. So, for determining superior functional form, all estimated functional forms should be compared using econometrics criteria like parsimony, simple interpretation, calculation simplicity, goodness of fit, power of modeling and forecasting (Green, 1993 and Thompson, 1988). Accuracy in functional form selection could express production relationship more actually and avoids misspecification in inputs and output relationships showing (Hossienzad and Salami, 2002).

Present study estimated three functional forms include Quadratic, Translog and Leontief for determination of irrigation water economic value. Variables in the mentioned functional forms include production quantity *(Y)*, irrigation water consumption in cubic meter *(wat)*, seed consumption in Kg *(sed)*, labor in day work *(lab)*, pesticides consumption in liter *(pes)* and fertilizers consumption in Kg *(fer)*. Quadratic functional form could be specified as below:

$$\begin{aligned} y &= \alpha_0 + \alpha_{wat} wat + \alpha_{sed} sed \\ &+ \alpha_{lab} lab + \alpha_{pes} pes + \alpha_{fer} fer \\ &+ 0.5\beta_{wat}(wat)^2 + 0.5\beta_{sed}(sed)^2 \\ &+ 0.5\beta_{lab}(lab)^2 + 0.5\beta_{pes}(pes)^2 \\ &+ 0.5\beta_{fer}(fer)^2 + \beta_{watsed} watsed \\ &+ \beta_{watlab} watlab + \beta_{watpes} watpes \\ &+ \beta_{watfer} watfer + \beta_{sedlab} sedlab \\ &+ \beta_{sedpes} sedpes + \beta_{sedfer} sedfer \\ &+ \beta_{labpes} labpes + \beta_{labfer} labfer + \beta_{pesfer} pesfer \quad (2) \end{aligned}$$

The value of marginal product of irrigation water could be specified according to the quadratic functional form as below:

$$\begin{aligned} VMP_{wat} &= P_y . MP_{wat} \\ MP_{wat} &= \alpha_{wat} + \beta_{wat} . wat + \beta_{watsed} sed \\ &+ \beta_{watlab} lab + \beta_{watpes} pes + \beta_{watfer} fer \quad (3) \end{aligned}$$

Considering mentioned inputs, the Translog

functional form could be specified as below:

$$\ln y = \alpha_0 + \alpha_{wat} \ln wat + \alpha_{sed} \ln sed + \alpha_{lab} \ln lab + \alpha_{pes} \ln pes + \alpha_{fer} \ln fer + 0.5\beta_{wat}(\ln wat)^2 + 0.5\beta_{sed}(\ln sed)^2 + 0.5\beta_{lab}(\ln lab)^2 + 0.5\beta_{pes}(\ln pes)^2 + 0.5\beta_{fer}(\ln fer)^2 + \beta_{watsed} \ln wat \ln sed + \beta_{watlab} \ln wat \ln lab + \beta_{watpes} \ln wat \ln pes + \beta_{watfer} \ln wat \ln fer + \beta_{sedlab} \ln sed \ln lab + \beta_{sedpes} \ln sed \ln pes + \beta_{sedfer} \ln sed \ln fer + \beta_{labpes} \ln lab \ln pes + \beta_{labfer} \ln lab \ln fer + \beta_{pesfer} \ln pes \ln fer \quad (4)$$

From which VMPwat could be determined.

$$VMP_{wat} = P_y.MP_{wat}$$

$$MP_{wat} = \frac{\partial \ln y}{\partial \ln wat}.\frac{y}{wat} = (\alpha_{wat} + \beta_{wat}.\ln wat + \beta_{watsed} \ln sed + \beta_{watlab} \ln lab + \beta_{watpes} \ln pes + \beta_{watfer} \ln fer).(\frac{y}{wat}) \quad (5)$$

The third functional form which is estimated in present study is Leontief.

$$y = \alpha_0 + \beta_{wat} wat^{0.5} + \beta_{sed} sed^{0.5} + \beta_{lab} lab^{0.5} + \beta_{pes} pes^{0.5} + \beta_{fer} fer^{0.5} + 0.5\beta_{watwat} wat + 0.5\beta_{sedsed} sed + 0.5\beta_{lablab} lab + 0.5\beta_{pespes} pes + 0.5\beta_{ferfer} fer + \beta_{watsed} wat^{0.5} sed^{0.5} + \beta_{watlab} wat^{0.5} lab^{0.5} + \beta_{watpes} wat^{0.5} pes^{0.5} + \beta_{watfer} wat^{0.5} fer^{0.5} + \beta_{sedlab} sed^{0.5} lab^{0.5} + \beta_{sedpes} sed^{0.5} pes^{0.5} + \beta_{sedfer} sed^{0.5} fer^{0.5} + \beta_{labpes} lab^{0.5} pes^{0.5} + \beta_{labfer} lab^{0.5} fer^{0.5} + \beta_{pesfer} pes^{0.5} fer^{0.5} \quad (6)$$

Irrigation water economic value based on this functional form could be determined as below:

$$VMP_{wat} = P_y.MP_{wat}$$

$$MP_{wat} = 0.5\beta_{wat} wat^{-0.5} + 0.5\beta_{watwat} + 0.5\beta_{watsed} wat^{-0.5} sed^{0.5} + 0.5\beta_{watlab} wat^{-0.5} lab^{0.5} + 0.5\beta_{watpes} wat^{-0.5} pes^{0.5} + 0.5\beta_{watfer} wat^{-0.5} fer^{0.5} \quad (7)$$

Supply cost of irrigation water calculated base on two general approaches includes accounting and engineering economics. In accounting approach, industrial accounting techniques have been used. Annual depreciation calculated as annual investment costs which summed with operational and maintenance costs and outcome divided by output quantity (water). In engineering economics approach, total costs of investment, substitution, operation and maintenance calculated for project life cycle and by constructing costs liquidity operation table, supply cost of irrigation water was determined. The main source of difference in two mentioned approach is considering money time value which is considered only in engineering economics approach. One of the methods of supply costs determination in engineering economics approach is average cost method which is used in present study. Average cost method calculated average cost for each volume unit of irrigation water. One of the advantages of applying this method is that average cost of a water unit (supply costs) per fixed costs (investment) and variable costs (operation and maintenance) could be calculated.

Requested data sets for estimating economic value of irrigation water has been acquired through surveys and filling questionnaires in Golestan province of Iran during 2010-2011. Investigated crops include irrigated wheat, soybean and tomato that the number of questionnaire of these crops was 154, 104 and 96 farms, respectively. Also, for calculating supply costs of irrigation water, regional water company data and Iran's water resource management company database were used.

RESULTS AND DISCUSSION

For each of investigated crops, results of different functional forms estimation were presented, separately. Determination of superior functional form done base on two criteria include normality of residuals and the number of significant coefficients. After determining superior functional form for each crop production function, economic value of irrigation water was calculated. Estimating three functional forms for irrigated wheat production using cross section data of 154 farms showed below summarized results in Table 3.

P-value amounts for calculated JB statistics in quadratic and Leontief functional forms showed that null hypothesis of residuals normal distribution have been rejected. Hence, these two functional forms could not be used for determining economic value of irrigated water in

Table 3: Comparing different estimated functional forms for irrigated wheat.

Functional form	Coefficients*	Significant coefficients*	Jurque-Bera Statistics**	p-value
Quadratic	21	11	93.46	0.000
Translog	21	8	4.59	0.1
Leontief	21	7	32.82	0.000

* Number ** Residuals Normality test

Table 4: Results of Translog functional form estimation for irrigated wheat production.

Variable	Coefficients	Standard Error	t-statistic	p-value
α_0	2.57	7.981	0.32	0.748
α_{wat}	-1.53	4.492	-0.34	0.734
α_{sed}	1.14	3.505	0.33	0.746
α_{lab}	-1.23	0.499	-2.47	0.015
α_{pes}	0.37	0.171	2.17	0.032
α_{fer}	2.14	1.53	1.4	0.164
β_{wat}	0.72	1.312	0.55	0.585
β_{sed}	-1.44	0.935	-1.54	0.126
β_{lab}	0.01	0.026	0.49	0.622
β_{pes}	-0.01	0.007	-1.34	0.182
β_{fer}	-0.21	0.192	-1.11	0.267
β_{watsed}	0.11	1.007	0.11	0.914
β_{watlab}	0.3	0.152	1.96	0.052
β_{watpes}	-0.2	0.068	-2.97	0.004
β_{watfer}	-0.85	0.515	-1.65	0.101
β_{sedlab}	-0.2	0.134	-1.51	0.133
β_{sedpes}	0.3	0.095	3.12	0.002
β_{sedfer}	1.14	0.459	2.49	0.014
β_{labpes}	-0.04	0.015	-2.69	0.008
β_{labfer}	-0.04	0.053	-0.69	0.493
β_{pesfer}	-0.04	0.025	-1.52	0.131

Adjusted $R^2 = 0.98$ D-W statistic = 2.02

wheat production and Translog functional form chosen as the superior one. Table 4 contains results of Translog functional form estimation.

Using Translog functional form economic value of irrigation water in wheat production equals 1624 IRR per cubic meter.

Estimating three functional forms for irrigated soybean production using cross section data of 104 farms revealed below summarized results in Table 5.

According to the JB statistics and its p-value, quadratic functional form residuals did not follow normal distribution. So from two functional forms of Translog and Leontief, the last one chosen based on the number of significant coefficients. Results of Leontief functional form estimation reported in table 6.

Economic value of irrigation water in soybean production calculated based on above estimation results. Mentioned value equals 2084 IRR per cubic meter.

Finally, estimating three functional forms for irrigated tomato production using cross section data of 96 farms revealed below summarized

Table 5: Comparing different estimated functional forms for irrigated soybean.

Functional form	Coefficients*	Significant coefficients*	Jurque-Bera Statistics**	p-value
Quadratic	21	13	23.09	0
Translog	21	4	3.93	0.13
Leontief	21	11	8.1	0.12

* Number ** Residuals Normality test

Table 6: Results of Leontief functional form estimation for irrigated soybean production.

Variable	Coefficients	Standard Error	t-statistic	p-value
α_0	-3002	908.8	-3.3	0.001
β_{wat}	-189	86.6	-2.18	0.032
β_{sed}	1756	771.3	2.28	0.025
β_{lab}	1355	296.2	4.57	0.000
β_{pes}	-1035	415.1	-2.49	0.015
β_{fert}	-25	83.2	-0.3	0.765
β_{watwat}	-12	7.4	-1.68	0.097
$\beta_{sedsedt}$	-259	462.7	-0.56	0.577
$\beta_{lablabt}$	-142	97.1	-1.46	0.147
$\beta_{pespest}$	369	320.5	1.15	0.253
β_{ferfer}	38	5.9	6.43	0.000
β_{watsed}	88	54.3	1.62	0.109
β_{watlab}	-20	27.9	-0.7	0.485
β_{watpes}	70	52.1	1.35	0.181
β_{watfer}	27	7.5	3.58	0.001
β_{sedlab}	-32	249.8	-0.13	0.899
β_{sedpes}	-699	435.7	-1.61	0.112
β_{sedfer}	-310	62.3	-4.97	0.000
β_{labpes}	383	120.1	3.19	0.002
β_{labfer}	56	29.6	1.88	0.063
β_{pesfer}	-94	79.1	-1.19	0.239

Adjusted R^2 = 0.98 D-W statistic = 1.85

results in Table 7.

Considering the rejection of null hypothesis in JB normality test for two quadratic and Translog functional forms, only, Leontief functional form could be used for determining economic value of irrigated water in tomato production that estimation results of this form reported in table 8.

Using above functional form, the economic value of cubic meter irrigation water in tomato production equals 3250 IRR. Considering acreages of these three products in Golestan province, weighted average economic value of irrigation water in mentioned region calculated which equals to 1795 IRR per cubic meter.

Supply costs of irrigation water in surface water resource for each of Golestan province dams calculated using data of initial investments costs, operation and maintenance costs and annual adjustable water volume of dams. Usually, operation and maintenance costs calculated base on 0.6 percentages of initial investments costs for a dam and 1.3 percentages for irrigation network. Initial investments costs updated using civil index. Also, annual investments costs of dams calculated using discount factor of 7 percentages and 50 years time horizon. After calculating equivalent of annual investments costs for all dams, this amount summed with operation and maintenance costs. Dividing this outcome by annual adjustable water volume, price of irrigation water per cubic meter gained. Table (9) contains supply costs of irrigation water per Golestan's province dams.

Annual adjustable water volume (A.A.V.) calculated by considering 90% transfer and distribution efficiency.

For calculating A.A.V. in irrigation and drainage network 81% of dam's A.A.V. applied.

For calculating supply cost of groundwater,

Table 7: Comparing different estimated functional forms for irrigated tomato.

Functional form	Coefficients*	Significant coefficients*	Jurque-Bera Statistics**	p-value
Quadratic	21	8	14.75	0.000
Translog	21	7	16.06	0.000
Leontief	21	15	1.61	0.44

* Number ** Residuals Normality test

Table 8: Results of Leontief functional form estimation for irrigated tomato production.

Variable	Coefficients	Standard Error	t-statistic	p-value
α_0	16555.2	9653.4	1.71	0.091
β_{wat}	-659.4	290.4	-2.27	0.026
β_{sed}	90070.6	30016	3	0.004
β_{lab}	3566.2	2091.2	1.71	0.093
β_{pes}	-15577.1	10103	-1.54	0.128
β_{fert}	-2669	853.1	-3.13	0.003
β_{watwat}	6.3	9.3	0.68	0.50
$\beta_{sedsedt}$	-20554.4	26177.2	-0.79	0.435
$\beta_{lablabt}$	-1428.1	680.4	-2.1	0.039
$\beta_{pespest}$	4398.5	2224.9	1.98	0.052
β_{ferfer}	505.8	121.9	4.15	0.000
β_{watsed}	471.7	570.5	0.83	0.411
β_{watlab}	207.6	69.4	2.99	0.004
β_{watpes}	-401.7	93.8	-4.28	0.000
β_{watfer}	37.3	23.9	-1.56	0.123
β_{sedlab}	-18293.3	6269.2	-2.92	0.005
β_{sedpes}	22589.7	4194.8	5.39	0.000
β_{sedfer}	544.8	2160.5	0.25	0.802
β_{labpes}	4125.5	1828.6	2.26	0.027
β_{labfer}	-137.1	126.3	-1.09	0.281
β_{pesfer}	-1132.4	535	-2.12	0.038

Adjusted $R^2 = 0.94$ D-W statistic = 1.99

Table 9: Supply costs calculation of irrigation water per dams in Golestan province.

Condition	Dams	Investments costs (Million Rials)		A.A.V (MCM)	Price of irrigation water (Rial/m^3)
		Dam	Irrigation network		
Existing dams	Voshmgir	126878	448232	91.31	109
	Kosar	36551	17852	6.73	426
	Golestan 1	477906	158643	116.17	321
	Golestan 2	413652	247673	32.28	1005
	Alagol	109863	7106	83.4	103
Planned dams	Chayli	1006795	444264	204	387
	Normab	1903178	521900	115	1298
	Kivdoval	311138	938060	54.2	450

Annual adjustable water volume (A.A.V.) calculated by considering 90% transfer and distribution efficiency.
For calculating A.A.V. in irrigation and drainage network 81% of dam's A.A.V. applied.

considering wells as the main source of groundwater in Golestan province, for investigating supply costs of groundwater, equivalent of annual investments cost calculated and divided by well's output water volume. In calculating annual investments cost, discount factor equals 8% and 35 years time horizon considered. Using sample data of a well in investigated region, average supply costs of groundwater in Golestan province equals 1695 IRR per cubic meter. 25 percentages of total irrigation water use in Golestan province provides by surface water and 75 percentages supplied by groundwater resources. So, weighted average supply costs of irrigation water in Golestan province equals 1399 IRR per cubic meter.

CONCLUSION

After determining economic value and supply cost of irrigation water in Golestan province, different pricing scenarios investigated in order to provide proper framework for policy makers and planners in water sector.

1) Scenario of existing non-beneficial irrigation water supply company

If water regional companies would work in a non-beneficial framework which means pricing irrigation water equal to its supply cost then this company should price irrigation water 1399 per cubic meter. While, water tariffs in modern and semi-modern irrigation networks, is 20 to120 IRR per cubic meter, now. So, regional water company should increase irrigation water tariff to its supply cost.

2) Scenario of equating water tariff to its economic value

Economic value shows consumers' willingness to pay for water. Each additional cubic meter of irrigation water added 1795 IRR to farmer's income in average. In means farmers will pay 1795 IRR for a cubic meter of irrigation water. If there are circumstances in which water scarcity raised in investigated region, pricing irrigation water equal to its economic value would be a good policy for optimal allocation of irrigation water.

3) Scenario of pricing based on equity of value and cost

Considering irrigation water as an economic commodity, water supplier act efficiently only when for each output unit (water) marginal cost equals marginal revenue or benefit. Results of present study showed that supply cost of cubic meter irrigation water equals 1399 which is less than its economic value (1795 IRR). So, improvement of water demand and supply management could be achieved by using price policies.

Results of present study provide good framework for regional managers in order to improve water resource management considering sustainable development rules in their region. It is necessary to study economic value and supply cost of irrigation water in different provinces of Iran. Mentioned studies provide good policy implications in Iran's water sector.

REFERENCES

1- Chambers, R.G. (1988). *Applied production analysis: A dual approach*, Cambridge University Press. USA.

2- Deacaluwe, B., Patry A., & Savard L. (2004). when water is no longer heaven sent: comparative pricing analysis in an AGE model Retrieved from *http://econpapers.hhs.se*

3- Debertin, D.L. (1997). Agricultural production economic, (M. Mosa Nejad & R. Najarzadeh, Trans), institution of economic research, Tarbiat Modares University.

4- Green, W.H. (1993). *Econometric analysis*, Macmillan publishing company, New York.

5- Guerrero, B., Wright, A., Hudson, D., Johnson, J., & Amosson, S. (2010). The Economic Value of Irrigation in the Texas Panhandle, Selected *Paper prepared for presentation at the Southern Agricultural Economics Association Annual Meeting,* Orlando, FL, February 6-9, 2010.

6- Guohau, X.U. (1986). Irrigation water charges in China, Report on the expert consultation on irrigation water charges, FAO, Rome, Italy.

7- He, L., & Tyner, W. (2004). Improving irrigation water allocation efficiency using alternative policy option in Egypt, Retrieved from http:// econpapers.hhs.se

8- Hosseinzad, J., & Salami, H. (2002). Select a production function to estimate the economic value of agricultural water, the case study of wheat production, *the quarterly of agricultural economic and development*, 48.

9- Huang, Q., Rozelle, S., & Howitt, R. (2006). Irrigation Water Pricing Policy in China. American Agricultural Economics Association Annual Meeting.

10- Hussain, I., Sial, M. H., Hussain, Z., & Akram, W. (2009). Economic Value of Irrigation Water: Evidence from a Punjab Canal. *The Lahore Journal of Economic,* 14(1), 69-84.

11- Hussain, R.Z., & Young, R.A. (1985). Estimates of the economic value productivity of irrigation water in Pakistan from farm surveys, *Water Resources Bulletin* (26.6): 1021-1027.

12- Judge, G.R.C. Hill, W.E. Griffiths. H. Lutkepohl, & Lee, T.C. (1988). *Introduction to the theory and practice of econometrics.* John Wiley and Sons, New York

13- Renzeti, S., & Dupont, D. (1999). An Assessment of the impact of charging for provincial water use permits. *Paper presentation, 6th conference of the international water and resource consortium Hawaii, Agecon.*

14- Sadeghi, A., Mohayidin, M. G., Hussein, M. A., & Baheiraei, A. (2010). Determining the economic value of water in production of wheat in Iran. *Australian Journal of Basic and Applied Science,* 4(6), 1391-1402.

15- Seagraves, J.A., & Easter, K.W. (1983). Pricing irrigation water in developing countries, *Water Resources Bulletin,* 19.4. 663-672.

16- Singh, K. (2007). Rational pricing of water as an instrument of improving water use efficiency in the agricultural sector: a case study in Gujarat, India *International Journal of Water Resources Development*, 23, 679–690.

17- Thompson, C.D. (1988). Choice of flexible functional forms: Review and appraisal, *Western Journal of Agricultural Economics,* 13, 169-183.

18- Young, R. A. (2005). *Determining the economic value of water: concepts and methods,* Washington, DC, USA.

9

Ex-ante Impact Study on the Integrated Natural Resources Management in the Middle East and North Africa Region Rangelands Rehabilitation Project in Iran

Habib Shahbazi [1*], Hamidreza Alipour [2] and Akram Abbasifar [3]

Abstract

The purpose of the Integrated Natural Resources Management program in the Middle East and North Africa Region (MENARID) is to bring national investment projects in the field of Integrated Natural Resource Management (INRM). The impact evaluation study tests whether the selected technology has significantly contributed to restore and maintain of the ecosystem functions and productivity, and whether has improved the economic and social well-being of the targeted communities. We chose Kamkooyeh Village in the Behabad County where a "Village Development Group" (VDG) has been established by following participatory and capacity building approaches such as social mobilization and micro credit mechanism. A set of complementary activities and interventions were recommended and implemented in the site by consultation and active participation of the local communities and beneficiaries. One of the proposed interventions is rangelands rehabilitation project (RRP). The expected benefits and costs of the RRP were predicted through financial valuation and an ex-ante evaluation of socioeconomic impacts. The financial and social benefit-cost ratio of RRP was estimated to be about 0.97 and 2.15, respectively.

Keywords:
Impact Study, MENARID, Rangelands rehabilitation, ICARDA, Kamkooyeh Village, Iran

[1] *Assistant professor, Department of Agricultural Economics, Sayyed Jamaleddin Asadabadi University (SJAU), Hamedan, Iran*
[2] *Department of Management, Rasht Branch, Islamic Azad University, Rasht, Iran*
[3] *Research Assistant, Consumer and Producer Protection Organization, Ministry of Industry, Mine and Business, Tehran, Iran.*
* *Corresponding author's email: Habib_susa@yahoo.com*

INTRODUCTION

The purpose of the Integrated Natural Resources Management (INRM) program in the Middle East and North Africa Region (MENARID) is to bring national investment projects in the field of INRM. Within the scope of MENARID, ICARDA planned to support a full and details impact study that captures the adoption of all INRM technologies implemented by the Iran MENARID project titled "Institutional strengthening and coherence for Integrated Natural Resources Management", and further assess a few high potential technologies selected in close collaboration with the local project team, and for which primary and secondary data are available. Such a study should also explore the gender aspect of the technologies analyzed. This approach should contribute to decisions related to up scaling of the activities within and outside of the target project area as wells as learning for policy and decision making in integrated NRM development under similar conditions in the future.

The impact evaluation study should test the following objects: The technology, in the specific context in which it has been implemented, has significantly contributed to restore and maintain of the ecosystem functions and productivity, and has improved the economic and social well-being of the targeted communities. This hypothesis should be tested in the project districts that are predominantly covered by the assessed technology.

The impact study focuses on attribution and contribution of these technologies in bringing about desired changes, and also covers its relevance, effectiveness, efficiency, impact and reliability. To achieve this goal, it:

i. determines extent and depth of the use of the technologies/practices by the target beneficiaries,

ii. measures the economic and environmental benefits of using this technology; and

iii. Undertakes a benefit-cost analysis of the technology, considering the full cost of the technology (costs of utilization, costs of promotion, subsidies and etc.).

We chose Kamkooyeh Village in the Behabad County (Iran) where a "Village Development Group" (VDG) has been established by following participatory and capacity building approaches such as social mobilization and micro credit mechanism. A set of complementary activities and interventions were recommended and implemented in the site by consultation and active participation of the local communities and beneficiaries. The proposed interventions include Rangelands rehabilitation project (RRP) by different practices like seeding and plantation in hilly landscapes/terrains

Since Kamkooyeh village is located in a hyper arid region (near to desert and with mean annual rainfall of about 153.9 mm) so water saving and water use optimization is a very important and necessary issue. This issue is more important in agriculture sector because it consumes more than 90% of the available water resources. One of the solutions for conserving water in soil profile, reducing runoff, soil conservation and as a result reducing soil erosion rate, forage production, increasing carbon sequestration and improving plant biodiversity, an intervention including sowing and seeding of natural vegetation cover has been done by one of the community member in his own lands (based on the customary right) in upstream of Jannat Watershed. This practice covers about 10 hectares and has been implemented by three labors and also one day work of 30 persons (members of village development group).

The seeding practice should be done before snow fall to give more opportunity for germination of seeds and also to be covered by soil because of sheep and goats walk but there was a small problem (delay) in providing seeds. On the other hand, this spring has less precipitation because of drought condition, but there is a hope that next year, the plants could grow better.

The sowing intervention had two aspects. One to test different sowing methods to find out which one is better for this area and on the other hand to convince local beneficiaries that they can rehabilitate and also improve rangelands condition by these type of intervention which is useful for their animal grazing and improving their own livelihood condition.

Rural development policy impact study could evaluate from economic, ecosystem and social aspects. These appraisals are categories to two approaches of ex-ante and ex-post evaluations (Shahbazi, 2013, Hosseini and Shahbazi, 2013 and 2014). One of evaluation methods is benefit-cost analysis approaches. Rangelands Rehabilitation Project (RRP) in Kamkooeh village is not operational yet. So, RRP impact study is ex-ante evaluation. This executive MENARID policy in Iran will be evaluated from economic, ecosystem and social aspects by ex-ante benefit-cost analysis approach. Benefit-cost analysis is method for project relative advantage according to optimal and effective allocation of resources.

Benefit-cost analysis is a method for evaluating relative advantage of investment projects in terms of efficient allocation of resources. Benefit-cost analysis aims to improve the efficiency of resources for economic prosperity in other words; the aim of the evaluating is to help select the best type of desired decision to the efficient use of resources (Wieck, 1993).

The first theoretical framework of benefit-cost analysis is bringing up by French engineer and economist, Dupit in 1844. He used the concept of consumer surplus. After that date the benefit-cost analysis has played a vital role in the economic well-being (Vreeker *et al.*, 2002). The first practical application of benefit-cost analysis was in 1930 for the development of water resources in the US (Pakzad, 1993). There are many definitions for benefit-cost analysis. Boardman *et al.* (2005) believed that benefit-cost is balance for measuring. Reh (2015) believed benefit-cost analysis is method for finding all the costs and benefits of a project as quantity value. In developing countries due to lack of capital resources, the best alternative investment allocation is a critical issue (Khalili, 1995).

Accordingly, many studies have focused on the evaluation of rural development projects. In Iran Malek Hosseini and Mirakzade (2014), assessed economic impacts of development projects on rural area (Kermansha province irrigation and drainage network). Salehi (2002) showed that the watershed project had positive impact on rural migration reducing and employment increasing. The effect of watershed projects on rural development, Efati (2000); Mansourian and Mohammadi Golrang (2003) also conducted a study. Hosseini Tavasol *et al.* (2007) studied the economic development impact of rural dam in Khuzestan, assessed the impact of villager's contribution in rural development and successful implementation of rural projects. Other studies in this filed are conducted by Khobfekr (1999) and Dadrasi Sabzevar (2007). Abbasifar (2008) assessed the economic and environmental impact of foresting projects in Iran. She also used the benefit-cost analysis. All of these studies have been used ex-post evaluation approach.

In other countries, Tolliver *et al.* (2011) study was to quantify the investment and maintenance needs of the county and local roads that serve as agricultural logistics routes in North Dakota. They found that the estimated resurfacing costs per mile of major agricultural distribution routes is 40% greater than the estimated resurfacing cost per mile on non-agricultural routes. They also discovered the average annual cost to resurface and maintain paved agricultural roads are $18,300. Jahren *et al.* (2005) conducted a study of Minnesota rural roads for the Minnesota Department of Transportation (DOT). The objective of the study was to identify the methods and costs of maintaining and upgrading a gravel road. They concluded that the historical costs to maintain both gravel and bituminous roads were between $1,500 and $2,500 per mile.

Anderson and Sessions (1991) used Mixed Integer linear Programming (MIP) to analyze the intermittent road management problem. They compute the total costs and open road segments if opening and closing costs are not considered simultaneously with transport and road maintenance costs. The total costs are 13% higher than the optimal solution that considers all four costs simultaneously. Baumet *et al.* (1986) estimated the benefits of keeping groups of existing roads in the county road system. Hanson *et al.* (1985) describe the variable costs of the predominant types of vehicles operating on Iowa rural county roads. Helmberger *et al.* (1990) develop a method to assess the economic impact of a rural road management study.

Sebealy *et al.* (2003) evaluate the impact of agricultural equipment on the actual response of low-volume roads in South Dakota. Other studies in this filed are conducted by Byrd and Gildestad. (2001).

In this section, first of all, the objectives of the impact study were mentioned then the study area was introduced. After that, the implemented projects (interventions) in the study area were introduced and were evaluated. In the next section, the methodology of the impact study will be explained. In section three, calculation of the environmental, economic and social benefits of the practice and B/C analysis (financial, economic and social) would be explained. In section four, conclusion and recommendation would be explained.

MATERIALS AND METHODS

We chose Kamkooyeh Village in the Behabad County in Yazd Province (Figure 1) where a set of complementary activities and interventions were recommended and implemented in the site by consultation and active participation of the local communities and beneficiaries including Rangelands Rehabilitation Project (RRP).

Due to the short time elapsed since the start of the interventions, (the lifetime of the largest intervention or project is approximately one year) so nobody expects to observe the resulting impacts in such a short period. Therefore, an ex-ante study was done to predict the potential impacts in the future.

In this project, we should consider all of the intervention or measures simultaneously while predicting the potential impacts. Accordingly, the impacts of this intervention (technology) on maintenance, restoration, or improvement of productivity, ecosystem functions, and social welfare of local communities were evaluated separately by considering the following Steps:

Ecosystem functioning (environmental, biophysical and biological) benefits

Ecosystem functioning (environmental, biophysical and biological) benefits of "Rangelands Rehabilitation Project (RRP)" are including: including "water conservation and increasing water infiltration", and as a result, "increasing ground water recharge" plus "reducing flood hazard" plus "soil erosion control" (because of low runoff in hilly landscapes) and on the other hand, improving natural vegetation cover and as a result, more "Carbon & CO_2 sequestration" plus "biodiversity value" plus "soil erosion control" (because of existing a better protection cover for the top soil by the improved canopy cover).

Calculation of "increasing water infiltration" benefits

Calculation of "increasing water infiltration" benefits in terms of "increasing ground water recharge" benefits plus "reducing flood hazard" benefits plus "soil erosion control" benefits is as the followings:

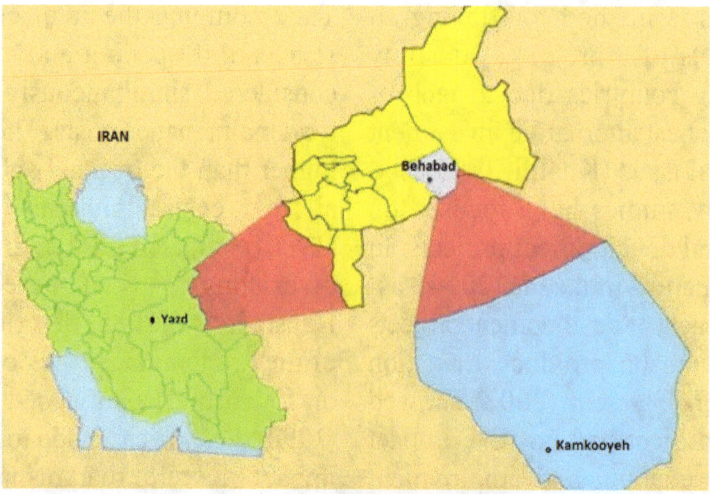

Figure1. Kamkooyeh Situation

$$V_{IWI}= V_{IGWR}+ V_{RFH}+ V_{SEC} \quad (1)$$

Where V_{IWI} is "increasing water infiltration" benefits V_{IGWR}, V_{RFH} and V_{SEC} are "increasing ground water recharge" benefits" plus "reducing flood hazard" benefits plus "soil erosion control" benefits, respectively. After field visit, by using a rainfall-runoff model (like *SCS*-Curve Number method) and considering canopy cover percentage of rangelands, before and after rangelands rehabilitation measures, the model will shows the impact of this activity on runoff control and also by using soil erosion model (like *PSIAC* or *USLE*), the impact of this activity on soil erosion control, would be determined and its benefits would be calculated.

Calculation of "CO_2 sequestration" benefits

Calculation of "CO_2 sequestration" benefits is as the followings:

$$V_{CS}= R_{AV}\times A_i\times C_i\times T_{con} \quad (2)$$

Where V_{CS} is "CO_2 sequestration" benefits because of rangeland rehabilitation measures, R_{AV} is carbon sequestration (ton/ha), A_i is area of rehabilitated rangelands (ha), C_i is the ratio of the increased biomass because of rangeland rehabilitation measures and Tcon is average tax on CO_2 dispersion-average countries (IRR/ton).

Calculation of "biodiversity value (V_{BS})" benefits

For calculation of "biodiversity value (V_{BS})" benefits of the improved rangeland (after rehabilitation measures), De Groot *et al.* (2012) table would be used. They have calculated the value of 10 different ecosystem services (in US $) in 2012.

Calculation of "soil erosion reduction" benefits

Calculation of "soil erosion reduction" benefits is as the followings:

$$V_{EE}=Q_I\times (M_N\times P_N+M_P\times P_P+M_K\times P_K) \quad (3)$$

Where V_{EE} is "erosion reduction" benefits, Q_I is quantity of sediment (m^3), M_N is percentage of Nitrogen in sediment, PN is market price of Nitrogen fertilizer, M_P is percentage of Phosphate in sediment, P_P is market price of Phosphate fertilizer and M_K is percentage of Potassium in sediment, P_P is market price of Potassium fertilizer.

Sum of the benefits from water conservation and "increasing water infiltration", "CO_2 sequestration", "biodiversity value" and "soil erosion reduction" are ecosystem functioning (environmental, biophysical and biological) benefits of *RRP*. It means, $V_{RRP}=V_{IWI} + V_{CS} + V_{EE} + V_{BS}$.

Economic (improving of productivity) benefits

Economic (improving of productivity) benefits of RRP, that is, Calculation of "improving of productivity" benefits is as the followings:

$$E_{RRP}=Q_F\times P_F \quad (4)$$

Where E_{RRP} is economic (improving of productivity) benefits of *RRP* because of improved vegetation cover after rangeland rehabilitation measures, Q_F is the amount of the increased livestock feed (ton), P_F is livestock fodder price (IRR/ton).

Social benefit/well-being

Social benefit/well-being of RRP could be determined by using the following equation:

$$SV_{RRP}=[1+((IN_{nmd}/IN_{pre\text{-}nmd})\times \gamma_{RRP})]\ [V_{RRP}+ E_{RRP} + EM_{RRP}] \quad (5)$$

Where SV_{RRP} is social benefit of *RRP*. V_{RRP}, E_{RRP} and EM_{RRP} are ecosystem functioning, economic benefits and employment value (temporary job opportunity for local community of *RRP*, respectively.

In addition to environmental and economic benefits of *RRP*, there are other impacts such as increasing active participation of local communities, improved group working manner, social mobilization which will cause improvement in social capital, therefore the rate of social participation could be calculated by using the $[(IN_{mnd}/ IN_{pre\text{-}mnd})\times\gamma_{RRP}]$ equation.

Actually, by considering this coefficient (rate), social benefits of *RRP* would be increased. In this equation, IN_{mnd} is average income of the village after implementation of the *MENARID* project, $IN_{pre\text{-}mnd}$ is average income of the village before implementation of the *MENARID* project and γ_I is the technology acceptance rate for *RRP*.

Calculation of "*EMRRP*" benefits

Calculation of "*EMRRP*" benefits is as the followings [1]:

$$EM_{RRP}=Q_{EM1}\times P_{w1} \quad (6)$$

Where EM_{RRP} is employment value of RRP activity, Q_{EM1} is number of the employed labors, Pw1 is labor daily wage in that region.

Benefit-cost analysis

Benefit-cost analysis of *RRP* is as follows:

Net annual equivalent uniform benefits of RRP could be calculated according to SVRRP for life period of the projects (for example, 20 years) and then Net annual uniform cost of RRP interventions for useful life period of the projects. Benefit-cost analysis would be determined based on the following equation:

$$B/C=\left[SV_{RRP}-EM_{RRP}\right]\Big/\left[C_{RRP}\left[\frac{r(1+r)^n}{(1+r)^n-1}\right]+K_{RRP}\right] \quad (7)$$

Where SV_{RRP} is annual social benefits of RRP intervention and EM_{RRP} is employment value of *RRP* intervention. C_{RRP} is net annual uniform cost (operational & overhead expenses) of *RRP* intervention, n is the life time of the project and *r* is the annual discount rate. Also K_{RRP} is the annual working capital cost (operational and maintenance costs).

Data

Some data such as water price in the study area, production price and agricultural production quantity are collected by questioners from local beneficiaries in the village, some data such as soil and carbon sequestration properties are collected from baseline studies, some data such as budget of projects (the allocated budgets) are collected form provincial project team. As it was mentioned before, the required data for doing this impact study were collected from different sources. First of all, those data that could be extracted from existing reports (base line study, the filled questioners by the *MENARID* team), field visit, were organized.

RESULTS

In this section, calculation of the environmental, economic and social benefits of the *RRP* practice will be explained. Benefit-cost analysis (financial, economic and social) of the projects would be presented in the last section.

Evaluation of ecosystem and socio-economic benefits of the RRP

In total, this practice (rangelands rehabilitation measures) has been implemented in 12 ha of rangelands. Its environmental functions includes: water conservation and "increasing water infiltration", and as a result, "increasing ground water recharge" plus "reducing flood hazard" plus "soil erosion control" and on the other hand, improving natural vegetation cover and as a result, more "Carbon & CO_2 sequestration" plus "biodiversity value" which their values will be explained in below.

Increasing ground water recharge and reducing surface runoff (flood hazard control):

Before explaining the calculation methodology, it is necessary to mention that:

- Based on the baseline survey, canopy cover of rangelands was 14.5% that it should be improved by 20% and to become 34.5%
- Soil characteristics (such as surface texture, infiltration rate and drainage condition) of the area has been assessed in medium rate and its infiltration rate is varying 3.81 to 7.62 millimeters (mm) per hours.
- After seeding and planting measures, vegetation cover condition has been changed from poor to moderate and *SCS – CN* (Curve Number) reduced from 79 to 69 therefore surface runoff from rehabilitated rangelands has been decreased (because of more infiltration rate).

Calculation of monetary value of ground water recharge

First of all, it is necessary to determine change in surface runoff depth after seeding and sowing measures:

$$Q=\frac{(P-0.2S)^2}{P+0.8S}$$

[1] *Although in economic and project evaluations literature, employment of installing projects is a cost item, but in social evaluations, this item is social benefit because of income development of farmer. For accurate evaluations benefits in comparison of cost, we subtracted employment benefits from social benefit (see equation 7).*

Where, Q is surface runoff depth (in inch), P is Precipitation depth (in inch), S is the maximum potential retention (after runoff begins). The equation [$S= 2540/CN - 25.4$] are used for calculation of S.

It was mentioned that after implementation of *RRP, CN* of the rangeland would be changed from 79 to 69. Therefore, S_1 (before the project) and S_2 (after the project) would be calculated as:

$S_1(CN=79)=(2540/79)-25.4= 6.75$

$S_2(CN=69)=(2540/796925.4=11.41$

Since change in the runoff depth is important so for simplifying the calculation, it could be assumed that precipitation depth *(P)* is equal to 1 inch.

$$Q_1 = \frac{(1-0.2\times 6.75)^2}{1+0.8\times 6.75} = 0.16$$

$$Q_2 = \frac{(1-0.2\times 11.41)^2}{1+0.8\times 11.41} = 0.02$$

$$Q_1 - Q_2 = 0.14$$

It means because of Rangelands rehabilitation measures; runoff depth would be decreased about 0.14 inches. This runoff depth is equal to 0.364 cm (36.4×10^{-4} m). On the other hand, decrease in runoff depth means increase in ground water recharge in rate of 36.4×10^{-4} m per unit area of ground surface. By considering 12 ha of rangelands under *RRP* measures (equal to 120000 m^2) in total 436.8 m3 of water would be infiltrated in ground water. This volume of water is equal to 436800 liters. Since discharge of the village *Qanat* is about 17 lit/sec so infiltration time becomes 25694.1 seconds (equal to 7.14 hours).

Monetary value of this function in case that unit price of irrigation water as 250000 IRR, it would be 1784300 IRR and unit price of irrigation water as 300000 IRR, it would be 2191200 IRR.

The function of reducing in soil erosion rate

Soil erosion causes sedimentation problems in downstream. In this section change in sediment yield as an indicator for evaluation of impact of *RRP* on soil erosion control, would be determined. Based on the MPSIAC model, sediment yield could be estimated by the following equation:

$Q_S= 18.6\ e^{0.0353R}$

Where Qs is sediment yield (cubic meter of sediment in each square kilometer of rangelands) and R is degree of sedimentation which will be derived from 9 causative factors (sum of their scores).

For calculating changes in Qs (or ΔQs), it is necessary to calculate changes in R (or ΔR), before and after implementation of *RRP*. It is necessary to mention that among 9 causative factors of soil erosion, only runoff factor (Y_4), Land cover factor (Y_6) and Land use factor (Y_7) could be change therefore it would be as:

$Y_4=0.2\ X_4$ *(Runoff)*

$X_4=$ *Runoff depth (mm)* $\times\ 0.03$

$Y_6=0.2\ X_6$ *(Land cover)*

$X_6=$ *Area percentage of bare lands*

$Y_7=20 - 0.2\ X_7$ *(Land use)*

$X_7=$ *Canopy cover percentage*

For determining of ΔR, it is necessary to determine ΔY_4, ΔY_6 and ΔY_7.

First:

$\Delta Y4=0.2\ \Delta X4$

Where, ΔX_4 is Runoff depth (mm) $\times$ 0.03. Then,

$\Delta Y_4=0.2\ [364mm\times 0.03]=0.22$

Second:

$\Delta Y_6=0.2\ \Delta X_6$

Where ΔX_6 is percentage of bare lands before the project minus percentage of bare lands after the project. Based on the baseline survey, canopy cover was 14.5% therefore percentage of bare lands before the project becomes:

$100 - 14.5=85.5\%$

The target of the *RRP* is improving the natural vegetation cover to 34.5% so its percentage of bare lands after the project becomes:

$100 - 34.5=65.5\%$

Then,

$\Delta X_6=85.5 - 65.5=20\%$

$\Delta Y_6=0.2\times 0.2=0.04$

Third: $\Delta Y_7=20 - 0.2\ \Delta X_7$

Where, ΔX_7 is primary canopy cover percentage minus secondary canopy cover percentage. Then,

$\Delta X_7=14.5 - 34.5=-20\%$

$\Delta Y_7=- 0.2\times (-20\%) + 0.04$

Finally,

$\Delta R=\Delta Y_4 + \Delta Y_6 + \Delta Y_7=0.22 + 0.04 + 0.04=0.3$

Now sediment yield changes could be estimated.

$$\Delta Q_s = 18.6\, e^{0.0353\Delta R}$$

$$\Delta Q_s = 18.8\, \mathrm{m}^3/\mathrm{km}^2$$

It means in each 100 ha, 18.8 m^3 of the sediment would be decreased. Therefore, in the rehabilitated rangelands (with area of 12 ha), about 2.26 m^3 of the sediments would be decreased.

Monetary value of this function would be evaluated based on cost of cleaning of the reservoir pool. Since each year about 2.26 cubic meters of the sediments would be decreased and its cleaning cost is equal to wage of one day labor so reduce in sediment yield worth equal to one day labor wage which is about 350000 to 400000 IRR [2].

The function of increase in carbon sequestration

To evaluate the monetary value of carbon sequestration the equation 2.9 could be used.

In Kamkooyeh area, in each 1153.9 ha of rangelands, 112652.7 Kg of carbon could be sequestrated per year. In other words, in this region, in each hectare of rangelands, 97.7 Kg (equal to 0.098 ton) of carbon could be sequestrated. Based on European Union records in year 2014, Tax for carbon emission is about 20 Euros per ton. Since increased canopy cover is about 20% so it could be calculated as:

0.098 ton × 12 ha × 20% × (20€ × 43000 IRR)= 202272IRR

(In the above equation, 43000 is exchange rate of Euro and IRR in year 2014)

Therefore, the monetary value of carbon sequestration in 12 hectares of rehabilitated rangelands would about 202300 IRR.

Economic benefits of RRP

This value would be calculated based on valuing added dried forage. Canopy cover of natural vegetation was 14.5% before the project and its produced dried forage in hectare was 193 Kg. The objective of implementation of RRP is improving canopy cover in such a way to reach 35.5% (it means 20% increase in canopy cover percentage) therefore, in each hectare of rangelands, 38.6 Kg of dried forage would be increased:

193 Kg/ha × 20%=38.6 Kg

Therefore, in total 12 hectares of the rehabilitated rangelands, 463.2 Kg of dried forage would be increased.

By considering unit price of dried forage in the region is about 9000 IRR in year 2013 so added monetary value in this part would be calculated as:

463.2 Kg × 9000 IRR=4168800 IRR

Therefore, after seeding and plantation measures in 12 hectares of rangelands, in total 463.2 Kg more dried forage would be produced which its value is about 4168800 IRR.

The employment value of RRP

In seeding and plantation measures RRP, one person with annual wage of 3500000 IRR and 9 labors (man/day) with wage of 400000 IRR have been worked so 7100000 IRR is its employment benefits.

9 days × 400000 IRR/day=3600000 IRR

1 person × 3500000 IRR/year=3500000 IRR

3600000 + 3500000=7100000 IRR

Social benefit of RRP.

Calculation of acceptance rate of the project

It was supposed that RRP implemented at 90 hectares of rangelands but in practice only 12 hectares of it has been implemented. Therefore its acceptance rate is as:

Acceptance rate=12 / 90=0.133

By considering the average income of the villagers as 41910000 IRR and assuming that after implementation of the MENARID project, the average income of the villagers could reach to the average income of Yazd province (110312101 IRR) or the average income of the country (101281362 IRR) then the monetary value of participation of the villagers would be determined as:

Based on the records of the Central Bank of Iran and also Center for Census data of Iran in year 2012:

- average monthly income of rural households at national level=101281362 IRR
- average monthly income of rural households at Yazd province=110312101 IRR

[2] *Each euro is equivalent to 43,000 IRR.*

A1- by considering national level: *101281362 / 4191000 × 0.133=0.35*

A2- by considering Yazd province level: *110312101/41910000 × 0.133=0.32*

It means the coefficient of added benefits, in the case that average monthly income of rural households at Yazd province is the target, would be 0.35 and in the case that average monthly income of rural households at national level is the target, would be 0.32.

Therefore, by considering the above mentioned rates, social benefits would be increased because of people participation in this activity and also improved social capital. The summary of the results of Eco systemic, economic and social benefits of *RRP* is shown in the Table 1.

Benefit-cost analysis of *RRP*

For calculation of benefit-cost of the implemented projects, equation 7 has been used. In this section both financial and social Benefit-cost analysis has been done.

Before benefit-cost analysis, implantation cost of the project including GEF, UNDP, Iran's Government, MENARID and Local Community Costs (Budgets) has been shown in Table 2.

Financial benefit-cost analysis

In this section only financial and economic benefits of the RRP has been considered. It means financial benefits of RRP in comparison to their costs would be analyzed. If lifetime of the projects considered as 20 years and annual discount rate assumed to be 12 percent, annual uniform cost of *RRP* has been presented in Table 3. In this table it is assumed that annual cost (current cost) of the project to be equal 30 percent of their financial and economic benefits. As it is clear in the Table 3, benefit-Cost ratio of RRP is less than 1 (about 0.97) and therefore financially (economically) is not acceptable.

Its sensitivity analysis for change in discount rate as 11 and 13 percent has been done which the result shows 1 percent increase in discount rate, benefit-cost ratio will be decreased to 0.92 and 1 percent decrease in discount rate, benefit-cost ratio will be increased to 0.1.01.

Social benefit-cost analysis

In this section only social benefit of the RRP has been considered. It means social benefits of RRP in comparison to their costs would be analyzed. If lifetime of the projects considered as 20 years and annual discount rate assumed to be 12 percent, annual uniform cost of RRP has been presented in Table 3. In this table it is assumed that annual cost (current cost) of the project to be equal 30 percent of their social benefits. As it is clear in the Table 3, in benefit-cost ratio of RRP is more than 1 (2.15) and therefore economically and socially are acceptable which is very impressive.

Its sensitivity analysis for change in discount rate as 11 and 13 percent has been done which the result shows 1 percent increase in discount rate, benefit-cost ratio will be decreased to 2.10

Table 1: Benefits of Rangelands rehabilitation Project (RRP) (10000 IRR)

Benefits	Water	Unit Price	Average
	25	30	
Ground Water Recharge	178.4	219.1	198.8
Decrease in Soil Erosion	40		40
Increased in Carbon Sequestration	20.2		20.2
Economic Values	416.9		416.9
Employment Creation Values	710.0		710.0
Ecosystem Benefits	238.7	279.4	259.0
Social Benefits (without considering participation values)	1365.5	1406.2	1385.9
Social Benefits (with considering participation values based on Yazd province condition)	1802.5	1856.2	1829.4
Social Benefits (with considering participation values based on Iran condition)	1843.5	1898.4	1870.9
Total average			**1850.2**

Table 2: Operational and overhead expenses in the Kamkooyeh village (GEF, Government of Iran & UNDP) in 10000 IRR

Stakeholders	Expenses type		Level	Operational and overhead
GEF	Official	Management	National	100
			Provincial	14
		Official & Equipment	National	2
			Provincial	2
	Travel		National	DW
			Provincial	160
	Capacity building	Meetings & workshops	National	-
			Provincial	InC
		Consultancy service	National	-
Government of Iran	Operational		Provincial	600
			Provincial	-
				20
	Official	Management	National	180
			Provincial	4
	Operational	Official & Equipment	National	8
Local Community			Provincial	1200
			Provincial	-
	Operational		Cash	L1
Total			In Kind	**2290**

DW: Distance Work
InC: Included in consultancy expenses
L1: 11 ha of seedling and mass plantation

and 1 percent decrease in discount rate, benefit-cost ratio will be increased to 2.20.

DISCUSSION AND RECOMMENDATIONS

According to the findings from calculations, the followings recommendations are proposed for better implementation of the projects:

I. In arid and semi-arid region which are facing with water shortage issue, flood hazard and soil erosion, using rangeland rehabilitation could improve water conservation and increasing water infiltration", and as a result, "increasing ground water recharge" plus "reducing flood hazard" plus "soil erosion control" (because of low runoff in hilly landscapes) and on the other hand, improving natural vegetation cover and as a result, more "Carbon & CO_2 sequestration" plus "biodiversity value" plus "soil erosion control" (because of existing a better protection cover for the top soil by the improved canopy and finally could have significant impact on socioeconomic condition of villagers.

II. Absolute attention to aspects of the economic benefit and sectoral viewpoint is a serious threat for the success of the projects. As the results show, socio-economic evaluation can lead to better and more accurate evaluation of the project impacts. Hence, integrated planning is more effective and efficient tool in design and implementation of the projects.

III. There is a risk that MENARID team are being more involved in the details of the implemented projects (hardware work) and paperwork so there is less attention to software works (in-

Table 3: Summary of benefit, cost and B/C ratio of RRP intervention

Type of Evaluation	Benefits (in 10000 IRR)	Cost (in 10000 IRR)	Annual uniform cost (in 10000 IRR)			B/C		
			Discount Rate			Discount Rate		
			12%	13%	11%	12	13	11
Financial	416.9	2290	431.6	451.1	412.6	0.97	0.92	1.01
Social	1850.2	2290	861.6	881.1	842.6	2.15	2.10	2.20

stitutional coherence and strengthening to achieve the goals of INRM).

IV. Integrated management/planning should provide a full coordination among all stakeholders. In some cases, there is a lack of coordination among the key stakeholders. As an example, in the project of irrigation water supply, the responsibilities of MENARID team have been well-fulfilled, however, those responsibilities by the local government and the villagers remains to be accomplished. As a result, during the field visit it was observed that the pipcline system was not under operation.

ACKNOWLEDGMENTS

This paper is derived from Institutional strengthening and coherence for Integrated Natural Resources Management (GEF-MENARID-Iran), Ex-ante Impact Study on the MENARID Intervention/Practices in Behabad Site, Kamkooyeh Village Project by kind financial support of ICARDA.

REFERENCES

1- Abbasifar, A. (2008). *The Financial and Economic Evaluation o Kheiroodkenar Forest whit Emphasis on Environmental Benefits,* Unpublished thesis. Departemant of Agricultural Economics. University of Tehran. Iran.

2- Andeson, J., & Session, J. (1991). *Managing low volume road Systems for Intermittent use*, Technical Report in Transportation Research Board#1291. US.

3- Baumel, C. P., Hamlett, C. A., & Pautsch, G. (1986). *The Economics of reducing the county road system.* Technical Report in Office of Secretary Transportation. Department of Transportation. Iowa State University. US.

4- Boardman, A. E., Greenberg, D. H., Vining, A. R., & Weimer, D. L. (2005). *Cost Benefit Analysis: Concepts and Practice (3rd Edition).* Prentice Hall Press. US.

5- Byrd, W. A., & Gildestad, B. (2001). *The Socio-Economic Impact of Mine Action in Afghanistan. A Cost - Benefit Analysis.* Technical Report in World Bank.

6- Dadrasi Sabzevar, A. (2007). Economic evaluation of watershed projects on agriculture, case study: Watershed of Mehr Sabzwvar. Paper presented at 6th Agricultural Economics Conferance. Mashhad. Iran.

7- DeGroot, R., Brander, L., Van der Ploeg, S., Costanza, R., Bernard, F., Braat, L., Christie, M., Crossmang, N., Ghermandi, A., Hein, L., Hussai. S., Kumar, P., McVittie, A., Portela, R., Rodriguez, L. C., Brinkm, P. T., & Van Beukering, P. (2012), Global Estimates of The Value of Ecosystems and Their Services in Monetary Units, *Ecosystem Services*, 1 (1),50–6.

8- Efati. (2000). The impacts of watershed Projects on Improving The Socioeconomic Status Villages' Watershed Basin. *Articles proceedings the role of Constructing Jahad in Agrcultural and Rural Development.* Reserch Centers and Rural Issuies. Iran.

9- Hanson, S. D., Hamlett, C. A., Pautsch, G., & Baumel, C. P. (1985). Vehicle travel costs on Paved, Granular, and Earth Surfaced County Roads. *Journal of Transportation Research Forum,* 26(1), 401-406.

10- Helmberger, P. J., Fruin, J., & Halbach, D. (1990). *Net benefits of rural road management strategies.* Unpublished thesis University of Minnesota. Minneeapolis. US.

11- Hosseini, S. S., & Shahbazi, H. (2013). Assessment of R&D role for Agricultural Supply and Demand Gap Adjustment. *Iranian Journal of Agricultural Economics and Development,* 84, 178-203.

12- Hosseini, S. S., & Shahbazi, H. (2014). Assessment of Iran's optimal Agricultural R&D Expenditure. *Iranian Journal of Agricultural Economics and Development Research*, 45(1), 23-40.

13- Hossini Tavasol, M., Kohandel, M., & Mortazavi farizhendy, G. (2007). *Prevent the Negative effects of Dam Construction and Use of the new Opportunities in the Areas of executive Strategies Related to natural Resources Case Study: Dam of Karon 3.* Presented at 1st National Confence of Dams and Hydrolic COnstravtion, Islamic Azad University, Karaj. Iran.

14- Jahern, C. T., Smith, D., Rukashaza-Mukome, M., White, D., & Johnson, G. (2005). *Economics of Upgrading an aggregate Road.* Technical Report in Research Services Division. Minnesota Department of Transportation. Minnesota. US.

15- Khalili, S. (1995). *Economic Evaluating of Songon Copper Mine Extraction*. Unpublished thesis. Shahid Beheshti University. Iran.

16- Khobfekr, H. A. (1999). *Assessment economic social and technical the watershed activities in Sistan of province.* Technical Report in soil consevation and Watershed Managment Research Center Press. Iran.

17- Malek Hosseini, A., & Mirrakzade, A. A. (2014). Economic Evaluation of rural Development (Case Study of Kermansha Province Irrigation and Drainage Network). *Iranina Journal of Agricultural Economics*

and Development Research, 45(1), 183-193.
18- Mansourian, N., & Mohammadi Golrang, B. (2008). Investigation economic-Social Community Watershed Projects in Iran: case study: the Basin of Kameh Watershed of Khorasan Razavi. Presented at 6th Agricultural Economics. Mashhad. Iran.
19-Pakzad, F. (1993). Economic evaluating of Investment Projects. Iran: Hirmand Press.
20- Reh, F. J. (2015). Cost Benefit Analysis, Retrieved from http:// management.about.com/ cs/money /a /cost benefit.htm. US.
21- Salehi, A. (2002). *Evaluation of social and economic impacts of the projects watershed province (play flooding and artificial feeding)*. Technical Report in Department of Rural Studies. Organization of Agricultural Jahad. Isfahan. Iran.
22- Sebaaly, P. E., Siddharten, R., & Huft, D. (2003). *Impact of heavy Vehicles on low Volume Roads.* Technical Report in Transportation Research Board#1819. US.
23- Shahbazi, H. (2013). *Economic evaluation of Iran's Agricultural R&D.* Unpublished dissertation Departmant of Agricultural Economics, University of Teharn. Iran.
24- Shahbazi, H., Mirghasemi, S. A., Abbasifar, A., Amjadi, A., & Pakravan, M. R. (2014). *Ex-ante Impact Study on the Menarid Intervention/Practices in Behabad Site, Kamkooyeh Village.* Technical report in International Center for Agricultural Research in the Dry Areas. Amman. Jordan.
25- Summary selection of Kamkooyeh base line study (2012), Iran's MENARID Team Report. Technical Report in Forests, Range and Watershed Management Organization. Tehran. Iran.
26- Tolliver, D. A., Dybing, A., Lu, P., & Lee, E. (2011). Modeling Investments in County and Local Roads to Support Agricultural Logistics. *Journal of transportation Forum*, 50(2), 101-115.
27- Vreeker, R., Nijkamp, P., & Ter Wekkle, C. (2002). *A Multicriteria Decision support methodology for evaluating airport expansion plans*. Technical Report in Transportation Research Board#3224. US.
28- Weick, e. (1993). *Cost- Benefit analysis and its possible application to the EARP Process*. Technincal Report in ESASINC. Retrieved from www.cyberus.ca. Canada.

10

Estimation of Water Demand of Agriculture Sector by Water-Yield Function (Case Study: Sistan Province)

Zahra Ghaffari Moghadam

Abstract

In this study, yield-water and profit function was used to obtain water demand function in agriculture sector. The results showed that the ratio of actual to potential Evapotranspiration had positive, significant effect on the ratio of actual to potential yield for wheat and barley. Also, square ratio of actual to potential Evapotranspiration had negative, significant relationship. After estimation of product function, water demand function for agriculture sector was obtained by profit function. Price elasticity of water demand for agriculture was -1.10 being less than -1 and showing that price policies can be an important factor in the control of non-optimum use of their valuable inputs.

Keywords:
Barely, Demand function, Profit function, Water- yield production function, Wheat

Institute of Agricultural Center at the University of Zabol.
** Corresponding author's email: ghafari_m_gh@yahoo.com*

INTRODUCTION

All living species need water for their survival. In addition, water is of vital importance in industrial, agricultural, metabolism, and other applications. The highest water consumption in Iran is in agriculture sector up to as high as about 90% (Amirnejad, 2006). As Iran is placed in a dry and semi-dry region, it suffers from water deficiency and it is necessary to consider the issue of water and especially water consumed in agriculture sector.

The two large components of agricultural water demand are irrigation demand and stock water demand. In this study, the former component was only considered. The most common methods for estimating agricultural demand are regression analysis and mathematical programming (Hooker and Alexander, 1998). This study uses non-linear programming models to estimate agricultural demand functions for irrigation districts in Sistan area. These programming models are used to predict optimal farm plans based on assumed input costs, product prices, and water-yield production functions. While programming models are one approach to estimate irrigation demands, an analogous approach in microeconomic theory would be to formulate the problem as a profit maximization problem and to derive a continuous factor demand for irrigation water using duality theory. The programming approaches have the advantage of explicit assumption about the interaction between inputs and outputs, and different types of technologies.

Mahan (1997) used nonlinear programming models to estimate agricultural demand function for irrigation districts in Southern Alberta. Wang (2005) used Irrigation Water Planning Model (IWPM) for deriving benefit functions of irrigation water which maximized the total profit of irrigated crop productions. When field survey data are limited, the economic benefit functions for irrigation water use may be derived using this model. In this study, water demand in agricultural sector was estimated by crop-water function and profit function in Sistan area.

MATERIALS AND METHODS

Crop production functions can be identified as having four categories: evapotranspiration and transpiration models, simulation models, estimated models, and hybrid models (Montazer and Riazy, 2008). Evapotranspiration models utilize linear yield-evapotranspiration relationships. Although evapotranspiration and transpiration models capture important aspects of crop-water relationships, they have limited ability to capture the impacts of non-water inputs and are of limited use for policy analysis. The simulation models simulate the crop production process in detail, while hybrid models combine aspects of the other three types. Among these model types, estimated production functions are more flexible than other types of models, and polynomial or quadratic functions are most widely used. In the present study, estimated models were used to estimate the production function. All irrigated crops in the region have their own water-crop functions. These functions are estimated by regression methods.

The quadratic polynomial form of crop production functions proposed by Dinar and Letey (1996) is expressed as follows

$$Y_a/Y_m=f(w)=a_0+a_1w+a_7w^2 \quad (1)$$

Where, *Ya* is the actual crop yield (metric tons (mt)/ha); *Ym* is the maximum potential crop yield (mt/ha); $w = WA/ETm$, is the ratio of the total available water *WA* (mm) to maximum seasonal potential evapotranspiration of the crop *ETm* (mm); the total water available to a crop includes effective rainfall, water irrigation and soil moisture (Mahan *et al.*, 2002). Assuming that the crop areas and irrigation efficiency are constant throughout the growing season, we have

$$WA_{j,cp}=SM_{j,cp}+EP_{j,cp}+EI_{j,cp} \quad (2)$$

Where, $WA_{j,cp}$ is the total water available to a crop field during whole growing season (mm); $SM_{j,cp}$ is soil moisture in the root zone at the beginning of the crop growing season (mm); $EP_{j,cp,t}$ is effective precipitation (mm) and $EI_{j,cp}$ is effective water applied to a crop field during period t (mm). Actual evapotranspiration is estimated based on the sum of spring soil moisture, effective precipitation during the growing season and the level of effective irrigation during the growing season that depends on timing and irrigation technology (Mahan, 1997).

Effective precipitation is zero for both wheat and barley in Sistan and since there was no information about the amount of moisture in the soil of this region, it was ignored in the calculations and it was assumed to be hidden in the effective irrigation.

Actual evapotranspiration was effective irrigation during the growing season. Potential evaporation was calculated by the use of monthly meteorological data for 26 years and CROPWAT software package and Penman-Manteith method. In Penman-Manteith method, potential evaporation is calculated on the basis of a daily or monthly period. In the present study, monthly evaporation was calculated as follows

$$Et_a = K_c \times ET_o \quad (3)$$

Where, ET_a is the amount of actual evaporation, K_c is crop coefficient which is different for various crops, and ET_o is potential evaporation (Alizadeh, 1995).

Demand function is estimated by profit function. Therefore, given the production function, the total profit of irrigation water use at demand site j was calculated based on the empirical seasonal crop-water yields functions, which is expressed as follows (Wang *et al.*, 2006)

$$\widetilde{B}_j = \sum pcp_{j,cp}.Ya_{j,cp}.AF_{j,cp} - \sum vc_{j,cp}.AF_{j,cp} \quad (4)$$

Where, $pcp_{j,cp}$ is the crop price, $cc_{j,cp}$ is the cultivation cost, and $vc_{j,cp}$ is the variable cost. The objective function is specified as

$$\begin{aligned} &\max \quad \widetilde{\beta}_j \\ &s.t: \quad \sum_{cp} AF_{j,cp} \le A_j \qquad \forall j \\ &AF^l_{j,cp} \le AF_{j,cp} \le AF^u_{j,cp} \qquad \forall j, cp \\ &\sum_{cp} EI_{j,cp}.AF_{j,cp} \le Q \qquad \forall j \end{aligned} \quad (5)$$

Where, A_j is land area (ha) in region j, $AF^u_{j,cp}$ and $AF^l_{j,cp}$ are minimum and maximum area limits for each crop, respectively, $EI_{j,cp}$ is effective irrigation during growing season (m^3/h), and Q is total water volume limit in site j. The demand function is estimated by solving model for various levels of effective irrigation water availability and plotting the corresponding shadow prices. The shadow prices of the water availability constraint measures the change in the objective function caused by a slight relaxation of the resource constraint. It can be interpreted as measuring the marginal value of water. Algebraically, the shadow price can expressed as follows (Mahan *et al.*, 2002)

$$MVW_j = \Delta\pi_j / \Delta Q_j \quad (6)$$

Where, MVW_{ij} denotes the marginal value of water (IRR/m^3) in site j, $\Delta\pi_{ij}$ is the change in profit (IRR) for a slight change in Q in region j, and ΔQ_j denotes the change in the total quantity of effective irrigation water in region j.

RESULTS AND DISCUSSION

Sistan area in the north of Sistan and Baluchestan Province is located to the southeast of Iran between the longitudes of 60°15' and 60°50' and latitudes of 30°50' and 31°28' (Kehkha *et al.*, 2004). The Hirmand river, which originates in Afghanistan, is the main water provider for this region and second most important border river of Iran among 26 border rivers (Ministry of Agriculture, 2000).

The average rainfall of the Sistan region is about 100 mm/yr as compared with an evaporation rate of about 4000 mm/yr. With about 245 000 ha of cultivable land, Zabol is the agricultural centre of the province (Ministry of Agriculture, 2000). The cultivation area depends on the Hirmand flow. The most important crops are wheat, barley, watermelon, melon, alfalfa, clover, and sorghum in the order of importance.

The agricultural water is solely supplied by the branches of the Hirmand river as well as the Sistan and Parian rivers. According Ministry of Agriculture (2000), about 82 500 ha of districts are irrigated by the Sistan river called Chah-Nimeh Irrigation Network. Ministry of Agriculture (2000), also, reports that annually about 1 250 million m^3 of the total incoming water from the Sistan river and the canals of the common Parian river and also Chah-Nimeh water reservoir are used for agriculture. This varies from 900 million m^3 to 1500 million m^3 per year. About 177 million m^3 of 1250 million m^3 are supplied by the Chah-Nimeh reservoir (Kehkhah *et al.*, 2004).

Areas studied in this research included Shipap and Poshtab in Sistan region. Their water requirement for agriculture is met by the reservoirs of Chah-Nimeh. They were selected for the study because they have same water supply sources. Also only wheat and barley were chose to avoid large model given the fact that they have the largest area under cultivation and the importance in the region.

Water-yield function estimated for wheat and barley was as follows

$$Ya/Ym = -0.97 + 0.16W - 0.067W^2$$
$$R^2 = 0.38 \qquad D.W = 1.7 \qquad F = 3 \qquad \text{ForWheat} \quad (7)$$

$$Ya/Ym = -0.6 + 0.1W - 0.0037W^2$$
$$R^2 = 0.40 \qquad D.W = 2 \qquad F = 3 \qquad \text{For barely} \quad (8)$$

Equation (7) indicates water-yield function for wheat and Equation (8) shows water-yield function for barely. The results indicate that the ratio of actual evaporation to potential evaporation has positive and significant effect on the ratio of actual yield to potential yield and the square ratio of actual evaporation to potential evaporation has negative, significant relationship.

Water demand function of agricultural sector

Demand function was estimated by profit function according to Equation (4). The benefit functions of agricultural demand sites are estimated by Irrigation Water Planning Model (IWPM) with different annual water availabilities. These programming models are used to predict optimal farm plans based on assumed input costs, product prices, and water-yield production function.

After the estimation of profit function, irrigation water demand of agriculture sector can be estimated by shadow prices obtained from profit function. The shadow price of the water availability constraint measures the change in the objective function caused by a slight relaxation of the resource constraint. It can be interpreted as measuring the marginal value of water at the corresponding level of resource availability. Therefore, the right side of water resources limitation was changed and a shadow price was obtained for each change in the limitation of water resources. The shado price was obtained as 1.50 (thousand IRR) for water. One unit increase in water amount, the profit is increased by 1.50 thousand IRR.

Also, 50 price-quantity pairs were estimated and then irrigation water demand was obtained by fitting regression equations to these demand schedules. Functional form was estimated including translog. In the model, the demand function is a function of price. Irrigation water demand is indicated in Graph 1 and Equation 9.

$$Q = 1.45 \times 10^{11} P^{-1.10} \qquad (9)$$

The results showed that the quantity of water demand is inversely related to the price of water so that higher water price results in lower demand for water.

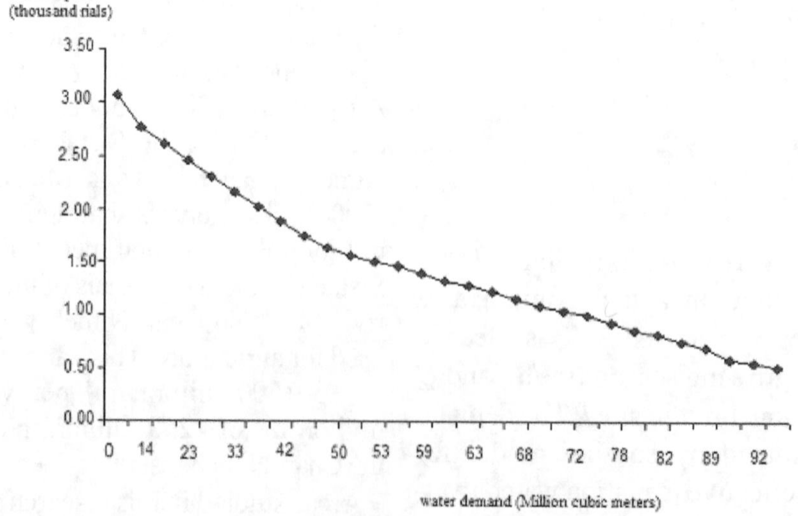

Figure 1: Irrigation water demand

The coefficient estimated for water price in demand function shows water price elasticity. The elasticity of water consumption in agriculture is -1.10. The elasticity of >1 showed high importance of pricing policy in the controlled use of this input in order to optimize the use.

REFERENCES

1- Alizadeh, A.(1995). *Applied Principles of Hydrology*, University of Imam Reza. Mashhad.

2- Amirnejad, H. (2006). Natural Resources Economics. Avay Maseih

3- Dinar, A., & Letey, J. (1996). *Modeling economic management and policy issues of water in irrigated agriculture,* Praeger Publishers.

4- Hooker, M., & Alexander, W. E. (1998). Estimating the demand for irrigation water in the central valley of California. *Journal of the American Water Resources Association*, 34 (3), 497– 505

5- Kehkha, A., Cacho, O., & Hardaker, B. (2004). *Water resource management focusing on drought mitigation in Iran: the case of the sistan region.* Paper presented at the 48thAnnual conference of the Australian Agricultural and Resource Economic Society, Melbourne February 11th to 13th.

6- Mahan, R.C., Horbulykb, T.M. & Rowse, J.G. (2002). Market mechanisms and the efficient allocation of surface water resources in southern Alberta. *Socio-Economic Planning Sciences.* 36: 25–49.

7- Mahan, R.C. (1997). *Efficient allocation of surface water resources in southern Alberta.* Unpublished dissertation, University of Calgary, Canada.

8- Ministry of Agriculture. (2000). Zabol cropping pattern a proposed plan. Unpublished working paper. Tehran. Iran.

9- Montazer, A., & Riazi, H. (2008). Optimization of water allocation in Qazvin irrigation command area. *Journal of Applied Irrigation Science*, 43(2), 129-143

10- Wang, L. (2005). Cooperative water resources allocation among competing users. unpublished dissertation Department of Systems Design Engineering, University of Waterloo.

11- Wang, L., Fang, L., & Hipel, K.W. (2006). *Basin-wide cooperative water resources allocation: Part II. Water and net benefits reallocation.* Technical Report No. 216-SM-280806, Department of Systems Design Engineering, University of Waterloo, Waterloo, Ontario, Canada.

11

Investigating the Effect of Variations in Irrigation Water Price on Cropping Pattern and Gross Margin under Uncertainty (Case Study: Khorasan Razavi)

Mostafa Mardani *, Saman Ziaee, Elham Kalbali and Samira Soltani

Keywords:
Optimization, Uncertainty, Water price, Khorasan Razavi Province

Abstract

Water shortage crisis is an issue that has led to drastic changes in different agricultural policies, especially in arid and semi-arid areas. Uncertainty in the amount of resources, e.g. water, used for agricultural production entails risk for farmers' income and cropping pattern changes. In the present study, the robust optimization model was used for optimal allocation of arable lands of Khorasan Razavi Province under uncertainty. During the allocation, the effect of water input price variations on total gross margin and cropping pattern was considered. It was found that under certain data, both parameters of total gross margin and total acreage are more than uncertain data. Given that water price variations resulted in tangible changes in wheat acreage, it is recommended to adopt appropriate policies to reduce its production risk.

Department of Agricultural Economics, University of Zabol, Bonjar Street, Zabol City, Sistan and Balochestan Province, Iran.
** Corresponding author's email: Mostafa.korg@yahoo.com*

INTRODUCTION

Changing a farm's ongoing costs is a powerful tool for orienting the optimal cropping pattern with regional macroeconomic policies. The variations in prices can be caused by the changes in production costs of inputs including the very limited input of water. Uncertainty in the amount of resources, e.g. water, used for agricultural production entails risk for farmers' income and also cropping pattern changes. Using a flexible model to determine the cropping pattern under uncertainty reduces the probability of system failure, where failure is defined as not meeting a given demand or other system constraint. Hazell and Norton (1986) believe that the distinction between risk and uncertainty is not beneficial in the mathematical programming because the available information for income distribution is usually limited to relatively small time series samples and subjective expectations held by farmer (farmer's mental background) and only estimation of possible income outcomes and probabilities related to each of these cases can be achieved (Chizari *et al.*, 2005).

Several studies have been conducted to optimize allocation of arable lands in different parts of the world (Sharma *et al.*, 2006 and Soltani *et al.*, 2008). Pfeiffer and Lin (2014) examined the effects of energy prices on groundwater extraction using an econometric model of a farmer's irrigation water pumping decision that accounts for both intensive and extensive margins. They found that energy prices had a significant effect on both the intensive and extensive margins and that the increase in energy prices would affect crop selection decisions, crop acreage allocation decisions, and farmers' demand for water. Various models have been used in research including goal programming (Sabouhi and Soltani, 2008; Sharma and Jana, 2009), fuzzy goal programming (Rastgaripoor and Sabouhi, 2009; Kohansal and Zare, 2008), multi-objective programming (Yeh and Labadie, 2003) and/or multi-objective fuzzy programming (Zeng *et al.*, 2010). In Iran, many studies have been done on determining optimal cropping patterns of agricultural products (El-Shishiny, 1988). In some of these studies, fuzzy models (Bagheri and Moazzazi, 2010) have been used to determine the optimal pattern. In these studies, the researchers chose different ways to apply existing risks in agricultural productions and calculated the existing risks in this sector.

The studied area is Khorasan Razavi Province with an area of 116,349 km^2. Agricultural section of Khorasan Razavi Province, as one of the largest and most important suppliers of agricultural products having a vast capacity of more than 1,081,130 hectares of various crops, has a decisive position in the national and provincial economy and an important role in providing critical needs of the community, food security, supply of material requirements of industries, and job creation (Biswas and Pal, 2005). The model used in the present study is to determine the optimal cropping pattern through robust optimization. One of the obvious advantages of robust optimization model is the flexibility in the application of uncertainty with regard to the social, economic and political conditions. Imposition of a type of optimal cropping pattern is avoided in this model and appropriate cropping pattern of each region is provided according to its specific circumstances.

MATERIALS AND METHODS

Table 1 lists the signs and indicators used in the present study.

Robust optimization model

One of the mathematical programming assumptions under certainty is that all parameters (input data) are fully known and determined. In practice, this assumption is unrealistic since most predicted or measured parameters are associated with uncertainty. To limit the uncertainty, a reliable system that can be designed. Soyster (1999) offered the following linear programming model to find answer for all uncertain data belonging to a convex set:

$$\begin{aligned} &\textit{Maximize} \quad cx \\ &\textit{subject to} \quad \sum_{j=1}^{n} \tilde{A}_{\cdot j} x_j \le b, \qquad \forall \tilde{A}_{\cdot j} \in K_j \\ &\quad j = 1,\dots,n, \qquad X \ge 0, \ \forall j. \end{aligned} \tag{1}$$

$\tilde{A}_{\cdot j}$ determines the jth column of the constraints

Table 1: The list of signs and indicators used in the present study

Sets	
j: the set related to product *j*,	$j \in \{1,2,...,J\}$
s: the set related to season *s*,	$s \in \{1,2,...,S\}$
i: the set related to month *i*,	$i \in \{1,2,...,S\}$
t: the set related to the type of fertilizer *t*	$t \in \{1,2,...,T\}$
e: the set related to basic products *e*	$e \in \{1,2,...,E\}$
Parameters	
p_{si} : price of water in month *i* of season *s*	
A_e : Minimum acreage requirement for basic products of *e*	
$C_{sj(e)}$: The amount of gross margin (without deducting the cost of water) per hectare of crop *i* or *e* in season *s*	
$W_{sij(e)}$: The amount of water requirement to produce one hectare of crop *i* or *e* in season *s* (m³/hectare)	
$f_{tj(e)}$: The amount of fertilizer type *t* requirement to produce crop *i* or *e*	
Random Data	
$\tilde{W}_{si}$: The average amount of water available in month *i* of season *s*	
$\tilde{A}$: Available acreage for all crops under study	
$\tilde{L}$: The number of available workforce	
$\tilde{F}_t$: The total amount of available fertilizer of *t* type	
Decision variable	
$x_{sj(e)}$: acreage of crop *i* in season *s*	
Model Variables	
Γ_i degree of conservation control parameter (uncertainty):	
$\tilde{\eta}_i$: Random variable for constraint *i*	
b_i : Right side values of constraint *i*	
ε :Level of specific uncertainty	
$\bar{a}_i$: Nominal value of uncertain data in constraint *i*	
$\hat{a}$: nominal value of uncertain data multiplied by the level of specific uncertainty	
$\tilde{a}_i$: random value of uncertain data in constraint *i*	
Ω_i : degree of uncertainty control parameter (for quadratic programming)	
P : probability of each constraint violation from its boun	

matrix and it is assumed that the uncertainty column belongs to the known and convex set of *K*. This model defines hard constraints (i.e. all such constraints must be provided) for all subsets of K_j. Thus, the optimal solution in this situation is likely to lose its optimality compared to the problem in certain circumstances because of the conservation model against uncertainty. The growth of conservation to ensure system reliability in uncertainty increases the operating and maintaining costs of the systems. On the contrary, flexibility in economic models leads to various options to choose from as compared to the rigid models such as Soyster's model (Soyster, 1973). To solve this problem, a different method is introduced to incorporate uncertainty in the problem. Uncertain variables are expressed as random perturbations for uncorrelated variables (Bertsimas and Sim, 2004):

$$\tilde{a}_{ij} = \bar{a}_{ij} + \tilde{\eta}_{ij}\varepsilon\bar{a}_{ij} = \bar{a}_{ij} + \tilde{\eta}_{ij}\hat{a}_{ij} \quad (2)$$

where, $\bar{a}_{ij}$ determines the nominal value of the uncertain parameter, $\varepsilon > 0$ specifies the determined uncertainty level, $\tilde{\eta}_{ij}$ is random variables distributed symmetrically in [-1, 1], $\hat{a}_{ij}$ is obtained by multiplying the nominal value of the variable $\bar{a}_{ij}$ e and the specified level of confidence, ε. Thus, the variable $\tilde{a}_{ij}$ has symmetric and bounded distribution in the constraint of $\lfloor \bar{a}_{ij} - \hat{a}_{ij}, \bar{a}_{ij} + \hat{a}_{ij} \rfloor$. Model (1) was modified after adding two additional variables of y and z as follows (9):

$$\begin{aligned} &\text{Maximize} \quad cx \\ &\text{subject to} \quad \sum_j \bar{a}_{ij}x_j + \sum_{j\in J_i} \hat{a}_{ij}y_{ij} + \Omega_i\sqrt{\sum_{j\in J_i} \hat{a}_{ij}^2 z_{ij}^2} \le b_i, \quad \forall i \\ &-y_{ij} \le x_j - z_{ij} \le y_{ij}, \quad \forall i, j \in J_j \\ &l \le X \le u, \\ &y \ge 0 \end{aligned} \quad (3)$$

where, J_i is a subset of the uncertain data indicator in constraint i, and Ω_i is a conservation control parameter for constraint i. Although this method caused conservation control, quadratic programming problems which had computational complexity had to be used. To overcome this problem, a new method was developed where model (1) remained linear and the degree of conservation was also controlled (Oliveira *et al.,* 2003). Consider the following optimization problem:

$$\begin{aligned} &\text{Maximize} \quad cx \\ &\text{subject to} \quad \sum_{j=1}^{n} \tilde{a}_{\cdot j}x_j \le b, \quad \forall i, j \in J_j \\ &l \le X \le u. \end{aligned} \quad (4)$$

Uncertainty in this model is the same as that in model (3). J_i is a subset of indicators related to the uncertain parameter of $\tilde{a}_{ij}$, which is determined for every constraint of i. It is assumed that all $\tilde{a}_{ij}$ are independent, symmetric, and bounded in the range of [-1, 1]. To control the degree of conservation, a parameter (Γ_i) is defined that a real number in the range of $[0, J_i]$ and can be attributed to it. Here, model (4) is rephrased in an optimization form by parameters which control the degree of conservation that improves the reliability of systems under uncertainty (Oliveira *et al.,* 2003):

$$\begin{aligned} &\text{Maximize} \quad cx \\ &\text{subject to} \quad \sum_i \bar{a}_{ij}x_j + \max_{\{S_i\cup\{t_i\}|S_i\subseteq J_i, |S_i|=[\Gamma_i], t_i\in J_i\setminus S_i\}} \\ &\left\{\sum_{j\in S_i} \hat{a}_{ij}y_j + (\Gamma_i - [\Gamma_i])\hat{a}_{it_i}y_t\right\} \le b_i, \quad \forall i \\ &-y_j \le x_j \le y_j, \quad \forall j \in J_j \\ &l \le X \le u, \\ &y \ge 0 \end{aligned} \quad (5)$$

where for each j, we have $y_j=|x_j^*|$. In model (5), $\sum \bar{a}_{ij}x_j \le b_i$ represents the *ith* constraint under certainty. In robust optimization model, there is an additional maximizing term (second term) as compared to the previous condition. This maximizing term ensures reliability of the model against uncertainty by the degree of conservation control parameter of (Γ_i). Certainty level of the model against uncertainty depends on the value of Γ_i parameters. Whenever Γ_i=0, maximizing term is deleted from the model and the constraint under uncertainty is converted to constraint under certainty. Whenever $\Gamma_i=\lfloor j_i \rfloor$, the model protection against uncertainty reaches its peak and is completely done. In this method, an evaluation is performed between system conservation against uncertainty (Γ_i) and system capacity (x_i). In other words, the more the system conservation is increased against uncertainty, the less the capacity of the system becomes.

Model (5) can be solved by linear programming without computational complexity. Also, the maximizing term can be calculated out of this model. Γ_i is a suitable tool for investigating system power against uncertain parameters or its inability against these parameters. There are different values for Γ_i depending on the probability of *ith* constraint violation from its bound and also the number of uncertain parameters in that constraint. Inserting in equation (5) as the optimal solution, the probability of *ith* constraint violation from its bound is defined as follows:

$$pr\left(\sum_j \tilde{a}_{ij}x^*_j > b_i\right) \le B(n, \Gamma_i) \quad (6)$$

such that:

$$B(n,\Gamma_i) \le (1-\mu)C(n,\lfloor v \rfloor) + \sum_{l=\lfloor v \rfloor+1} C(n,l)$$

where, $n = |K_i|,\ v = (\Gamma_i + n)/2,\ \mu = v - \lfloor v \rfloor$ (7)

and also,

$$C(n,l) = \begin{cases} \frac{1}{2^n}, & \text{if } l = 0 \text{ or } l = n \\ \frac{1}{\sqrt{2\pi}}\sqrt{\frac{n}{(n-l)n}}\exp\left(n\log\left(\frac{n}{2(n-l)}\right) + l\log\left(\frac{n-l}{l}\right)\right), & \text{otherwise} \end{cases} \quad (8)$$

To calculate Γ_i, a desirable level of probability of *ith* constraint violation from its bound is considered. Regarding the number of uncertain parameters in that constraint, equation (8) is used for its computation. Assuming that equation (7) is tight, equation (8) can be directly used to calculate Γ_i.

Objective

The objective of the present study was to maximize gross margin by cropping pattern optimization. To investigate the effect of water price variations on optimal cropping pattern, water-related costs are individually included in the objective function. The objective function is as follows:

$$Max: \quad Z = \sum_{s=1}^{S}\sum_{j=1}^{J}(c_{sj(e)} - p_{si}w_{sij(e)})x_{sj(e)} \qquad \forall i \quad (9)$$

where, z is the total gross margin, $C_{sj(e)}$ is gross margin (without deducting the cost of irrigation) from one hectare of product j or e in season s, P_{si} is the price of water in month i of season s, $W_{sij(e)}$ is the amount of water required to produce one hectare of crop j or e in month of season s and $x_{sj(e)}$ is the acreage of product j e or in season s.

Uncertain Data

In a cropping pattern model, uncertainty takes place due to the fluctuations of some parameters in the model to optimize it. Uncertain data here are: average amount of available water ($\tilde{W}_{si}$), total acreage of crops ($\tilde{A}$), number of available workforce ($\tilde{L}$) and the total amount of fertilizer type t ($\tilde{F}_t$). According to equations (2) and (6), random form of each uncertain parameter can be written as follows:

$$\tilde{W}_{si} = \overline{W}_{si} + \tilde{\eta}_{si}\hat{W}_{si}, \qquad \forall s,\ \forall i \quad (10)$$

$$\tilde{A} = \overline{A} + \tilde{\eta}_3\hat{A} \quad (11)$$

$$\tilde{L} = \overline{L} + \tilde{\eta}_4\hat{L} \quad (12)$$

$$F_t = \overline{F}_t + \tilde{\eta}_{t+4}\hat{F}_t, \qquad \forall t \quad (13)$$

Equation (10) corresponds to uncertain parameters of an average amount of available water where $\tilde{W}_{si}$ is the total amount of available water in month i of season s, $\overline{W}_{si}$ is the nominal value, and is equal to 10% of the amount of $\overline{W}_{si}$ ($\varepsilon = 0.1$). Equation (11) corresponds to uncertain parameters of the total acreage in which $\tilde{A}$ is the total acreage of crops, $\overline{A}$ is the nominal total acreage and $\tilde{A}$ is equal to 10% of $\overline{A}$. Similarly, equations (12) and (13) can be similarly interpreted. $\tilde{\eta}_{si}$, $\tilde{\eta}$, $\tilde{\eta}_4$ and $\tilde{\eta}_{t+4}$ are random variables in [-1, 1].

Constraints

The total consumed water for all products in season s should not be more than the total available water in month i of season s.

$$\sum_{s=1}^{S}\sum_{j=1}^{J} w_{si}x_{sj(e)} \le \sum_{i=1}^{I}\tilde{W}_{si} \quad (14)$$

Available water in month i of season s is considered as uncertain parameters. In other words, the available water in each month is considered as an uncertain parameter. So, there are six uncertain parameters in one season. So, the random constraint for the first season (from October to the end of March) is as follows:

$$\sum_{j=1}^{J} w_{1i}x_{1j(e)} \le \sum_{i=1}^{6}(\overline{W}_{1i} + \tilde{\eta}_{1i}\hat{W}_{1i}) \quad (15)$$

In addition, this equation for the second season (from April to late September) is written as follows:

$$\sum_{j=1}^{J} w_{2i}x_{21j(e)} \le \sum_{i=7}^{12}(\overline{W}_{2i} + \tilde{\eta}_{2i}\hat{W}_{2i}) \quad (16)$$

Table 2: Descriptions of Data

	Wheat	Barely	Corn	Sugar beet	Cotton	Tomatoes
Gross margin (IRR*/ ha)	3593541	2937292	14162471	9177122	6105133	22263984
Water requirement (m^3/ha)	4983	4183	11283	15200	12550	13517
Phosphate (Kg /ha)	179	148	210	255	231	323
Nitrogen (Kg/ha)	246	204	197	371	323	408
Potash (Kg/ha)	19	12	44	80	31	18
Workforce (Man-day/Ha)	26.87	20.12	53.46	116.46	69.2	209.36

*Note: US $ 1≈ 32000 IRR
Source: Organization of Agriculture Jihad (2011)

Random constraints of (15) and (16) can be rewritten using the definition of the degree of conservation control parameters of $\Gamma_1 \geq 1$ and $\Gamma_2 \geq 1$, respectively and according to the model (5) as follows:

$$\sum_{j=1}^{J} w_{1i}x_{1j(e)} - \sum_{i=1}^{6}\overline{W}_{1i} + (\Gamma_1 - 1)\sum_{i=1}^{6}\left|-\hat{W}_{1i}\right| \leq 0 \tag{17}$$

$$\sum_{j=1}^{J} w_{2i}x_{2j(e)} - \sum_{i=7}^{12}\overline{W}_{2i} + (\Gamma_2 - 1)\sum_{i=7}^{12}\left|-\hat{W}_{2i}\right| \leq 0 \tag{18}$$

and it can be rewritten to the following form for $\Gamma_1 < 1$ and $\Gamma_2 < 1$:

$$\sum_{j=1}^{J} w_{1i}x_{1j(e)} - \sum_{i=1}^{6}\overline{W}_{1i} + \Gamma_1\sum_{i=1}^{6}\left|-\hat{W}_{1i}\right| \leq 0 \tag{19}$$

$$\sum_{j=1}^{J} w_{2i}x_{2j(e)} - \sum_{i=7}^{12}\overline{W}_{2i} + \Gamma_2\sum_{i=7}^{12}\left|-\hat{W}_{2i}\right| \leq 0 \tag{20}$$

Other constraints used in this study are also converted to the degree of conservation control parameters. The final forms of these constraints are as follows:

$$\sum_{j=1}^{J} x_{sj(e)} - \overline{A} + \Gamma_3\left|-\hat{A}\right|, \quad \forall S \qquad \Gamma_3 \in [0,1] \tag{21}$$

$$\sum_{j=1}^{J} l_j x_{sj(e)} - \overline{L} + \Gamma_4\left|-\hat{L}\right| \leq 0, \quad \forall s \qquad \Gamma_4 \in [0,1] \tag{22}$$

$$\sum_{t=1}^{T}\sum_{j=1}^{J} f_{tj}x_{sj} - \overline{F}_t + \Gamma_{t+4}\left|-\hat{F}_{tj}\right| \leq 0, \quad \forall t \qquad \Gamma_5, \Gamma_6, \Gamma_7 \in [0,1] \tag{23}$$

The constraints (21), (22) and (23) are related to arable land, workforce and chemical fertilizer, respectively. The constraint (23) with parameters Γ_5, Γ_6, and Γ_7 is for the three types of potash, nitrogen and phosphate fertilizers, respectively. Another constraint is related to the land acreage of the crops which are regarded as essential to meet the domestic needs. This constraint is defined as follows:

$$\sum_{e=1}^{E} x_{sj(e)} \geq A_e, \quad \forall S \tag{24}$$

where, A_e represents the minimum acreage required for essential products of e. Monte Carlo simulation method is used to evaluate the model for which 1000 random numbers (with a predetermined probability distribution), for each of generated uncertain data and optimal cropping

Table 3: Current and optimal cropping pattern with different water prices under certainty (p=1)

				Water Price (IRR/m^3)			
	Product	Variable	Current	120.5	250	640.2	1000
Acreage (ha)	Wheat	x_{11}	284573	332509	33250	-	-
	Barely	x_{12}	142946	29384	29384	364869	-
	Corn	x_{23}	23315	85846	85846	67838	7322
	Sugar beet	x_{24}	24443	-	-	-	-
	Cotton	x_{25}	45280	5200	5200	5200	5200
	Tomato	x_{26}	14491	30498	30498	45531	94048
	Total	A	535048	483437	184178	483438	106570

Table 4: Optimal cropping pattern with different water prices under uncertainty with the probability of p=0.1

			Water Price (IRR/m³)			
	Product	Variable	120.5	250	640.2	1000
Acreage (ha)	Wheat	x_{11}	288232	288232	-	-
	Barely	x_{12}	29840	29840	320651	-
	Corn	x_{23}	85580	85580	69970	16787
	Sugar beet	x_{24}	-	-	-	-
	Cotton	x_{25}	5200	5200	5200	5200
	Tomato	x_{26}	26241	26241	39272	83668
	Total	A	435093	435093	435093	105655

pattern model, are resolved for different levels of the probability of each constraint violation from its bound (p). The final value obtained from these numbers for the constraints in which there is uncertain parameter, is compared to the final value of original data. For each series of random numbers generated on each constraint, if the intended constraint is provided, "one" and otherwise "zero" is attributed to them. Finally, the percentage obtained from total number of "one" in this process is determinant of the functionality and flexibility of the proposed model in providing optimal solutions.

The data used in this study were obtained from Organization of Agriculture Jihad of Khorasan Razavi Province in 2011 (Jehad-Agriculture Organization, 2011).

RESULTS AND DISCUSSION

In this study, the optimization issue consists of decision variables to determine the acreage of six crops: wheat, barley, corn, sugar beet, cotton and tomato. Cotton is one of the crops for which an acreage of at least A_e=200 hectares is considered (to meet the needs of some spinning firms). The issue was optimized under uncertainty and with each constraint violation from its bound at the level of 10% (p=0.1), and also under certainty (p=1), for different water prices of (120.5, 250, 640.2, and 1000 IRR[1] /m³). According to equation (8), for p=0.1, to are {4.34, 4.34, 1, 1, 1, 1, 1}, respectively. Required data are summarized in Table 2.

Table 3 shows the existing cropping plan and the results of the optimization model under certainty (p=1) to determine the optimal cropping pattern.

The total cultivated area in the current model shows 11% decrease in average water price (120.5 million) as compared to the optimal cropping pattern in the studied area. The optimal cropping pattern is reported at three different irrigation water prices in Table 3. Average water price in this area is 120.5 IRR and other prices are presented by an expert in Organization of Agriculture Jihad of Khorasan Razavi for the management of water resources of the province from supply management to demand management.

As seen, tangible changes are not observed in cropping pattern in spite of the rising of water price from 120.5 to 250 IRR/m³. Crop acreages, such as tomato which has more gross margin than other products, were greatly increased as water price was increased from 250 to 640.2

Table 5: Total gross margin related to the model at different conditions (unit - billion IRR)

Description	Water Price (IRR/m³)			
	120.5	250	640.2	1000
Total gross margin under certainty	326	285	158	102
Total gross margin under uncertainty	295	258	143	92

[1] *32000 Iranian Rials (IRR)≈US $ 1*

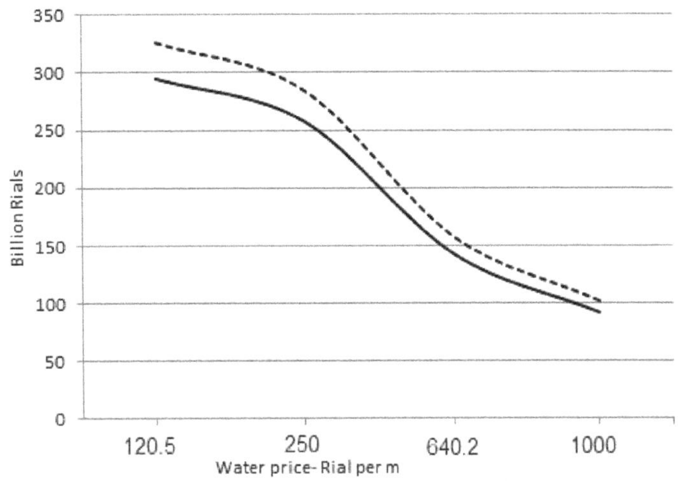

Figure 1: Total gross margin at different prices and conditions

IRR/m³ (to compensate higher water costs). The increase in water price from 640.2 to 1000 IRR/m³ resulted in sharp reduction of the acreage of all products. This reduction is lower in products with lower gross margin. In addition, this price increase caused the reduction of the total acreage of 483,437 to 106,570 hectares.

In addition, it can be seen that sugar beet is placed in the model in the current cropping pattern; however, it is recommended in none of the optimal patterns of cultivation. The main reason can be seen in Table 2. As it can be seen in Table 3, technical coefficients of the sources needed to produce this crop are high (especially water requirement) and their benefits are low. Also, according to experts, given that the bulk of the production in South Khorasan Province is to meet the needs of the province, there is no need to restrict the minimum planting area.

Results in Table 3 are related to the conditions in which all data are known and determined. In contrast, the results under uncertainty are different. Table 4 shows the results of robust optimization model with possibility level of . It is considered that in addition to the previous optimal cropping pattern (issue under certainty), acreage of crops with higher gross margin increases with the increase in water price under uncertainty. The increase in water price from 640.2 to 1000 IRR/m³ causes sharp reduction of the total acreage from 435,093 to 105,655 ha.

The main difference between these two issues

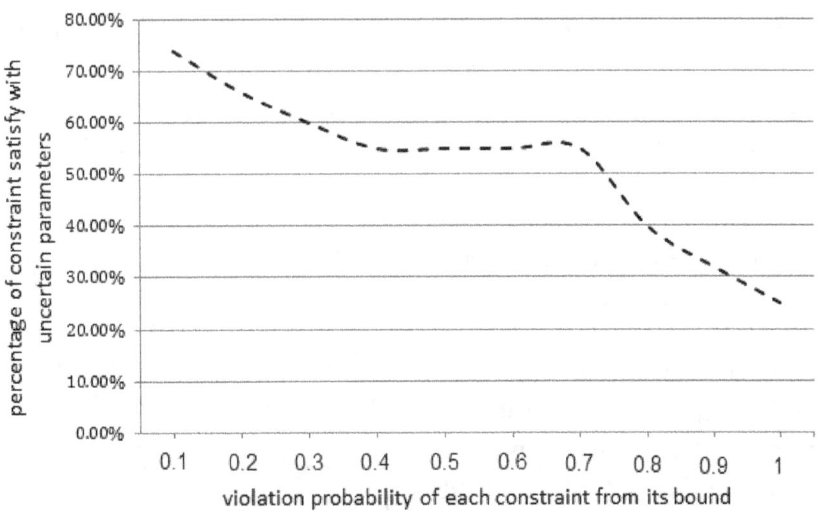

Figure 2: Determination of constraints supply percentage with uncertain parameter using Monte Carlo simulation (random numbers with a normal distribution)

(under certainty and uncertainty) is that the total acreage of crops under uncertain data with the probability level of at fixed prices is less than certain data Also, two crops, tomato and wheat, have less acreage under uncertainty than under certainty at different water prices due to higher gross margin and therefore, higher risk. For example, tomato acreage at a fixed price of 120.5 IRR/m^3 water is 30,498 ha under certainty and 26,241 ha under uncertainty (P=0.1). Also, it is seen that wheat price variation from 120.5 to 640.2 is neglected in cropping pattern under both certainty and uncertainty. However, barley acreage was increased sharply by price variations. The most stable acreage under different conditions and prices is related to corn.

Table 5 is related to the total gross margin of optimal cropping production at different water prices (under certainty and uncertainty with the probability of).

It is seen that the total gross margin is higher under certainty than under uncertainty at the probability level of , at fixed prices and for each cubic meter of water. For example, at a fixed price of 250 IRR total gross margin is equal to 285 billion IRR under certainty and 258 billion IRR under uncertainty. Also, gross margin decreases by the increase in water price at the constant condition (Figure 1). With respect to the subjects covered in this section, it can be observed that the issue of uncertainty in optimal cropping pattern is very important. Therefore, using models like the robust optimization triggers various options under different economic, social and political conditions for agricultural decision- and policy-makers.

Figure 2 shows the results of Monte Carlo simulations for optimal cropping pattern model. Generating 1000 random numbers for each level of constraint violation probability from its bound (P) (normal distribution with a convergence of 99.99 %). It is seen that the percentage of constraint provisions having uncertain parameters increases due to the increase in system protection in optimal cropping pattern against uncertain data (reduction of the degree of constraint violation probability from its bound (P). In the probability of 0.1 (maximum system protection), over 70% of random numbers satisfied the constraints. This amount of constraint provision in 1000 random numbers (in normal distribution with the convergence of 99.99%) is very justifiable and appropriate for each uncertain parameter. Also, it is noted that with the elimination of the system protection against uncertain data, less than 30% of random numbers is satisfied in uncertain constraints. This figure shows the ability of optimization through conservative control parameters.

CONCLUSION

In this study, robust optimization model was used to assess the uncertainty issue in the allocation of arable lands. Moreover, the effect of the variations of water price on gross margin irrigation and acreage of crops was studied. It was found that the total gross margin is higher under certainty than under uncertainty (at constant prices). With the increase in water price at constant conditions, the amount of gross margin reduces. Considering that the change in water price causes tangible changes in wheat acreage and giving that this product is strategic in domestic production, it is proposed to adopt appropriate policies to reduce risk and to prevent the acreage loss of the crop.

A desirable level of income for farmers and also a controllable use of water can be obtained using water price control policies at different conditions.

ACKNOWLEDGEMENT

The authors would like to thank the staff of Organization of Agriculture Jihad. Further, we would like to thank the reviewers of this manuscript for their very helpful advice.

REFERENCES

1- Jehad-Keshavarzi Organization Office of Statistics and Information Technology unpublished results (2011). Statistical Report of Irrigation Systems. Soil and Water Management Department, Iran.

2-Bagheri, M., & Moazzazi, F. (2010). Determining the optimal cropping pattern: Possibility programming application. *Agricultural Economics Research,* 1, 81-53.

3- Ben-Tal, A., & Nemirovski, A. (2000). Robust solutions of linear programming problems contami-

nated with uncertain data. *Journal of Mathematical Programming,* 88, 411–424.
4- Ben-Tal, A., & Nemirovski, A. (1999). Robust solutions of uncertain linear programs. *Journal of Operations Research, Letters,* 25, 1–13.
5- Bertsimas D. & Sim M. (2004). The price of robustness. *Journal of Operations Research,* 52 (1), 35–53.
6- Biswas, A., & Pal, B.B. (2005). Application of fuzzy goal programming technique to land use planning in agricultural system. *Journal of Omega,* 33(5), 391-398.
7- Chizari, A., Sharzchie, Gh., & Keramat Zadeh, A. (2005). Determining the economic value of water with ideal programming approach (Case Study: Barez and Shirvan Dam). *Economic Research*, 71, 66-39.
8- El-Shishiny, H. (1988). A goal programming model for planning the development of newly reclaimed lands. *Journal of Agricultural Systems*, 26(4), 245-261.
9- Hassan Shahi, M. (2007). Optimal pattern (economic) of crops under risk (application of objective models - MOTAD, Advanced MOTAD, linear programming and quadratic), case study of Arsanjan city. *Economics Research*, 43(1), 1-18.
10- Hazell, P.B.R. & Norton R.D. (1986). *Mathematical programming for economic analysis in agriculture,* Colliei MacMillan publisher, London, U.K., Pp. 387.
11- Kohansal, M. & Zare, A. (2008). Determining the optimal cropping pattern consistent with sustainable agriculture using fractional fuzzy programming with multiple objectives. A Case Study in Northern Khorasan Province, *Agricultural Economics and Development*, 62, 33-1.
12- Oliveira, F., Volpi, N.M.P., & Sanquetta, C.R. (2003). Goal programming in a planning problem. *Journal of Applied Mathematics and Computation*, 140(1), 165-178.
13- Pfeiffer, L., & Lin, C.Y.C. (2014). The effects of energy prices on groundwater extraction in agriculture in the High Plains Aquifer. *American Journal of Agricultural Economics*, 4(8), 213-222.
14- Rastgaripoor, F., & Sabouhi, M. (2009). Determining the cropping pattern using gray fuzzy programming (case study of Ghochan County). *Journal of Agriculture and Natural Resources Science and Technology,* 48, 413-405.
15- Sabouhi, M., & Soltani, Gh. (2008). Optimal cropping patterns in the basin area with emphasis on social benefits and net imports of virtual water: A case study of Khorasan region. *Journal of Agriculture and Natural Resources Science and Technology,* 43, 313-297.
16- Sharma, D. K., & Jana, R.K.(2009). Fuzzy goal programming based genetic algorithm approach to nutrient management for rice crop planning. *International Journal of Production Economics*, 121(1), 224-232.
17- Sharma, D.K., Ghosh, D., & Alade, J.A. (2006).A fuzzy goal programming approach for regional rural development planning. *Journal of Applied Mathematics and Computation*, 176(1), 141-149.
18- Soltani, Gh., Zibaee, M., & Kahkha, A. (2008). *Application of Mathematical programming in agriculture.* Research Organization, Agricultural Extension and Education. Tehran.
19- Soyster, A.L. (1973). Convex programming with set-inclusive constraints and applications to inexact linear programming. *Journal of Operations Research,* 21, 1154–1157.
20- Yeh, C.H., & Labadie, J.W. (2003).Multiobjective planning for agricultural decision: An application. *Agricultural Journal of Planning and Management*, 23, 336-343.
21- Zeng, X., Kang, S., Li, F., Zhang, L., & Guo, P. (2010). Fuzzy multi-objective linear programming applying to crop area planning. *Agricultural Water Management,* 7(23), 233-241.

12

Feasibility of Farmers' Participation in Optimal Irrigation Management System (Case of Guilan Province)

Mohammad Taleghani

Abstract

Farmers' participation in the maintenance and operation of water facilities has a positive impact on their attitude and internal trends and incentives to have closer cooperation with Regional Water Organization. The transfer of irrigation management to farmers enhances the productivity of water facilities and ultimately, results in efficient water use in agriculture. In order to evaluate the effect of participation and transfer of irrigation management to farmers, four hypotheses have been proposed. The study was an applied research according to the goal and was a descriptive survey as data collection method. Judgmental sampling was used to take samples from the population. A researcher-made questionnaire was employed for data collection. It was conducted among 120 people in Guilan Regional Water Organization engaged in agricultural sections. The questionnaire's validity was confirmed by content validity and its total reliability was estimated by Cronbach's alpha as to be 0.86 using SPSS software. The effects of participation on farmers' commitment, cooperation, and satisfaction and the legitimacy of Regional Water Organization of Guilan province were evaluated at 0.01 level, and no evidence was found to reject the hypotheses.

Keywords:
Farmers' participation, Commitment, Cooperation, Satisfaction, Legitimacy and irrigation system

Associate Professor, Faculty of Management and Accounting, Department of Industrial Management, Rasht Branch, Islamic Azad University, Rasht, Iran.
** Corresponding author's email: m.taleghani454@yahoo.com*

INTRODUCTION

The great challenge for the current century is to use less water to produce more food, particularly in countries with more limited water and land resources. The effective and sustainable use of water for agriculture has become a global priority requiring urgent and immediate solutions in view of intensifying competition. The current solution adopted by many countries, especially in Iran as one of the developing countries, is to create more and more irrigation facilities. However, these irrigation schemes have not achieved to high level of success as most of them are managed by government where farmers' participation is non-existent. Farmers' participation in the maintenance and operation of water facilities and irrigation management transfer to farmers are among the topics that have attracted the interests of officials and those involved in the water industry in recent decades (Omid *et al.,* 2012; Sato *et al.*, 2007; Tanaka and Sattto, 2005). Numerous studies indicate that most of the regional water companies, especially during the construction phase and the operational phase, do not care much for the wishes and opinions of farmers who are in essence the major water consumers and do not use their full potentials and power in the management of water utilities. Unfortunately, such an approach to water industry management results in the lack of maintenance and utilization of water facilities and has provided the grounds for early depreciation, while farmers' effective involvement in the management of water utilities can play an important role in solving these problems and helping regional water companies in the provision of maintenance and operation costs and the implementation of regulations for efficient water use in agriculture. Although, numerous studies have been carried out on the effects of participation on different items, there is shortage in research on the factors influencing participation. Thus, this article was organized to evaluate the effects of farmers' participation in the optimal management of irrigation systems in Guilan Province.

Bolfikova *et al.* (2009) suggest that "participation shall be exactly defined as ... who, what, when and what kind of involvement aspects". Participation is people's mental and emotional involvement in group situations that encourages them to achieve the group objectives and to share responsibility and assist the group. Mental and emotional involvement, motivation to assist and acceptance of the responsibility are considered as three important ideas of participation (Davis and Nyvastorm, 2013; Toosi, 2014).

In the Indian localization movement, participation is defined as social and creative process by those who determine and meet their needs and interests. This is not a passive participation in activities designed by others and not simply the exploitation of the results of the economic and social activities; rather, it is indeed pioneering in making decisions about the type of work, how to work and then doing the work" (Ackley and Marsden, 2011). Participation can include all members of a working group or just appears between two persons. When all members of a working group are involved in one issue, they will have more information and provide different ideas (Toosi, 2014). Involvement and participation in decision making and implementation leads to the satisfaction of the members of the group. Michell (2013) believes that people, who are involved in management decisions, are satisfied with the decisions and are likely to support Z theory. By participation, people not only satisfy their emotional needs but also achieve to their higher levels of needs and return to their original human nature i.e. their social life – and in addition to natural growth, they increase their labor productivity (Moshabaki, 1998).

Rosenbach and Sashkin (2014) state satisfaction as a key factor for increasing the consent and believes that it is a causal relationship between them. Lower and Hickman also showed that service provision by a participatory approach has much higher quality than in authoritarian style (Moheb Ali, 2014). Participation is a complex process with dual strategic significance; in other words, participation is a strategy to improve the socio-economic system and requires social actors to understand the conflicting values and interests and to negotiate in order to improve this system (Rahnavard, 1998).

Development and promotion of intellectual,

innovation and empowerment of individuals depend on the use of intellectual resources. In this way, participation in decisions making strengthens all levels of organization. In general, participation will entail the following results:

- Creating a "commitment"
- Increasing the quality of work
- Flourishing new ideas
- Reinforcing motivation and increasing satisfaction
- Giving the feeling of ownership to participants
- Increasing efficiency and job satisfaction, and reducing losses and the costs and conflicts
- Breaking the culture of silence
- Providing the required backgrounds to "change"
- Destroying marginalization
- Leading to better acceptance of the decisions
- Creating "responsibility" among participants for doing work (Mehdizade *et al.*, 2013).

The advocates of participation regard cooperative management as the best method with an emphasis on the results of studies. Henkel (2014) identifies the key advantages of participation for learning according to which the participants learn the functions of the organization and gain richer understanding of the dynamics and priorities of a larger system. Rosenbach and Sashkin (2014), also, believe that the implementation of participatory management is morally necessary. There are three important properties in the process of participation that make people accept the commitment: selection (the creation of opportunities for participation), to be revealed (doing job in relation to everyone, especially those who understand the importance of the subject) and irreversibility and the feeling that the limit should be overstepped (Tashmn and Ariel, 2009).

McLgan and Krystvnel (2012) believe that participatory methods have a significant impact on our achievements. Also participatory working methods, closely related to the replacement rate, increase production and improve performance. Some management writers know participation as a process in which people have the opportunity to actively participate in management processes affecting affairs (Towers, 2012). In addition, it is emphasized that participation plays an important role in the development of the atmosphere and style of management in an organization (Cornelius, 2011). Others know participation effective in the spirit and the quality of life (Foot and Caroline, 2009). Mehdizade *et al.* (2013) found out that the participative style could improve organizational performance by two variables including attitude and feeling which included effectiveness, efficiency, productivity and innovation. Participation in high level prepares people for four key elements: power, information, knowledge and rewards. Authority is divided through a variety of mechanisms in order to receive the data for decision making. Information on the activities of the organization are given to people as so they can appropriately participate in decision-making with sufficient understanding (Taleghani *et al.*, 2014). Knowledge allows people to do their complex affairs and thus, increase decision-making responsibilities through skill training, learning from each other about related affairs and education of the group working. Finally, reward systems are modified to provide incentives to increase the skills and commitment to the organization's overall effectiveness.

Hyman and Meisen (2014) noted the different aspects of participation in decision-making. They emphasized that TQM will be establish with individuals' collaborative partnership and their commitment. The effect of participation not only on commitment but also on other aspects of management has been studied. The available evidence about the impact of participation on satisfaction, productivity and conflict show that participation techniques have potentially useful results for these areas (Cowling and Philip, 2014). Despite the positive aspects of participation in management, some authors have considered the limitations of this management style which is briefly discussed below.

Participation is like a plant that does not grow easily in the human environment (UNDP, 1993). Powerful interest groups that are motivated by personal greed are making numerous barriers against the way of people's political and economic power. These barriers include legal and juridical systems, bureaucratic constraints, values and social traditions and improper distribution of assets (Toosi, 2014). Some researchers believe that participatory and group decision-making has some disadvantages as follows:

Working group is slower. Group spends more time to reach final decision and the waste of time means the waste of capital.

The team's efforts lead to reconcile that from the view of its effectiveness, is not always a good decision.

Group is under the influence of an individual or a small group.

Excessive dependence on the group decision may limit the management's ability to act quickly and decisively in the urgent cases (Michell, 2013).

In general, there are eight components affecting the implementation of the partnership: monitoring will disappear; the decision will take longer time; the group thinking will reduce the quality and efficiency; individualism will be lost; apathy will be a threat; rights and responsibilities are not equal and the management will be gone (McLgan and Krystvnel, 2012). Scientific resources of research considered various aspects of participation in the management as far as they were relevant to the present study. A review of their findings will shed light on how consistent these theoretical debates are with the results of the present study.

Humans are social creatures tending to live in groups for material needs (Golabi, 2012) because participation in group activities leads to increase team collaboration and plays an important role in satisfying them. With this description in mind, the present study proposes the following hypotheses:

H_1: There is a significant relationship between participation and farmers' commitment.

H_2: There is a significant relationship between participation and farmers' cooperation.

H_3: There is a significant relationship between participation and farmers' satisfaction.

H_4: There is a significant relationship between participation and legitimacy among farmers.

H_5: There is a significant difference between mean of commitment, cooperation, satisfaction and legitimacy.

MATERIALS AND METHOD

The research method according to goal was applied and based on data gathering, it was a descriptive survey. Research was conducted on farmers and Regional Water Organization of Guilan Province. Statistical population consisted of all staffs that work in Guilan Regional Water Organization and engaged in agricultural sections. Judgmental sampling method was used for taking samples from the population. The population size was 600 people but all of them were not engaged in agriculture section. So, after negotiating with supervisions of the related departments of the organization, 120 people were introduced who were working in Guilan Regional Water Organization and were engaged in agricultural sections. Questionnaires were distributed through judgment sampling and finally, 105 questionnaires were extracted and used.

The questionnaire used in this study consisted of five sections: (1) participation; (2) commitment; (3) cooperation; (4) satisfaction and (5) legitimacy. The survey instrument was composed of 16 items. Respondents were asked to indicate the extent to which they agreed or disagreed with each item by using five-point Likert type scale from "1=strongly disagree to 5=strongly agree". In order to ensure the content validity of the questionnaire, ten experts were interviewed about the relationship between the proposed questions and hypotheses. The reliability of the measurements in the survey was tested using Cronbach's alpha (α). The reliability coefficients (α) of different parts were as follows: participation (0.843), commitment (0.791), cooperation (0.814), satisfaction (0.703), and legitimacy among farmers (0.809).

The data were entered into SPSS software package for processing and the necessary calculations; thus, the needed information was extracted by drawing frequency and contingency distribution table; and the possibility of describing the population and analysis of data and hypothesis has provided. According to research properties and hypotheses, "Chi-square" was used to test the differences in the means of the constructs. Then, the significance relationship between dependent variables to each other and with independent variable was analyzed by using ANOVA and Tukey post-hoc procedure for means comparison.

RESULTS

The result of descriptive statistics showed that in terms of age, the highest frequency allocated to the age range of 30- 40 years. In terms of the land area, the highest frequency (78.1%) was belonged to land area of <2 Hectare. The study of statistical population table showed that the farmers of statistical population had very high level of commitment (45.7%) and cooperation (61%) and that their satisfaction with the organization was 66.7 percent. Furthermore, following the implementation of participation plant, the legitimacy of the Water Organization was increased to 62.9% among farmers.

Also the study of frequency of statistical sample in terms of participation showed that 59% of the statistical samples had high participation, 36.2% average participation and 4.8% low participation. Then, to determine whether there was evidence for confirming or rejecting research hypothesis, research hypotheses were tested and were ranked based on the influence of participation variable. Chi-square was used for hypothesis testing and according to Table 1 no evidence was observed to refute all four hypotheses at the 0.05 level.

This question to answer was which independent variables the participation had the highest relationship with and whether there was a significant relationship between the dependent variables. Table 1 show that not only was there a correlation between participation and dependent variables but also there was a correlation between the dependent variables. According to the above descriptions and Table 1, it was observed that participation had the highest correlation with the satisfaction and the lowest correlation with the commitment.

According to Table 1, Chi- Square with a confidence level of 95% show a significant relationship between dependent and independent variables (such as farmers' commitment, farmers' cooperation, farmers' satisfaction and legitimacy among farmers); and these variables are not independent. Therefore, the hypothesis H_1, H_2, H_3 and H_4 are accepted.

Also, there was a correlation between dependent variables commitment, cooperation, satisfaction and legitimacy, so that cooperation had a significant correlation with satisfaction (r_s=0.532), the farmers' cooperation had the highest correlation with farmers' satisfaction (r_s=0.636), the legitimacy had the highest correlation with farmers' satisfaction (r_s=0.527), the legitimacy had the highest correlation with farmers' cooperation (r_s=0.454), farmers' commitment had moderate correlation with the legitimacy of the organization (r_s=0.446), and farmers' cooperation had the highest correlation with farmers' commitment (r_s=0.612). Also, according to Table 2, there are

Table 1: Result of hypothesis testing

No.	Hypothesis	X^2	p-value	Result
H_1	There is a significant relationship between participation and farmers' commitment.	66.54	< 0.05	Accepted
H_2	There is a significant relationship between participation and farmers' cooperation with water organization.	63.44	< 0.05	Accepted
H_3	There is a significant relationship between participation and farmers' satisfaction.	73.11	< 0.05	Accepted
H_4	There is a significant relationship between participation and legitimacy among farmers.	86.07	< 0.05	Accepted

Table 2: Spearman Correlation between participation and dependent variables.

Dependent variable	Farmers' satisfaction	The legitimacy of the organization	Farmers' cooperation	Farmers' commitment	Participation
Farmers' satisfaction	1				
The legitimacy of the organizatior	0.527	1			
Farmers' cooperation	0.532	0.454	1		
Farmers' commitment	0.636	0.446	0.612	1	
Participation	0.808	0.783	0.744	0.716	1

Table 3: Results of Tukey HSD method

Variable	n	α=0.05	
		1	2
Commitment	105	.1762	-
Cooperation	105	-	3.4143
Satisfaction	105	-	3.4429
Legitimacy	105	-	3.4476

significant relationship between participation and farmers' satisfaction (r_s=0.808), the legitimacy of organization (r_s=0.783), farmers' cooperation (r_s=0.744), farmers' commitment (r_s =0.716).

Also, using analysis of variance (ANOVA), means were compared to determine whether there was a significant difference between the means and if there was, which one was due. Finding shows that at the confidence level of 95%, there is a significant differences between commitment, cooperation, satisfaction and legitimacy ($F=4.88$, $p<0.05$).

To answer this question which variables the difference result of, using the method of analysis of variance, "Tukey HSD", the difference was investigated and it was found that at the 95% confidence level commitment is significantly different from the average of other variables. Thus, according to Table 3, commitment was placed in a group and cooperation, satisfaction and legitimacy were classified in the other group.

CONCLUSIONS

The results of the study revealed a significant, positive relationship between participation and commitment of farmers and confirmed the theoretical discussions so that as participation increased, farmers showed greater sense of commitment to the maintenance and exploitation of water facilities and also efficient water use.

It was shown that 3.15% of farmers showed low to moderate commitment; in other words, they do not show interest in issues and events of the optimal management of irrigation system. Given the direct relationship of partnership and commitment, the root of the lack of interest should be sought in their unwillingness to participate.

Several factors play role in farmers' unwillingness to participation including: unfair distribution of water, dissatisfaction with contractors of pumping stations, the distance of agricultural lands to water divider valves, fatigue of irrigation channels, inappropriate behavior of water organization staff with farmers, not voting of farmers about current issues and future of pumping stations and not to go on sale of agricultural products after harvest. Based on the principles of scientific research, there is a relationship between participation and cooperation. As the results of the study showed, the relationship between farmers' participation and cooperation with water organization was significant and direct meaning that farmers' cooperation with water organization would increase with higher participation, and the coefficient of the correlation between these two variables was high. Farmers' cooperation with regional water companies about issues of irrigation may be done in different forms including the provision of all or part of the costs of maintenance and exploitation of irrigation channels, cooperation in canal dredging and weeding, distribution of water, settled disputes over water rights, the implementation of efficient agriculture water use. In general, regional water companies if know that there is a strong relationship between participation and cooperation will be able to perform their laws and programs with minimum cost and difficulty meanwhile they will provide financial and human resources needed for water installations and ultimately, increase organizational productivity. Also, the evaluation of research scientific resources emphasized the relationship between participation and satisfaction. The findings of this study corroborate the theoretical foundations. The results of the study showed that there was a direct, significant relationship between farmers' participation and satisfaction with water organization, and test of "Chi-square" also confirmed this point so that the correlation coefficient between farmers' participation and satisfaction was as high as rs =0.808. In other words, participation has the highest effect on farmers' satisfaction with organization water. Several factors may play role in farmers' satisfaction with water organization, some of which are mentioned below:

After implementation of participation plan, the employees of water organization established a close relationship with farmers, so that they tried to resolve the problems and issues related to the pumping stations in several public meetings with all farmers and private meetings with their representatives. Also in order to improve the condition of stations and the equitable distribution of water, joint decisions were made including participation in the selection of exploitation contractor, determination of the rate of delivery water, and the distribution of agricultural water.

Water requirements of agricultural land after the implementation of participation plan will be given at the disposal of farmers regularly, confidently, fairly and timely.

The participation system has been designed bottom-up in accordance with.

The organization of water helps farmers as much as possible in times of water scarcity and drought and other unexpected events

It will be seen that the above-mentioned actions are possible to be done in the form of participation plan. They had a positive impact on the attitude of the farmers and consumers of water of pumping stations and is effective in satisfying and their cooperation with the water organization. It is evident that awareness of the relationship between above variables will help regional water companies in implementing the program and achieving the organization's objectives. Based on the results, the relationship between participation and the legitimacy of water organization among farmers is significant and direct. Also, Table 2 shows that the correlation between two above variables is as high as rs =0.783. As previously mentioned, several factors may be involved in increasing the legitimacy of water organization among farmers of water consumer of pumping stations, some of which are as follows:

After the implementation of participation plan, Water Organization of Guilan Province had more powerful presence in related activities of pumping stations. Observing the principle of impartiality, the organization also provided technical and advisory services and monitored constantly on matters relating to the station. It was able to distribute water among farmers fairly and on-timely.

Before the implementation of the participation plan, most of the devices and facilities of pumping stations were old and often inefficient such as pumps, alternators, electrical panels, water transfer channels and... due to their undesirable and incorrect exploitation by farmers. After implementation of the project, most of defective channels and systems were repaired and re-launched under the management of water organization and farmers' cooperation and self-empowerment.

Understanding the existing conditions, Water Organization provides farmers with the required facilities such as mechanical shovel (if necessary).

The staff of water organization is involved in resolving problems related to pumping stations by listening to farmers' complaints in the water sector.

All these factors, in addition to several other factors have led farmers to understand the legitimacy and necessity of the presence of water organization as an adjusting and amplifying lever and to trust it. In principle, the ultimate goal of any research is to give recommendations to help overcome the existing problems. Also research results and its recommendations pave the way for researchers who want to study the same issue. Therefore it is necessary for any research to finally present some suggestions to direct future researchers and to solve the problems of related organizations and companies, at least partially. Based on the results of the present study, the following recommendations can be drawn:

According to the positive and strong relationship between participation, commitment and cooperation, regional water companies can increase farmers' sense of commitment and responsibility towards their water facilities by implementing participatory plans and delegating irrigation management to farmers and can make a profit in meeting the costs of maintenance and exploitation of water facilities overcoming the problems of shortage of human resources by using their cooperation. Also, implementation of participation plan will be followed by optimum consumption of agricultural water and increasing productivity of facilities and irrigation networks.

A pattern of participation can be fully successful if it is designed on the basis of farmers' opinions and demands and local culture of major consumers of water so that after the implementation of the plan it is also necessary for the coordination to be in accordance with farmers by holding successive meetings on the current problems of water facilities.

In order to increase the effectiveness of water users' participation and satisfaction, it is recommended that main water consumers of are studied the construction of water facilities and their suggestions are considered.

Farmers' participation in the ownership and exploitation of irrigation systems had positive effect on increasing their cooperation and commitment. To achieve this goal it is essential that regional water companies encourage farmers by implementation of promoting programs and assignment of long-term loans and share them in the ownership of water facilities.

It seems that participation is effective in the reduction of maintenance and exploitation costs, reduction of depreciation and the increase in irrigation efficiency and optimal use of agricultural water. We suggest that researchers interested in participation investigate the above-mentioned variables in special studies. Also, it seems appropriate that a comparative study is conducted to compare farmers' participation in the management of water facilities among selected two or more regional water companies. In addition, a comparative study of farmers' participation in the management of water facilities in the country with other countries will provide good results. The results of this type of researches can play an important role in identifying the strengths and weaknesses of farmers' participation and offer more useful methods for their participation.

ACKNOWLEDGEMENT

Author gratefully thanks the employees of Guilan Regional Water Organization and experts that participated in the survey, in particular, my dear colleague Mehran Mehdizade (Lecturer at Guilan Applied Science and Payame Noor Universitie).

REFERENCES

1- Ackley, P., & Marsden, D. (2011). *Approaches to participation in rural development* (M. Mahmudnizhad, Trans). Tehran: Centre for Research and Evaluation the rural issues of Ministry of the Construction Jihad.

2- Bolfikovaa, E., Hrehova, D., & Chrenovac, J. (2009). Analysis of satisfaction with a participation in organizations. *Serbian Journal of Management*, 4(1), 105- 115.

3- Cornelius, N. (2011). *Human resource management.* United Kingdom, UK: Thomso.

4- Cowling, A., & Philip, J. (2014). *The Essence of personnel management and industrial relations.* New York, NY: Prentice Hall.

5- Davis, K., & Nyvastorm, J. (2013). *Human behavior at work (organizational behavior)* (M. A. Toosi, Trans.). Tehran: Public Administration Training Center.

6- Foot, M., & Carolinc, H. (2009). *Introducing human resource management.* England: Longman.

7- Golabi, S. (2012). *The principles of sociology.* Tehran: Mitra Publication.

8- Henkel (2014). *Information on how to participate in the 2014 Annual General Meeting, How to vote by post and how to assign and use proxies (powers of representation)*. Annual General Meeting of Henkel AG & Co. KGaA, April 4

9- Hyman, J., & Meisen, B. (2014). *Managing employee involvement and participation.* London: SAGE Publications.

10- McLgan, P., & Krystvnel, (2012). *Era of partnership* (M. Islamieh, Trans.). Tehran: Cultural Research Bureau.

11- Mehdizade, M., Dorostkar Ahmadi, N., & Ramzanian, M. R. (2013). Designing conceptual model of participative decision-making as a strategy for improvement of employee's attitude, emotions, and job performance in work-life (case of: Rasht branches of Maskan Bank). *Transformation Management,* 5(9), 104-133.

12- Michell, T. (2013). *Management of organizations, introduction to behavior of the staff in organization.* (M. H. Nazarynezhad,Trans). Tehran: The Institute

of printing and Publications of Astan Quds Razavi.
13- Moheb Ali, D. (2014). *Scientific methods of improving the productivity of human resources*. Tehran: Public Administration Training Center.
14- Moshabaki, A. (1998). The position of participation in organizational productivity. *Journal of Management and Development Process*, 12(1), 13-28.
15- Omid, M. H., Akbari, M., Zarafshani, K., Eskandari, Gh. H., & Fami, H, Sh. (2012). Factors influencing the success of water user associations in Iran: A case of Moqan, Tajan, and Varamin. *Journal of Agricultural Science and Technology*, 14(1), 27-36.
16- Rahnavard, F. (1998). Supporters and opponents of participatory management. *Knowledge Management,* 11(41), 73-91.
17- Rosenbach, W. E., & Sashkin, M. (2014). *The leadership profile on becoming a better leader through leadership that matters.* Rosenbach & Associates, Inc.
18- Sato, M., Kono, S., & Onvichit, T. (2007). Principles and methods for participatory irrigation management and role charring between government and farmers. *The 4th Asian regional Conference and 10th International Seminar on Participatory Irrigation Management*, 2-5 May, Tehran, Iran, 283-297.
19- Taleghani, M., Asgari, M. H., & Modabernia, Y. (2014). *Industrial marketing*. Rasht: Rasht Branch, Islamic Azad University Publishing.
20- Tanaka, Y., & Sato, Y. (2005). Farmers managed irrigation districts in Japan: assessing how fairness may contribute to sustainability. *Agriculture Water Management,* 77(3), 196-209.
21- Tashmn, M., & Arieli, CH. (2009). *Innovation platform of victory.* Tehran: Rasa Cultural Institute.
22- Toosi, M. A. (2014). *Participation in the management and ownership.* Tehran: Public Administration Training Center.
23- Towers, B. (2012). *The handbook of human resource management*. Cambridge, Massachusetts: Blackwell.
24- United Nations Development Programme (1993). *Human Development Report 1993*. New York: Oxford University Press.

13

Analysis of Farms Performance Using Different Sources of Irrigation Water: A Case Study in a Semi-Arid Area

Hayat Lionboui [1], Tarik Benabdelouahab [1], Aziz Hasib [2] and Abdelali Boulli [2]*

Keywords:
Farm efficiency; Water valuation; Source of irrigation water; Irrigated perimeter of Tadla, Morocco

Abstract

Improving production efficiency is the main objective of government action to avoid efficiency losses and to increase the income of farmers. The aim of this study was to analyze performance levels of farms in the irrigated perimeter of Tadla in Morocco, according to the source of irrigation water. Thus, technical, allocative and economic efficiency were analyzed for farms in this area. To estimate the efficiency indices, the approach adopted is based on Data Envelopment Analysis method. Sixty samples of farms were chosen according to the mode of access to irrigation water. The results showed significant variability in technical, allocative and economic efficiency between the observed farms. The source of irrigation water is an important determinant of farm performance in the Tadla region. Thus, the average of economic efficiency varies between 45% and 83% according to the source of irrigation water. In terms of water valuation, farms that use only groundwater achieve a better value of irrigation water (2.19 MAD[1] /m^3) in comparison with those that combine between surface and groundwater and those which are limited to the use of surface water.

[1] National Institute of Agricultural Research, Research Center of Tadla, Beni Mellal, Morocco.
[2] Sultan Moulay Slimane University, Faculty of Science and Technology, Beni Mellal, Morocco.
* Corresponding author's email: lionbouihayat@gmail.com

[1] At the time of writing (mid-May 2015), MAD1.00 was approximately equivalent to £0.07, $US0.10, and €0.09.

INTRODUCTION

In the world, irrigated areas produce more than half of all foodstuffs and thus contribute hard to food security. This activity consumes about 72% of available water resources (Geerts and Raes, 2009; Seckler *et al.*, 1999). The importance of the irrigated area has been reiterated with the launch of the Green Morocco Plan's Strategy (GMP) in 2009. This strategy aims, among other things, to adapt Moroccan agricultural sector to climate change and to improve farmers' profits (MAPM, 2009). The interest, henceforth, will be focused not only on increasing production, but rather on an economically efficient production, that valorize, as well as possible, the available natural resources, especially water.

Across regions of Morocco, the Green Morocco Plan's Strategy was declined in regional agricultural programs. Tadla-Azilal region, equipped with an irrigated perimeter (Tadla perimeter), has benefited from a regional agricultural program focusing on the efficiency of production and water use, and on increasing of farmers' incomes. The agricultural development projects created under this strategy are based on the aspects of production, spatial distribution, water scarcity, land status and the integration of the maximum of agricultural value chains.

In the irrigated perimeter of Tadla, agricultural activities depends on climatic conditions, which makes water management as a technical and economical imperative and an excellent means for socio-economic development (Lionboui *et al.*, 2014). However, irrigation efficiency is still low with the dominance of traditional irrigation techniques (flood irrigation) which the rate of efficiency does not exceed 50% and which dominate 90% of the irrigated area. In addition, productivity does not yet reached the potential per irrigated hectare for some farms and the cubic meter of water is not sufficiently valued by crops. The valuation of water resources is an important element for sustainable water resource management (Edens and Graveland, 2014). Therefore, stakeholders in the study area are expected to improve production inputs efficiency, especially, irrigation water.

In this context, the study focuses on the analysis of the indices of technical, alocative and economic efficiency, as well as the valuation of irrigation water, through a sample of sixty farms selected according to the mode of access to irrigation water. The objective of the paper was to investigate the influence of different sources of irrigation water on Tadla farms' performance. We adopted a non-parametric approach based on linear mathematical programming through the "Data Envelopment Analysis (DEA)" method.

This paper is divided into three parts. The first part introduces the concepts and methodological approach. The second part is devoted to present the main results and the final section discusses the implications and recommendations emanating from the results of this work.

MATERIALS AND METHODS

Study area

The study area is the Tadla plain. It is located in central of Morocco (Figure1). The agricultural area is about 106,000 ha, including 98,300 ha of irrigated areas. The rainfall, in the study area, is mainly characterized by an annual irregularity and interannual variability, with an average amount of 330 mm/year (DRA, 2012).

Created in the 1940s, Tadla perimeter was among the first large irrigation schemes in the country and was aimed at benefiting small farmers and introducing modern farming techniques and industrial crops (Préfol, 1986). The irrigated perimeter of Tadla is divided into two sub-schemes by the Oum-Er-Rbia River flowing from east to west and taking its source in the Middle Atlas. The sub-scope of Beni-Amir is supplied from the Ahmed-Al-Hansali dam (capacity of 750 million m^3) built on the Oum-Er-Rbia river. On the left bank of the Oum-Er-Rbia River the Beni-Moussa sub-perimeter is fed by the waters of the Bin-El-Ouidane dam (capacity of 1.5 billionm^3). Also, Beni-Moussa and Beni-Amir Groundwater are used for agriculture in the Tadla plain. In addition to these two sources, much pumping is done from other superimposed layers, where the Turonian aquifer utilized by drilling 80 to 150

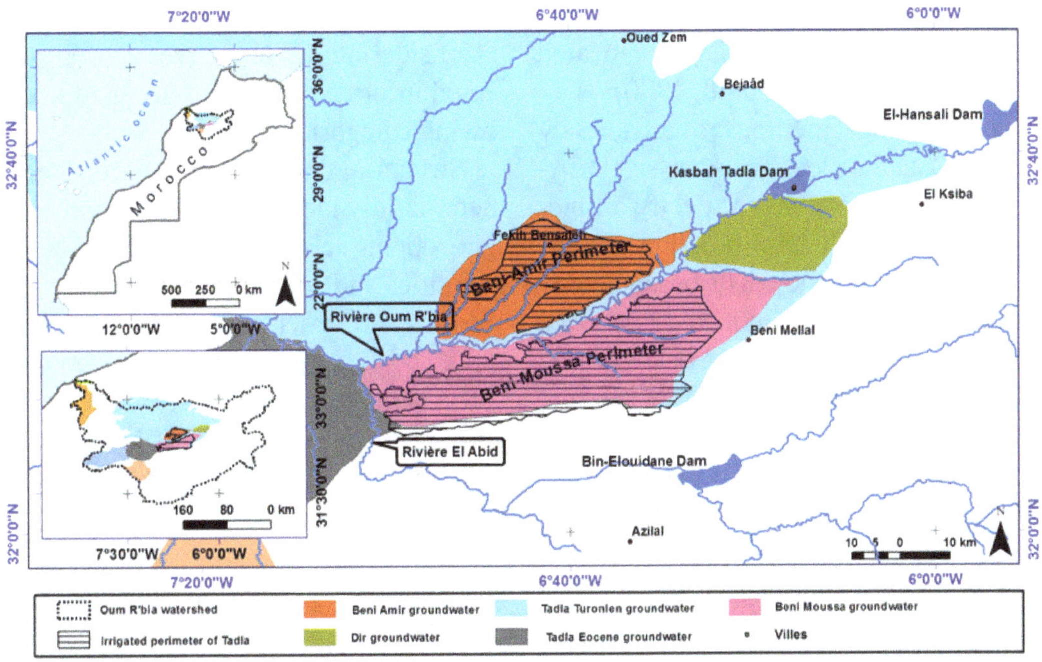

Figure 1: Irrigated perimeter of Tadla in Morocco

m depth (ABHOER, 2012).

On the concept of efficiency

Several indicators may be used to measure performance. In this work we focus on Technical Efficiency (TE), Allocative Efficiency (AE) and Economic Efficiency (EE).These three types of efficiencies were developed by Farrell (1957).

The first type measures the ability of one system to produce a maximum of output by using a limited quantity of input. This is the level of output that represents the production frontier (Barnes *et al.*, 2011). The technical efficiency index is given by the ratio of the actual volume produced on the potential volume (maximum). Thus, the technical inefficiency is the complement to one of the index calculated;

The second one consists of maximizing production using the least costly combination of inputs (Al-Feel and Al-Basheer, 2012). Indeed, it takes into account the prices of inputs and it is assumed that the optimal mix of inputs is one which ensures that the ratio of substitution marginal rates is equal to the ratio of prices. In other words, it is necessary that the producer equalizes the marginal value of outputs to marginal costs of inputs;

Finally, the last one takes into account the previous two types of efficiency (technical and allocative efficiency). In this sense, it is not enough to choose the most technically efficient output-input combination, but, it is necessary that this combination is also the one which occurred with the least cost. This complementarity allows providing meaningful criteria to assess the performance level of production systems. Two approaches are commonly used to measure efficiency: the non-parametric approach and the parametric approach.

In this work, we opted for a non-parametric approach based on mathematical programming through the "Data Envelopment Analysis (DEA)" method.

Data Envelopment Analysis (DEA)

Through the "Data Envelopment Analysis (DEA)" method we opted, in this work, for a non-parametric approach based on mathematical programming. This method enables us to compare all similar decision making units (DMU) in a given population, taking into account several dimensions simultaneously (Heinrichs *et al.*, 2013).

We collected data on input and output quantities of each DMU and we determined the efficiency frontier from the perspective of the best practice. The empirical application of this method consists in constructing a non-parametric piece-wise surface over the data through mathematical programming. The production frontier is estimated using a convex polyhedron enveloping all observations, which the most efficient are located on the frontier (Seiford and Thrall, 1990).

Compared to other methods of efficiency measurement, Data Envelopment Analysis (DEA) allows to avoid requirements to specify a functional form of the production function and restrictions on the inefficiency distribution (Yang *et al.,* 2015). DEA can be conducted under the assumption of constant returns to scale (CRS) or variable returns to scale (VRS). While the CRS assumption is only appropriate when all farms are operating at an optimal scale, the use of the VRS specification permits the calculation of efficiency devoid of scale effects (Coelli *et al.*, 1998). We assume that the farms involved in the study do not all operate at an optimal scale. Thus, we opted for data envelopment analysis model input-oriented under the assumption of variable returns to scale to analyze the efficiency of Tadla farms according to mode of access to irrigation water. This model allows determining input cost savings that can be achieved for each production unit of the sample if it is efficient.

Data used in the analysis of farms' performance

The data used are obtained from the farms survey conducted over the irrigated perimeter of Tadla during 2011/2012 cropping season (from September to August). Thus, we had applied a farm typology to understand the diversity of production potential at the regional level. Previous work on farmer strategies in terms of water management and agricultural production in the Tadla region has been useful to choose the typology. To analyze the diversity observed in the farmers behaviour, especially towards water management, we used the typology generated by Bacot (2001), based on the mode of access to irrigation water. This typology considers four types of farm types: a- farms with an easy access to groundwater and surface water (A_SGW); b- farms with an easy access to surface water and with limited access to groundwater (A_SW); c-farms with easy access to groundwater and not having access to surface water (A_GW); d- farms who don't have access to surface water and groundwater (R_A).

Considering the objectives of this study, collected information can be grouped into the following categories:

• Structural data of farms: farm size, employment, family labor and farm equipment's.

• Cropping selection, obtained yields, marketed quantities and selling prices.

• Use and costs of inputs for different cultures.

The sample includes sixty farms, chosen according to the source of irrigation water. The price data has considered into account 2012 as the reference year.

Data analysis by DEA method

In the Data Envelopment Analysis method (DEA), the efficient frontier is defined by the best performing practices of the sample used (Hoang, 2013). To define the efficiency frontier that serves as a reference for the measures of technical, allocative and economic efficiency, it is necessary to define the outputs and the inputs used in the production process. To calculate the scores of technical, allocative and economic efficiency, an aggregate of outputs and five aggregates of inputs were considered.

Outputs

The farms of this study are characterized by the diversity of their products (outputs). Therefore, the aggregation of outputs will be expressed in monetary value. Plant production is used in this study, as one aggregate of production. It includes the monetary value of all crops for each farm.

Inputs

The aggregation of production factors will be based on the quantities. Furthermore, we need to know the prices of these production factors (inputs) to calculate efficiency indices. Thus, production factors retained in this study are:

Irrigation water: It includes surface water delivered through the water supply network (0.32 MAD/m^3) and groundwater coming from wells or boreholes and designed for privately irrigation (0.6 MAD/m^3).

Sowing - Fertilization - Phytosanitary treatment:

This aggregate comprises quantities used in seeds, fertilizers and phytosanitary treatment.

Labor: Labor is expressed in Agricultural Work Unit (AWU) for each operation. The price of this aggregate is fixed in 50 MAD for each AWU.

Mechanization: This aggregate presents the different operations of agricultural preparation and maintenance of soil and harvesting. For each operation, the prices paid by farmers are available.

Agricultural land: the variation in land prices as well as the multitude of soil types in the irrigated perimeter of Tadla makes comparison between farms so delicate. To overcome this difficulty, we used the average rental price per hectare in the region according to the installed crop.

Water valuation indicator

The valuation of water resources by the agricultural sector is at the heart of the debate on the competitiveness and sustainability of this sector in the irrigated perimeter of Tadla. With the aim of preserving water, we need to know the farms which valuate better water. Through concrete examples of sixty farms in the irrigated perimeter of Tadla, we proceeded to assess indicators of water valuation for these farms relating to the mode of access to irrigation water resource.

The valuation of irrigation water has been discussed from the perspective of the community, that is to say in terms of value added per cubic meter of water used (VA/m^3). While gross margin appears to involve the enrichment of individual agents, the value added measures wealth creation for the community as a whole (including labor income). It represents the sum of labor remuneration, financial costs and taxes or subsidies, in addition to the gross profit of the agent (producer).

Therefore, the value added is not only an element of enrichment, but also a form of representation of the distribution of income to four fundamental agents of the national economy: households (labor remuneration), financial institutions (financial costs), government (taxes and subsidies) and non-financial companies (gross profit). The main crops that have been observed in the examined farms are: durum wheat, soft wheat, alfalfa, sesame, corn silage, sugar beets, olives and citrus.

RESULTS AND DISCUSSION

Farms efficiency

The economic or overall efficiency is jointly determined by the concepts of technical and allocative efficiency. The obtained results show certain homogeneity of efficiency scores in the observed farms. In terms of technical efficiency, the average score was 0.90 with amplitude ranging from 0.51 to 1. On the one hand, the results suggest that globally farmers could increase their production by almost 10% without increasing the level of inputs used. This indicates that these farmers do not use all of the available technological knowledge. On the other hand, 48.33% of observed farms have an index of

Table 1: Frequency distribution of technical, allocative and economic efficiency indices

Classes	Technical efficiency		Allocative efficiency		Economic efficiency	
	%	Cumulative	%	Cumulative	%	Cumulative
0.0 – 0.5	-	-	18.33	18.33	20.00	20.00
0.5 – 0.6	3.33	3.33	6.67	25.00	15.00	35.00
0.6 – 0,7	5.00	8.33	8.33	33.33	11.67	46.67
0.7 – 0.8	8.33	16.66	16.67	50	23.33	70.00
0.8 – 0.9	18.33	34.99	21.67	71.67	15.00	85.00
0.9 – 1.0	65.00	99.99	28.33	100.00	15.00	100.00
Average	0.90		0.75		0.69	
Standard Deviation	0.13		0.20		0.20	
Min	0.51		0.30		0.27	
Max	1.00		1.00		1.00	

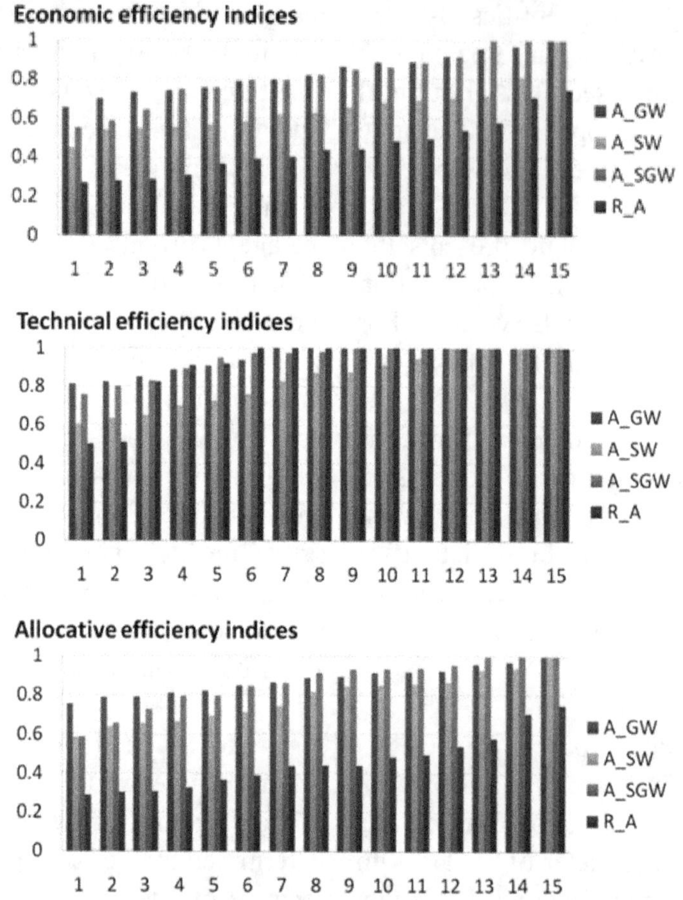

Figure 2: Efficiency indices distribution according to the mode of access to irrigation water

technical efficiency equal to 1%. This indicates that, these farms have mastered the technology available, and the improvement of productivity for this group of farm can be achieved only through technological progress (Table 1).

Concerning allocative efficiency, about 18.33% of farmers have efficiency scores lower than 0.5. This indicates that these farms do not optimally allocate production resources to produce a given goods level. For the latter, significant shortfalls can be recovered in terms of allocative efficiency. Nearly one farmer of ten has a score of allocative efficiency at around 1.

The economic efficiency combines simultaneously technical and allocative efficiency. The average index of economic efficiency is about 0.69. It should be noted that 20% of farms have an index of economic efficiency less than 0.50, which gives an idea about the potential scope for improving production and increasing farmers' income. Moreover, 8.33% of observed farms have an economic efficiency index equal to 1. This shows that the concerned farms have mastered the available technology and they are able to equalize the marginal value of their products to the marginal costs of production factors.

Farms efficiency according to mode of access to irrigation water

Indices of technical allocative and economic efficiency at the different types of farms in the irrigated perimeter of Tadla are presented in the following figure (Figure 2).

According to the figure below (Figure 3), we observe that the farms conducted in "A_GW" mode are technically the most efficient followed by those conducted in "A_SGW" mode with technical efficiency indices of 0.95 and 0.94 respectively on average. Then, the farms conducted in "A_SW" mode are in the third ranking with an index of technical efficiency equal to 0.84. Finally, farms conducted in "R_A" mode have the lowest technical efficiency index 0.82 on average. Furthermore, we note that there is no

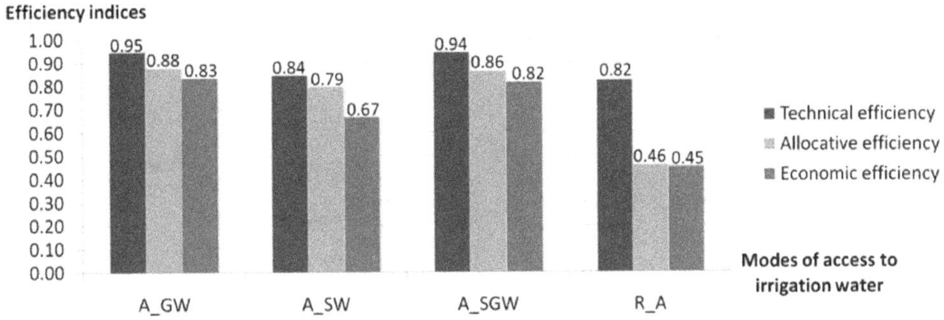

Figure 3: Average efficiency according to the mode of access to irrigation water

a big difference between the average indices of technical efficiency for the different types of farms. Therefore, there is no significant heterogeneity between farms types in mastery of available technology.

Except for farms conducted in "R_A" mode, which present an average allocative efficiency index of 0.46, we observe that there is not much difference between the average indices of allocative efficiency for the three other types of farms. Hence, we observe that farms conducted in "A_GW" mode are the most efficient in terms of resource allocation, with a score of 0.88, followed by those conducted in "A_SGW" mode, which have an index of allocative efficiency around 0.86. Then, the farms Conducted in "A_SW" mode come with an average allocative efficiency index equal to 0.79.

The results of this work show that there is no great difference between the average indices of economic efficiency for the first two types of farms. Farms conducted in "A_GW" mode, are the most economically efficient with an average index of 0.83. Farms conducted in "A_SGW" mode come in second place with an average economic efficiency index of 0.82. Both of these types of farms are characterized by the fact of having the ability to irrigate when considered appropriate to their crops and to designate the best suited irrigation technique for the context of the farm. Then farms conducted in "A_SW" mode come with an economic efficiency index of 0.67. This value can be explained by the irregularity of surface water supplies through the canals of the irrigation network. Also, this group of farms receives irrigation water in inappropriate times, in the opinion of farmers, and the majority of these farms are using a traditional irrigation called flood irrigation. At last, farms conducted in "R_A" mode come with an economic efficiency index of 0.45. These farms are directly dependent on rainfall which quantity differs according to the climatic years.

Efficiency and irrigation water valuation

In Morocco, the issue of water valuation is at the focus of interest of policymakers. Most studies that analyze this concept, consider water valuation indicator as the gross margin per cubic meter of water used (e.g. Sali and Monaco, 2014; Stanley, 2005; Waughray and Rodríguez, 1998). However, the interest of the farmer is not necessarily similar to that of the community, which is more interested in the overall income and in several development goals in the long term like the conservation of water resources. In this paper, water valuation indicator has been approached according to the community interest, that is to say, in terms of value added per cubic meter of irrigation water used by farm (Table 2).

The results show that while farms conducted in "A_SGW" mode realize the highest gross

Table 2: Water valuation and gross margin by mode of access to irrigation water

Farm Type	Water Valuation (MAD/m³)	Gross margin (MAD/ha)	Use of irrigation water (m³/ha)
A_GW	2.19	17,931.80	8,771.34
A_SW	1.83	11,637.33	7,305.08
A_SGW	1.98	18,002.24	9,859.97

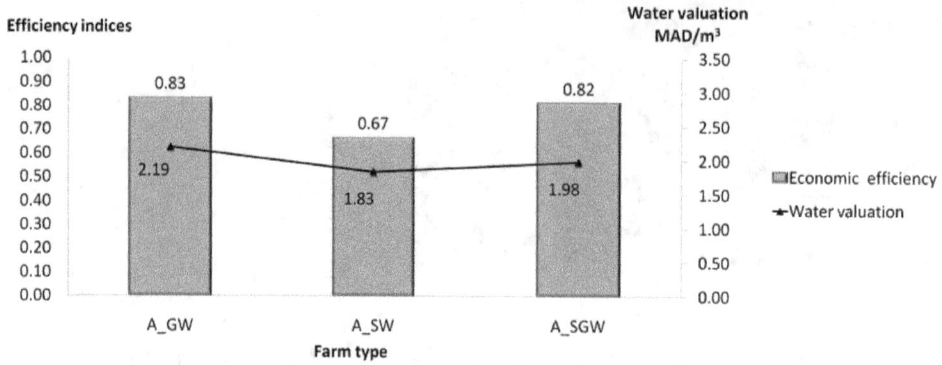

Figure 4: Economic efficiency and water valuation indicators by farm type

margin. In water valuation they come in second place (1.98 MAD/m^3) after farms conducted in "A_GW" mode (2.19 MAD/m^3). The interests of the community and those of farmers are divergent. It is therefore important, in this case, for policymakers to consider in addition to farms profitability, the valuation of scarce resources. Finally, farms conducted in "A_SW" mode are classified in last position, with an indicator of water valuation of 1.83 MAD/m^3.

In order to determine the relationship between water valuation and economic efficiency of farms in the irrigated perimeter of Tadla, following figure shows the average indices of economic efficiency and water valuation indicators for each farm type, according to the mode of access to irrigation water (Figure 4).

This figure shows, first, that even if there is no significant difference between farms conducted in "A_GW" mode and those conducted in "A_SGW" mode relating to economic efficiency, the last type of farms values less irrigation water than the first. This proves that the efficiency of farms and water valuation are two objectives that do not always converge. Farms conducted in "A_SGW" mode achieve more value added and they realize high yields because they have the ability to irrigate at the appropriate moment and with sufficient quantities of water relatively less expensive compared to those using only groundwater. Farms in this latter type, however, use less important quantities of irrigation water than the first, which allows them to achieve a good water valuation. Finally, we note that farms conducted in A_SW mode which are economically less efficient, have also the lowest indicator of water valuation (1.83 MAD/m^3).

CONCLUSIONS

The results of this research show that there is a considerable variation in technical, allocative and economic efficiency between farmers of the studied sample. In terms of technical efficiency, a substantial number of farmers have a high value. For these farms, the potential gains through available production technologies are minimal. Therefore, the adoption of new technologies is the effective way to increase their productivity. However, for farmers with a low level of technical efficiency, productivity could be increased by improving their expertise in the available technologies through mentoring and outreach.

One of the main guidelines of the Green Morocco Plan's Strategy (GMP) is one that concern the mechanism of a double integration of the actors. Vertical integration of upstream and downstream of the value chains activities, and horizontal integration which is more concerned about stakeholders and agricultural institutions. In terms of allocative efficiency, this mechanism could significantly reduce the inefficiency observed in the studied farms. Indeed, this mechanism is supposed not only able to circumvent the production constraints (financing, supply and management) but also, farmers can benefit of profit margins previously held by intermediaries in addition to gains due to economies of scale and the reduction of transaction costs.

In a region where the limiting factor is not the land, it is important to focus on the scarce resource, especially water in the case of irrigated perimeter of Tadla. Thus, the water valuation has been approached relating to the community interest, which is to say in terms of value added per cubic meter of water. It appears that irrigation water is better valued by farms which use only groundwater (2.19 MAD/m^3) followed by those that combine between surface and groundwater (1.98 MAD/m^3). For farms which are limited to the use of surface water, they are classified in last position, with an indicator of water valuation of 1.83 MAD/m^3. Also, the results suggest that efficiency does not always reflect water valuation by crops: an efficient farm for the farmer is not necessarily the one that values better irrigation water.

Accordingly of the results obtained in our study, and in order to remedy the failures in scores of technical, allocative and economic efficiencies, as well as in the water valuation indicators, the recommendations that could improve the efficiency and water valuation levels of Tadla Farms are:

- Take into account the water scarcity in the agricultural policies, and the valuation of this resource must remain at the heart of all strategies.
- Give more interest to extension services and organize awareness campaigns to circulate information between farmers.
- Allocate more importance to issues concerning the acquisition price, the quality and marketing channels of inputs.
- Take into account economically efficient farms by extension services which can serve as a reference model for inefficient farms.
- Allocate water resources to crops which value it at most.

In conclusion, on the one hand, farms, examined through this study have consistently required efforts to increase their efficiency levels. On the other hand, the preservation of water resources and especially their allocation should be subject of a public debate to reflect strategies and perceptions of local actors by involving them in a common prospective reflection.

ACKNOWLEDGEMENT

This work was supported by the National Institute for Agronomic Research of Morocco. The authors are indebted to the staff of the Regional Centre for Agronomic Research of Tadla for their collaboration.

REFERENCES

1- ABHOER (2012). Management Plan of Integrated Development of Water Resources. Hydraulic Basin Agency of Oum Er Rbia River.

2- Al-Feel, M.A., & Al-Basheer, A.A.R. (2012). Economic efficiency of wheat production in Gezira scheme, Sudan. *Journal of the Saudi Society of Agricultural Sciences,* 11, 1-5.

3- Bacot, M.L. (2001). Les stratégies d'utilisation conjuguée des eaux de surface et souterraines pour l'irrigation dans le périmètre irrigué de Tadla (Maroc).- Ecole nationale du génie de l'eau et de l'environnement de Strasbourg, France, Strasbourg.

4- Barnes, A.P., Rutherford, K.M.D., Langford, F.M., & Haskell, M.J. (2011). The effect of lameness prevalence on technical efficiency at the dairy farm level: An adjusted data envelopment analysis approach. *Journal of Dairy Science,* 94, 5449-5457.

5- Coelli, T., Prasada Rao, D.S., & Battese, G.E. (1998). An Introduction to Efficiency and Productivity Analysis, Springer Science & Business Media, Boston.

6- DRA (2012). (Regional Direction of Agriculture) Monography of Tadla region. Ministry of Agriculture and Marine Fisheries, Morocco.

7- Edens, B., & Graveland, C. (2014). Experimental valuation of Dutch water resources according to SNA and SEEA. *Water Resources and Economics,* 7, 66–81.

8- Farrell, M. J. (1957). The measurement of productif efficiency. *Journal of Royal Statistical Society,* 120, 253-281.

9- Geerts, S., and Raes, D. (2009). Deficit irrigation as on-farm strategy to maximize crop water productivity in dry areas. *Agricultural Water Management,* 96, 1275-1284.

10- Heinrichs, A.J., Jones, C.M., Gray, S.M., Heinrichs, P.A., Cornelisse, S.A., & Goodling, R.C. (2013). Identifying efficient dairy heifer producers using production costs and data envelopment analysis. *Journal of Dairy Science,* 96, 7355-7362.

11- Hoang, V.N. (2013). Analysis of productive performance of crop production systems: An integrated analytical framework. *Agricultural Systems,* 116, 16-24.

12- Lionboui, H., Fadlaoui, A., Elame, F., & Benabdelouahab, T. (2014). Water pricing impact on the economic valuation of water resources. *International Journal of Education and Research,* 2 (6), 147-166.

13- MAPM (2009). Ministry of Agriculture and Marine Fisheries, Regional Agricultural Plan of Tadla Azilal, Green Morocco Plan.

14- Préfol, P. (1986). Prodige de l'irrigation au

Maroc. Le développement exemplaire du Tadla,1936-1985. Nouvelles Editions Latines, Paris, France.
15- Sali, G., & Monaco, F. (2014). Irrigation Water Resource in a Rice-Growing Area: Economic Evaluation under Different Pricing Conditions. In "Agricultural Cooperative Management and Policy" (C. Zopounidis, N. Kalogeras, K. Mattas, G. van Dijk and G. Baourakis, eds.), pp. 23-40. Springer International Publishing.
16- Seckler, D., Barker, R., & Amarasinghe, U. (1999). Water scarcity in the twenty-first century. *Water Resources Development* 15, 29–42.
17- Seiford, L. M., & Thrall, R. M. (1990). Recent Developments in DEA: The Mathematical Programming Approach to Frontier Analysis. *Journal Econometrics,* 46, 7-38.
18- Stanley, C. B. (2005). Investigation of the economics of water as used by smallholder irrigation farmers in South Africa. University of Pretoria, Department of Agricultural Economics, Extension and Rural Development, South Africa.
19- Waughray, D. K., & Rodríguez, A. (1998). Valuing water as an economic good in dryland areas - balancing the need for food, environmental and financial security. In "World Congress of Natural Resource Economists", Venice.
20- Yang, L., Ouyang, H., Fang, K., Ye, L., & Zhang, J. (2015). Evaluation of regional environmental efficiencies in China based on super-efficiency-DEA. *Ecological Indicators,* 51, 13–19.

14

Assessment of Water Use Efficiency Indices in Selected Plains of Fars Province, Iran

Hamidreza Mirzaei Khalil [1], Mahdi Esfandiari [1], Javad Shahraki [2] and Morteza Yaghoubi [2]*

Abstract

This paper aimed to evaluate Water Use Efficiency (WUE) indices for crops and horticulture productions. In doing so, after gathering data from 164 farmers, we investigated water withdrawal cost indices and monetary return per cubic meter of water through two econometric and managerial approaches. Besides, water shadow pricesand dry matters of per cub meter of water were calculated. The results revealed that the average monetary return per cubic metre of water is 3875.4 IRR[1] (0.134 USD) and cost per cubic metre of well water equals 839.3 IRR (0.029 USD), four times more than the value of some current development projects of water resources. Finally, according to the findings, we suggested a corrected price trade term to adjust water monetary return indices. Control policies and government participation in funding of artificial recharge of undergorund aquifers projects were recommended as well.

Keywords:
Monetary return, Withdrawal Cost, Dry matter, Water use efficiency, Water shadow Price

[1] *Department of Agricultural Economics, Shahid Bahonar University of Kerman, Kerman, Iran.*
[2] *Department of Economics, University of Sistan and Baluchestan, Zahedan, Iran.*
* *Corresponding author's email: yaqubi@pgs.usb.ac.ir*

[1] *3875.4 Iranian Rials (IRR)≈US $ 1*

INTRODUCTION

Given the limited water in arid and semi-arid regions along with global warming (Hanjra and Qureshi, 2010) and the fact that almost 93 percent of renewable surface and underground water resources go into watering (Turner, 2004), Water Use Efficiency (WUE) is a serious criterion for sustainable production (Neal *et al.*, 2011). Currently, In areas with low rainfall and consecutive droughts, underground waters withdrawal has substantially increased which leads to dramatic loss of underground water levels and therefore, a water crisis in the near future is not unexpected (Hanjra and Qureshi, 2010). On way to deal with this crisis and its consequences, is determining Water Use Efficiency (WUE) and its indices e.g. monetary return, dry matter production, the cost of water per cubic meter of water and the cost of water per hectare. A precise evaluation of these indices, at least, help us for rational allocation of scarce water (Ward and Michelsen, 2002) and aid to avoid worsening the conditions through optimal management of water resources. In doing so, we can find a wide variety of research in this context. For instance, Medrano *et al.* (2015) had a review on water use efficiency of vineyards in semi-arid regions. They indicated cover crops in these areas have a desirable effect, but to avoid excessive water consumption, it is necessary to careful management by these cover crops. At a case study in the North China Plain Lu *et al.* (2016), investigated changes in water use efficiency in grain production in a long term. According to their findings, water footprint technique is better than WUE in evaluating the effectiveness of crop water use.Using a mathematical approach, Medellín-Azuara *et al.* (2010) evaluated the economic value of agricultural water under changing conditions. They concluded that the economic value of water at both the farm and aggregated level are similar. By adopting a mathematical programming, Sabuhi *et al.* (2012) also determined the efficiency of water consumption for rice producers in Fars province of Iran. Their results showed that rice producers had the potential of 65% reduction in water consumption. Using the same approach, Farija *et al.* (2009) investigated WUE of Tunisia greenhouse and its effective factors. The results indicatededuction and investment in irrigational technologies have positive effects on WUE, while the size of land has negative effects. Speelman *et al.* (2008) also analyzed WUE of farm's South Africa using DEA and it's effective factors. Their results demonstrated that factors such as method of irrigation, land ownership, land size, and crop selection are effective on WUE. Pala and Oweis, (2001) did a study on WEU and concluded that in regions with available water resources, agriculture production and WUE can be higher than complete irrigation through supplement irrigation. Tennakoon and Milroy (2003) investigated water use efficiency on irrigated cotton farms in Australia. They indicated that there is a huge potential to enhance the efficiency among the farmers. Wise *et al.* (2011) had a comparison of the biophysical and economic water-use efficiencies and finally found a better inform resource allocations. Some studies like, Berbel *et al.* (2000), de Andrade Resende Filho *et al.* (2015), Johansson (2000), Molden *et al.* (2010), Orprecio *et al.* (2016), Rogers *et al.* (2002), Speelman *et al.* (2009), Shahraki and Yaghoubi (2014), Tsur *et al.* (1997), Tang *et al.* (2015), also have analyzed water pricing and its link with efficiency and the costs of per cubic meter of water. However, to the best of authors' knowledge, no study can be found on investigating water use efficiency indices in Fars province of Iran. This is especially important because, this province is the pole of agricultural production and despite Iran severe water restrictions, increasing tendency of farmers to develop cultivation, demands for digging new wells and more intensive extraction of water resources of existent well are boosting. Therefore, the clarification of WUE indices are of policy priorities and to achieve more in saving water resources, determining water extraction costs are the principal aim of the present study.

MATERIALS AND METHODS

We collected data by interviewing 164 farmers used well water, through questionnaires in selected Fars Province plains (i.e., Shiraz,

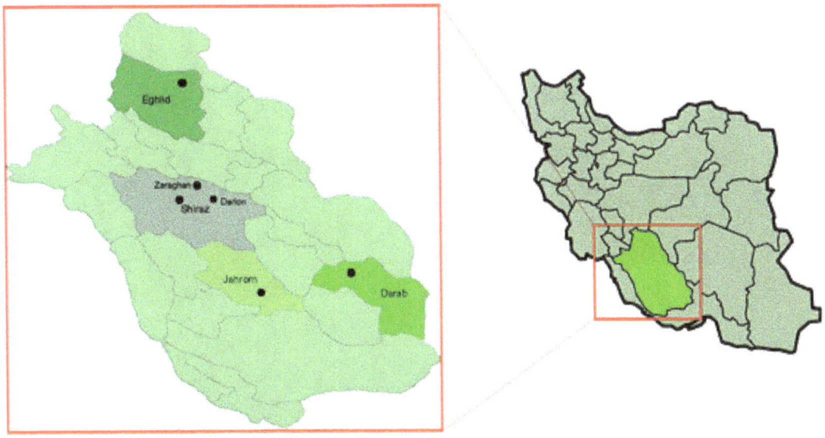

Figure 1: The geographical locations of the study area, Fars province, Iran

Zarghan, Jahrom, Darion, Darab and Eghlid) through integrating two purposive and multistage cluster sampling techniques. This province, with an area of 133,000 Kilometers, is located in the south of Iran (Figure 1).

In the first step, the amount of water withdrawal, water consumption and water requirements for each crop were determined. Next, in order to determine the monetary returns of water per cubic, an econometric approach was used. To evaluate WUE indices, we applied both econometrics and managerial methods. In Managerial method, we used a model of residual value. This relationship shows average production of a crop toward its water use. Furthermore, marginal and average productivity and the value of monetary return of water per cubic were determined. Important details of the equations and models are as follows:

Calculating the cost of water withdrawal at water wells

We used an econometric approach to calculate the costs. In this relation, not only the costs of water withdrawal of the well were considered, but also the level of costs related to distribution and transfer of water were taken into account as follows:

1- Cost of investment (wells, purchase and installation of pump motors and related equipments)

2- Operating costs (maintenance and management, fuel, repairs, etc.)

3- Transmission cost of water (pipes, construction of canal water, and other pressurized irrigation equipments from the location of the pump motor to the targeted field.)

Considering the life of well and equipments and an adequate interest rate, primary investment was converted into annual investment (Soltani, 1992). In doing so, first, the effect of inflation was neutralized. Next, the following equation was applied to convert fixed investment into annual investment.

$$A = P\left[\frac{i(1+i)^n}{(1+i)^n - 1}\right] \quad (1)$$

where, i represents discount rate (25%), P shows the current value of investment and *an* equals uniform investment during the life of well. Then, this cost was added up to annual variable costs (annual operation and maintenance of wells) and thereafter the total annual costs of well were attained. Through dividing annual aggregated cost to the level of water use of each production, we obtained monetary cost of each cubic meter of water, extracted from wells.

Evaluating monetary returns per cubic meter of water

To determine the crops with more efficiency and monetary return of production, we applied the following model:
residual value model:

$$y = \frac{TR - (VC - WYC)}{VW\ (m3)} \quad (2)$$

By equation (2), monetary return per cubic meter of water can be calculated for each of crops or for the whole farm.

where, Y represents monetary return per cubic

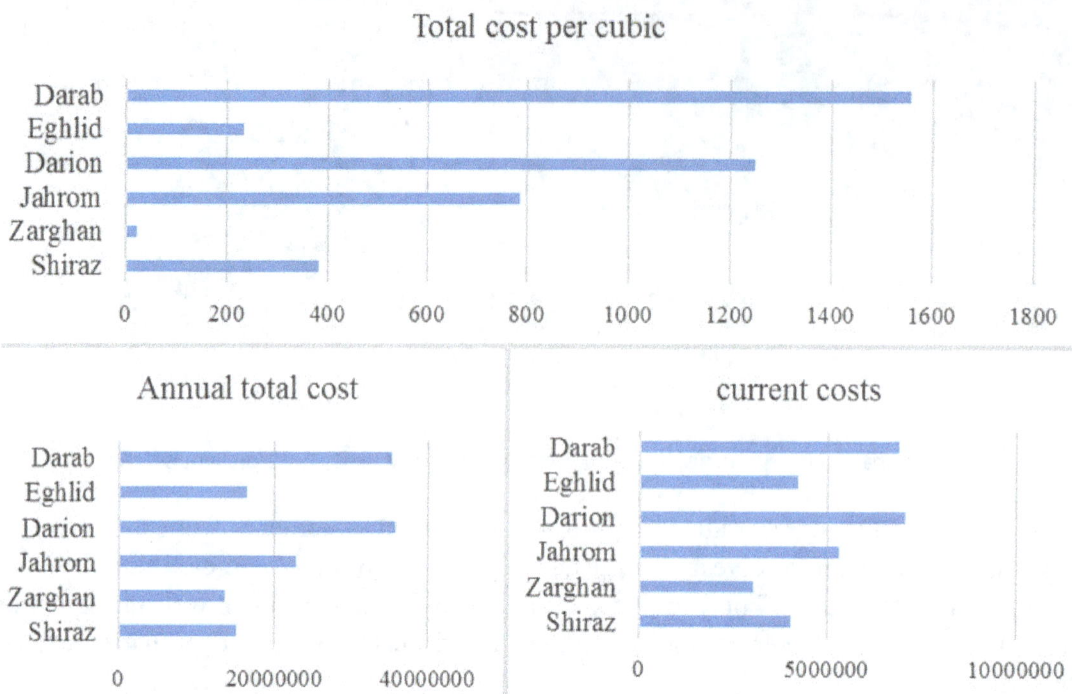

Figure 2: Clustered bar of the total cost, annual total cost and current cost

meter of water, TR is total revenue, VC shows annual production costs of crops, WYC equals annual water cost and VM indicates the volume of water use.

It is worth to note that the annual water cost of production costs was also considered. The above formula shows that how much of the monetary return connected to a production is due to water use. Since monetary return is related to the other inputs, equation (2) does not indicate monetary return of water clearly. In other word, it does not show the return per cubic meter of waterpurely.. However, its results can be usedfor comparative purposes.

Calculating dry matter of per cub meter of water use

This ratio is considered as one of the agricultural water efficiency indices, simply calculated by dividing the total amount of product (s) on water consumption.

Calculating value marginal production of water and determining optimal water use

Through estimating the following production functions, i.e., linear forms of Cobb-Douglas and transcendental function, we can assess the value marginal production of water i.e., its shadow price:

$$LnY = LnA + \beta 1 LnX1 + \beta 2 LnX2 + \ldots + \beta n LnXn \quad (3)$$

$$Y = A \prod_{i=1}^{n} X_i^{\alpha_i} e^{\beta_i x_i} \quad (4)$$

$$Vmp = Px1 \quad (5)$$

where, X is the amount of wheat's inputs, Y reperesents wheat's production and A shows constant coefficient. *VMP* is the value marginal production of last per cubic meter of water and Px1 shows the water price. Needless to say, VMP>Px1 reveals underused water and VMP<Px1 indicates overused water. To estimate above relations, two SPSS and Excel softwares were used.

RESULTS

The results of calculating water use efficiency indices (i.e. water withdrawal cost, monetary return and dry matter production per cubic meter of water) for different plains are represented below:

Evaluate and comparison the cost of withdrawal water:

Average annual current costs of water is 4644753 IRR [1] and the equivalent annual cost

[1] *1.00 IRR=0.0000344700 USD*

Table 1. Statistic characterize of annual withdrawal cost of water in Fars Province, Iran (2013)

	Current costs (IRR)	Annual total cost (IRR)	Total cost of per cubic (IRR)
Total	4644753	24666196	840
Shiraz	4037037	15164723	382
Zarghan	3056000	13712499	22
Jahrom	5326315	22767384	786
Darion	7065583	35772154	1249
Eghlid	4197727	16541180	234
Darab	6891903	35303239	1557

(EAC) - the annual deprcciation for investment in well - for each well equals 20021443 IRR. Total annual cost of each farm (well) is 24666196 IRR. The cost of per cubic meter of withdrawal water of underground resources equals 839.3 IRR, four times of some development projects of water resources.

Figure 2 intuitively shows the total cost, annual total cost and current cost of the study areas.

From these clustered charts, we can quickly become aware of the possibility of cost reduction. As these three bars show us, unlike specially Darion and Darab, Zarghan has the minimum level of the total cost, annual total cost and current cost. So then by relying on management practices, some costs of production can significantly be reduced and the production process would be more affordable.

According to Table 1, the cost of water withdrawal in Jahrom, Darion and Darab are nearly double in comparison with other plains. Evaluation of initial data also indicates making higher investment in these plains with the aim of withdrawal water from deep wells.

Calculating monetary return per cub meter of water:

Table 2 states some statistical characteristics of water monetary return in selected plains which shows high dispersion around the mean of the observations.

Monetary return of per cub meter of water of Shiraz, Zarghan, Jahrom, Darion, Eghlid and Darab are 2737.3, 4383.1, 4124, 3450, 2137.5 and 5267.5 IRR, respectively. The standard deviation of monetary return per cubic meter of water in these plain are 2184.1, 4105, 3448.7, 2906.3, 2227.7 and 7409 IRR. Therefore, the amount of standard deviation in selected plain is considerably less than standard deviation of the whole plains. This is a natural consequence. Because monetary return depends on inputs use and various combinations of inputs that these amounts are more uniform in comparison to the whole province.

Calculating the shadow price and production of dry matter per cub meter of water:

Considering wheat's base year price i.e., 2200 IRR/kg for transcendental and Cobb-Douglas functional forms, the water shadow price equal 280 and 310 IRR/m³, and by taking into account the price 2070 IRR/kg are 261 and 290 IRR/m³, respectively (Table 3). Marginal product and water price indicate most farmers are in the third region of production. This overuse reminds the necessity of promotional-educational programs. Therefore, besides consider technical progress in planning of cropping pattern policy makers, the WUE indices need to be taken into account as a basic planning cultivation.

Another water use efficiency index is the pro-

Table 2: Statistical characteristics monetary return per cub meter of water (2013)

	Description (IRR per cub meter)	f	Min	Max	Mean	SD
1	Monetary return crop and horticultural products in total	164	67	46697	3875	4748
2	Monetary return crop and horticultural products in Shiraz	28	211	9203	2737	2184
3	Monetary return crop and horticultural products in Zaraghan	25	263	13896	4383	4105
4	Monetary return crop and horticultural products in Jahrom	20	688	11908	4124	3448
5	Monetary return crop and horticultural products in Darion	24	598	12062	3450	2906
6	Monetary return crop and horticultural products in Eghlid	22	67	10951	2137	2228
7	Monetary return crop and horticultural products in Darab	45	187	46697	5267	7409

Table 3: The results of production function for evaluating the value of per cubic meter of water in wheat production (2013)

Wheat price (IRR)	Production elasticity of water (Cobb-Douglas form)	Production elasticity of water (transcendental form)	Value per cubic meter of water (transcendental form) (IRR)	Value per cubic meter of water (Cobb-Douglas form) (IRR)
2200	.000126	.00014	277.2	308
2070	.000126	.00014	261	290

Table 4: Some statistical characteristics of the dry matter production per cub meter of water

	Description	f	Min	Max	Mean	SD
1	The production of dry matter of crop and horticultural products in total (Ton)	162	0.1	32.8	3.3	3.7
2	The production of dry matter of crop and horticultural products in Shiraz (Ton)	27	0.2	7.1	2.7	1.9
3	The production of dry matter of crop and horticultural products in Zaraghan (Ton)	25	0.2	11.1	3.6	3.5
4	The production of dry matter of crop and horticultural products in Jahrom (Ton)	20	0.4	7.5	2.8	2.1
5	The production of dry matter of crop and horticultural products in Darion (Ton)	24	0.4	18.9	3.8	3.7
6	The production of dry matter of crop and horticultural products in Eghlid (Ton)	21	0.1	14.5	2.5	3.1
7	The production of dry matter of crop and horticultural products in Darab (Ton)	46	0.1	32.8	3.8	5.1

duction of dry matter. Table 4 shows statistical characteristics of production of dry matter per cub meter of water.

The results indicate the average of production of dry matter per cub meter of water and standard deviation are 3.3 and 3.7 respectively. Minimum and maximum of the production of dry matter per cub meter of water are 0.1 and 32.8 respectively. This shows high dispersion of observations around the mean. The results state total average of the production of dry matter per cub meter of water for Shiraz, Zarghan, Jahrom, Darion, Eghlid and Darab are 2.7, 3.6, 2.8, 3.8, 2.5 and 3.8 kg/m^3. As a result, the production of dry matter per cubic meter of water among the plains is almost equal.

CONCLUSION AND RECOMMENDATIONS

In determining water monetary return, applying econometric and mathematical programming approaches are more reliable than residual models. The results showed that the residual model and production function methods have a different monetary return. By finding factors caused these differences, we can provide the possibility of improving positive factors and decaying negative ones. Thus, factors that influence WUE indices change monetary return per cubic metre of water as well. Here, monetary return per cubic meter of water is influenced by inputs and outputs prices and their the level of use as well. Total factor productivity has a direct relationship with water use efficiency. As the prices of crops increase in a crop-year, water shadow price relatively will increase, quickly. Hence, correcting the price trade off relation, will correct the alleged water monetary productivity indices (monetary return).

Water withdrawal costs of critical plains are considerably increased. Therefore, if policies related to the control of underground water crisis do not apply, increasing direct withdrawal costs of water endangers economic use of underground water.

One of the agricultural features of Fars province is dependency to water resources on underground water. This causes that water distribution has drawbacks in comparison with other provinces that their water resources of agriculture attain through investigating in water resources development projects (dams and weir). Thus, due to the high potential withdrawal cost reduction, decline in direct costs of rinse and government participation in funding of artificial recharge of groundwater aquifers projects are necessary. According to the results, cost per cubic meter of water from underground water equals 839.3 IRR that is almost four times more than water resources development costs. Farmers in Fars province are forced to pay their withdrawal costs of water. While, for instance farmers of Payab dam, supply their water resources only with almost nothing pay (about 500000 IRR per hectare under cultivation). Therefore, amending the fair distribution of water is also indispensible.

High cost of underground water withdrawal suggests an inevitably endeavor to maintain the underground water resources and keep the underground water table. Notwithstanding the consequences of global warming, low raining, droughts and unequal distribution of rainfall, the possibility of development projects cannot be attained for most water resources. Therefore, development of artificial recharge schemes and strengthening the aquifers through performing watershed projects (maintain upstream vegetation cover) has also a high priority.

ACKNOWLEDGEMENT

We would like to express our appreciation to all those who help us for gathering data. Furthermore, we would also like to acknowledge with much appreciation the role of reviews for improving our manuscript through their useful comments.

REFERENCES

1- Berbel, J., & Gómez-Limón, J. A. (2000). The impact of water-pricing policy in Spain: an analysis of three irrigated areas. *Agricultural Water Management,* 43(2), 219–238.

2- de Andrade Resende Filho, M., Correa, J. S. O., & de Oliveira Torres, M. (2015). Water Pricing in Brazil: Successes, Failures, and New Approaches. *In Water Pricing Experiences and Innovations* (pp. 41–61). Springer.

3- Frija, A., Chebil, A., Speelman, S., Buysse, J., & Van Huylenbroeck, G. (2009). Water use and technical efficiencies in horticultural greenhouses in Tunisia. *Agricultural Water Management*, 96(11), 1509–1516.

4- Hanjra, M. A., & Qureshi, M. E. (2010). Global water crisis and future food security in an era of climate change. *Food Policy*, 35(5), 365–377.

5- Johansson, R. (2000). Pricing irrigation water: A literature survey. *World Bank Policy Research Working Paper*, (2449).

6- Lu, Y., Zhang, X., Chen, S., Shao, L., & Sun, H. (2016). Changes in water use efficiency and water footprint in grain production over the past 35 years: A case study in the North China Plain. *Journal of Cleaner Production,* 116, 71-79.

7- Medellín-Azuara, J., Harou, J. J., & Howitt, R. E. (2010). Estimating economic value of agricultural water under changing conditions and the effects of spatial aggregation. *Science of the Total Environment*, 408(23), 5639–5648.

8- Medrano, H., Tomás, M., Martorell, S., Escalona, J.-M., Pou, A., Fuentes, S., & Bota, J. (2015). Improving water use efficiency of vineyards in semi-arid regions. A review. *Agronomy for Sustainable Development*, 35(2), 499–517.

9- Molden, D., Oweis, T., Steduto, P., Bindraban, P., Hanjra, M. A., & Kijne, J. (2010). Improving agricultural water productivity: between optimism and caution. *Agricultural Water Management*, 97(4), 528–535.

10- Neal, J. S., Fulkerson, W. J., & Hacker, R.B. (2011). Differences in water use efficiency among annual forages used by the dairy industry under optimum and deficit irrigation. *Agricultural Water Management,* 98(5), 759-774.

11- Orprecio, J., Rola, A. C., Deutsch, W., Coxhead, I., & Sumbalan, A. (2016). *More efficient water pricing may avert a water crisis*. Makati City, Philippines: Philippine Institute for Development Studies.

12- Pala, M., & Oweis, T. (2001). Strategies for improved water use efficiency in dry areas. Natural Resource Management Program, ICARDA, Aleppo, Syria, Retrived from http://www. Ankara. Edu. /faculties/agriculture/seminar/abstract.html.

13- Rogers, P., De Silva, R., & Bhatia, R. (2002). Water is an economic good: How to use prices to promote equity, efficiency, and sustainability. *Water Policy,* 4(1), 1–17.

14- Sabuhi, M., Yaghoubi, M., & Esfandiari, M. (2012). A Survey on Water Use Efficiency of Rice Producers in Kamfirouz Region, Fars Province, Iran. *International Journal of Applied Operational Research*, 2(3), 45-53.

15- Shahraki, J., & Yaghoubi, M. (2014). Analysis of relative water use efficiency in Bahoukalat, Chabahar, Iran. *Journal of Operational Research and Its Applications*, 11(3 (42)), 93–108.

16- Soltani, Gh., B. Najafi & J. Torkamani. (1992). Management of agricultural units. Shiraz University Press, Iran.

17- Speelman, S., Buysse, J., Farolfi, S., Frija, A., D'Haese, M., & D'Haese, L. (2009). Estimating the impacts of water pricing on smallholder irrigators in North West Province, South Africa. *Agricultural*

Water Management, 96(11), 1560–1566.
18- Speelman, S., D'Haese, M., Buysse, J., & D'Haese, L. (2008). A measure for the efficiency of water use and its determinants, a case study of small-scale irrigation schemes in North-West Province, South Africa. *Agricultural Systems,* 98(1), 31–39.
19- Tang, J., Folmer, H., & Xue, J. (2015). Technical and allocative efficiency of irrigation water use in the Guanzhong Plain, China. *Food Policy*, 50, 43–52.
20- Tennakoon, S. B., & Milroy, S. P. (2003). Crop water use and water use efficiency on irrigated cotton farms in Australia. *Agricultural Water Management*, 61(3), 179–194.
21- Torkamani, J., Gh. Soltani and & H. Asadi. (1997). Pricing of water and investigating the value of the marginal efficiency of agricultural water. *Water Resources and Development*, 6 (1): 5-13.
22- Tsur, Y., & Dinar, A. (1997). The relative efficiency and implementation costs of alternative methods for pricing irrigation water. *The World Bank Economic Review,* 11(2), 243–262.
24- Turner, R.K. (2004). Economic valuation of water resources in agriculture: From the sectoral to a functional perspective of natural resource management (Vol. 27). Food & Agriculture Org.
25- Ward, F. A., & Michelsen, A. (2002). The economic value of water in agriculture: concepts and policy applications. *Water Policy*, 4(5), 423–446.
26- Ward, G.N., Jacobs, J.L. & Mc Dowell, A.M. (2001). Water use efficiencies of dry land brassica forage crops on contrasting soil types. Proceedings of the 10th Australian Agronomy conference, Hobart.
27- Wise, R. M., Dye, P. J., & Gush, M. B. (2011). A comparison of the biophysical and economic water-use efficiencies of indigenous and introduced forests in South Africa. *Forest Ecology and Management,* 262(6), 906-915.

15

Optimization Model of Hirmand River Basin Water Resources in the Agricultural Sector Using Stochastic Dynamic Programming under Uncertainty Conditions

Mahmoud Mohammadghasemi [1,2]***, Javad Shahraki*** [3*] ***and Mahmood Sabouhi Sabouni*** [4]

Abstract

In this study, water management allocated to the agricultural sector' was analyzed using stochastic dynamic programming under uncertainty conditions. The technical coefficients used in the study referred to the agricultural years, 2013-2014. They were obtained through the use of simple random sampling of 250 farmers in the region for crops wheat, barley, melon, watermelon and ruby grapes under the scenarios of drought, wet, normal, and water required in the most sensitive growth stages. Production function and profit function were obtained from the yield-water-product function of crops using Eviews software. Expected net profit of the system and optimal allocation of water were also calculated based on the GAMS economic analysis software. The results revealed that 14% of the cases over the past 30 years had wet years (high), 47% of the time and that 39% had experienced drought (low) and normal (average) years. In the best case, i.e. with high current levels, respectively at, 58, 67, 54, and 48% of water requirements for these crops and, in the worst case (with low current levels), 47, 35, 49, 53 and 48% of the water requirements provided during the most sensitive growth stages. Moreover, the results showed that the cultivation of the ruby grape was the best product with the highest expected profit in normal and rainfall conditions. In general, when the expected value of net profit is positive, managers would act optimistically and they would promise the optimal level of water provided to the farmers. Conversely, when the net value is negative they would prefer to be more conservative and would promise a lesser amount of water provided to the farmers. Hence, if the promised water to the farmer is not wasted, he will choose the loss incurred from a lesser harvest.

Keywords:
Expected value, Optimal allocation, Stochastic dynamic programming

[1] *PhD Student of Agricultural Economics, University of Sistan and Baluchestan*

[2] *Social and Economic Research Department, Sistan Agriculture and Natural Resources Research and Education Center, AREEO, Zabol, Iran*

[3] *Associate Professor of Economics, University of Sistan and Baluchestan*

[4] *Professor of Agricultural Economics, University of Ferdowsi Mashhad*

* *Corresponding author's email: j.shahraki@eco.usb.ac.ir*

INTRODUCTION

From three perspectives, water plays a key role in sustainable development. First, it is consumed as a final product. Second, water is an important input element in many businesses. Third, it has a key role in biological organisms on Earth (Divakar, 2011). Sustainable socio-economic development in countries with low water is limited to the availability of water and its reduced quality (Ghaffour *et al.*, 2013). In terms of water resource management, low water would provide high risk for different sectors of development programs (Salimfard and Mostafaee Dowlatabad, 2013). According to the latest estimates by UNESCO of the water cycle on Earth, it can be inferred that the average annual rainfall in Iran was 251 mm, having a significant difference with the average rainfall of each continent (Kessler and Van Drop, 1998). It could be compared with rainfall in semi-arid and desert-like countries of some continents (Lynch *et al.*, 2011). The average rainfall in all lands and Asia was 831 and 732 mm, respectively (Thompson and Powell, 1998). In Iran, the average annual rainfall was 413 billion m^3; however, the area of Iran is 1.1 % of all lands and 3.35% of the land area of Asia. Iran's rainfall volume comprises just 0.37% of all rainfall from the earth's lands and 1.29% of rainfall volume in Asia. Also, the average annual evaporation in Iran is estimated at about 70- 71% of annual rainfall. In this regard, just Africa and Australia, with 70% and 80% evaporation under undesirable rainfall conditions, respectively, are lower than Iran (Mihankhah, 2013). The Hirmand basin is located in the province of Sistan and Baluchistan. In all climate categories, the Sistan region had a hot and dry climate. Based on different calculation methods, its average annual temperature was 21° C, its annual rainfall was 61.4 mm, its relative humidity was 38%, and its potential evapotranspiration was 4196 mm (Ebrahimzade and Lashkaripoor, 2001). Of the total cultivated lands of the country, an area of 12 million hectares is located in Sistan and Baluchistan out of which 52.4% is located in the Sistan region (Mohamadghasemi, 2008). The purpose of this study was to estimate the optimization model of the Hirmand River Basin water resources in agriculture using stochastic dynamic programming under the conditions of uncertainty.

MATERIALS AND METHODS

The yield-water-product function was used to estimate crop production function. Under each irrigation condition, crops had their unique water-product functions estimated using regression methods (Tu *et al.*, 2003). This function expressed the relationship between the actual yield and the effective irrigation, so the second-degree polynomial function for estimating the crop-water function recommended by (Divakar, 2011) was as follows:

$$Y_a/Y_m = f(w) = a_0 + a_1 w + a_2 w^2 \quad (1)$$

where Y^a is the actual crop yield (tons / ha), Ym is the maximum potential yield (t/ha), $W=WA/ET_m$ is the ratio of total available water to the maximum potential seasonal evaporation in the crop, i.e. the ratio of actual evapotranspiration to the potential evapotranspiration.

Total water available for the crops was obtained through effective rainfall, irrigation and soil moisture (Mahan *et al.*, 2012) Hence,

$$WA_{j,cp} = SM_{j,cp} + EP_{j,cp} + EI_{j,cp} \quad (2)$$

where $WA_{j,cp}$ is the total available water for plants during the growing season, $SM_{j,cp}$ is soil moisture in the root zone at the beginning of the growing season, $EI_{j,cp}$ is effective rainfall, and $EI_{j,cp}$ is efficient water used for the crop. Depending on time and irrigation technology, actual evapotranspiration, is the sum of actual soil moisture, effective precipitation, and effective watering during the growing season (Mahan, 1997). In the studied region, the rate of effective rainfall for wheat, barley, melon, watermelon and ruby grapes was zero. Since there was no information available on soil moisture in the region, soil moisture was excluded from the calculations and it was assumed that it was hidden in the effective irrigation (Ghaffari Moghadam *et al.*, 2012).

Therefore, the actual evapotranspiration included effective irrigation during the growing season. Effective irrigation and potential evapotranspiration were also determined using monthly weather data for 26 years with Netwat-Cropwat Software and the Penman-Monteith method, respectively. In this method, based on the types of available data, potential evapotranspiration was calculated daily and monthly. Of course, in the present study, the monthly evapotranspiration was applied. For calculating the amount of actual evapotranspiration, the following formula was used (Abbott and Cohen, 2009):

$$ET_a = K_c \times ET_o \qquad (3)$$

where ET_a is the actual evapotranspiration, kc is the plant factor varying in different crops, and ET_o is the potential evapotranspiration.

Based on the water- crop yield, the total profit yield of irrigation water is expressed as follows (Dorfman , 1969):

$$\beta_j^{\sim} = \sum PCP_{j,cp} . Ya_{j,cp} . Af_{j,cp} - \sum Vc_{j,cp} . Af_{j,cp} \qquad (4)$$

where $PCP_{j,cp}$ is the crop price, $cc_{j,cp}$ is the cultivation cost, and $Vc_{j,cp}$ is the variable costs of crop production. The target function can be written as:

$$\begin{aligned} & max\beta_j^{\sim} \\ s.t: \quad & \sum_{cp} Af_{j,cp} \le A_j \\ & Af_{j,cp}^{i} \le Af_{j,cp} \le Af_{j,cp}^{u} \qquad (5) \\ & \sum_{CP} EI_{j,cp} . Af_{j,cp} \le Q \end{aligned}$$

where A_j is the total area under cultivation (ha) in the jth area, and

$Af^{u}_{j,cp}$ and $Af^{i}_{j,cp}$ are respectively the maximum and minimum levels of cultivated areas (ha), $EI_{j,cp}$ is the rate of effective irrigation required during the growing season (m^3/ha) and Q is the total amount of effective irrigation available in the jth region.

Following slight changes, the shadow price, which includes any variation in the target function, will be placed on the right side of resource limitations. It will be considered as an indicator for the ultimate value of water. In the form of an algebraic expression, the shadow price is expressed as follows: (Mahan *et al.*, 2012)

$$MVW_j = {}^{\Delta\pi_j}/_{\Delta Q_j} \qquad (6)$$

where MVW_j is the final value of water ($/$m^3$) in the jth region $\Delta\pi_j$, are changes in income (IRR) caused by slight variation of Q in the jth region, and ΔQ_j includes changes in the total amount of effective irrigation in the jth region. In their research on economic allocation of water resources in Sistan region, (Sharaki and Mohamadghasemi, 2014), used the dynamic optimization models, analyzed its effect on sustainable agricultural development of the area and chose two farming patterns for okra and cucurbit. Karim *et al.* (2012), studied the effects of cultivating ruby grapes on the economy of the Sistan farmers (Mohamadghasemi, 2008). Also analyzed the cost-benefit performance of agricultural crops (wheat, barley and triticale) in Sistan and Baluchistan. These crops included wheat, barley, melon, watermelon and ruby grapes which were planted under drought, wet, normal, and water requiring scenarios at the most critical growth stages. It is worth to note that, the most sensitive stage of water for horticultural and gardening crops was at the time when they enter the reproductive stage (flowering) (Ghasemi *et al.*, 2012).

Since the system manager always faces issues regarding water allocation between competing agricultural consumers (including various scenarios) and due to the fact that water supply tends be random in the future, the demand for water will also be estimated based on the needs of different scenarios and a logical period will be considered for all data (Zeng *et al.*, 2014).

In cases where the agricultural sector is informed that it has little water available, it will change its activities so that it would need less water. When there is uncertainty, the manager is supposed to create a plan in which, despite allocation of water efficiency, the system benefits increase, and in turn, the system risk reduces (Homauonifar and Rastgaripour, 2011). Hence, the random variable of water supply with Ptk (probability scenario k in time period t) was used to design a set of scenarios with branching structure. This model can be formulated as follows (7)

$$
\begin{aligned}
maxf &= \sum_{i=1}^{I}\sum_{t=1}^{T} NB_{it}\, W_{\,i\,it} - \sum_{i=1}^{m}\sum_{j=1}^{n}\sum_{k=1}^{r} p_{it}\; C_{it}\; D_{ik} \\
& s.t \\
& q_i \geq \sum_{i=1}^{m} (W_{it} - \ {}_{itk}) \quad \forall i,j \\
& W_{itmax} \geq W_{it} \geq D_{it} \forall i,t,k
\end{aligned} \tag{7}
$$

where F is the net system profit of the planning horizon, NB_{it} is the net income of i th crop per allocated water unit, W_{it} water promised for the product i, C_{it} is the farmer's losses per unit of water promised, but not delivered in period t, D_{itk} is water scarcity for the crop i under scenario k in period t (in other words, some of W_{it} was not delivered at q^{th}), q^{th} is random variable of water supply in period t, $W_{it}max$ is the amount of water allocated for i^{th} consumer at time t, P_{tk} is frequency probability of scenario k in period t, k are total number of scenarios and t is the most sensitive growth stage, i is the type of crop (i=1 wheat, i=2 barley, i=3 melon, i=4 watermelon, and i=5, ruby grape).

Model 7 expresses uncertainty in the amount of water supplied by the probability level of P_{tk}, but it considers the parameters of W_{it}, NB_{it}, C_{it} in their definite form. In the real world, however, these parameters may not be definite.

To solve this problem, the parameters of this model were considered periodically. The result of the model was as follows: (8)

$$
\begin{aligned}
maxf^{\pm} &= \sum_{i=1}^{m}\sum_{t=1}^{T} NB^{\pm}_{it}\, W^{\pm}{}_{it} - \sum_{i=1}^{m}\sum_{j=1}^{n}\sum_{k=1}^{r} P^{\pm}_{tk}\, C^{\pm}_{tk}\, D^{\pm}_{itk}) \\
s.tq_{th} &\geq \sum_{i=1}^{m} (W_{it}{}^{\pm} - D_{itk}{}^{\pm}) \quad \forall i,j \\
W^{\pm}_{imax} &\geq W^{\pm}_{it} \geq D^{\pm}_{it} \forall i,t,k
\end{aligned} \tag{8}
$$

where $f^{\pm}$ is the net profit of the system in the planning horizon, $NB_{it}^{\pm}$ is the farmer's profit resulted from cultivation of the i th crop in period t per unit of water allocation, and $W^{\pm}{}_{it0}$ is the water promised to the farmer for cultivation at time t, $C^{\pm}{}_{tk}h$ is the farmer's loss resulting from planting the crop per unit of water allocation promised but not delivered in period t, D_{itk} is water scarcity for the ith crop under the scenario k in period t (in other words, some of $W^{\pm}{}_{it}$ which was not delivered in time q^{th}), q^{th} is a random variable of water supply in period t, $W^{\pm}{}_{imax}$ is the maximum amount of water allocated for consumer i at time t, P_{tk} is the frequency probability of scenario k in period t.

Since $W^{\pm}{}_{it}$ is considered as a periodic parameter, equation 8 cannot be solved directly, so it needs to be oversimplified. To solve this problem y_{it} is defined as a decision variable (9):

$$
\begin{aligned}
W^{\pm}_{it} &= W^{-}_{it} + \Delta W_{it} y_{it} \\
\Delta W_{it} &= W^{+}_{it} - W^{-}_{it} \\
y_{it} &\in [0,1]
\end{aligned} \tag{9}
$$

in this equation, y_{it} is a decision variable used to define the optimal range $W^{\pm}{}_{it}$. ,when y_{mt} reaches its highest level, $y_{it}=1$. If the required water is delivered to the sectors, the system profit will reach its peak level. In case of losses, the reverse is true, too. When $y_{it}=0$ and the promised rate of water is delivered, the system profit will decrease dramatically, but if the promised water is supplied, it would have the least amount of loss for the system. Substituting model 9 for model 8, the following model is obtained: (10)

$$
\begin{aligned}
maxf^{\pm} &= \sum_{i=1}^{m}\sum_{t=1}^{T} NB^{\pm}_{it}\,(W^{-}_{it} + \Delta W_{it}\, y_{it}) - \sum_{i=1}^{m}\sum_{j=1}^{n}\sum_{k=1}^{r} P\, C^{-}_{tk}\, D_{itk}{}^{-}) s.tq^{+}_{th} \\
&\geq \sum_{i=1}^{m} (W^{-}_{it} + \Delta W_{it} y_{itop} - D_{itk}{}^{-}) \\
&\forall h, k = 1,2 \ldots k \;, t = 1,2, \ldots T \\
W^{+}_{imax} &\geq W^{-}_{it} + \Delta W_{it} y_{it} \geq D^{-}_{it} > 0 \forall i,t,k \\
0 &\leq y_{it} \leq 1
\end{aligned} \tag{10}
$$

when the $W^{\pm}{}_{it}$ interval is defined as the optimum case, model 10 is divided into two sub-models. After solving these two sub- models, the maximum and minimum rates of total system profit can be obtained. To obtain the highest profits of the whole system *(f^{+})* in model 11, the upper limit of interest *(NB^{+})* and the lower limit of losses *(C^{-})* of the farmer were considered. Model 11 can be formulated as follows: (11)

$$
\begin{aligned}
maxf^{+} &= \sum_{i=1}^{m}\sum_{t=1}^{T} NB^{+}_{it}\,(W^{-}_{it} + \Delta W_{it}\, y_{it}) - \sum_{i=1}^{m}\sum_{j=1}^{n}\sum_{k=1}^{r} P\, C^{-}_{tk}\, D_{itk}{}^{-}) \\
s.tq^{-}_{th} &\geq \sum_{i=1}^{m} (W^{-}_{it} + \Delta W_{it} y_{itop} - D_{itk}{}^{-}) \\
&\forall h, k = 1,2 \ldots k \;, t = 1,2, \ldots T \\
W^{\mp}_{imax} &\geq W^{-}_{it} + \Delta W_{it} y_{it} \geq D^{-}_{it} > 0 \forall i,t,k \\
0 &\leq y_{it} \leq 1 \qquad \forall i,t \\
& F_{opt}{}^{+}, y_{itop}, D_{itk}{}^{\pm}
\end{aligned} \tag{11}
$$

In the model (11), low farmer income (NB^{-}) and low loss of water consumption are indicated. Farmer (C +) was used for obtaining the lowest income limit of the system *(f^{-})*.

Model (12) is formulated as follows: (12)

$$
\begin{aligned}
maxf^{+} &= \sum_{i=1}^{m}\sum_{t=1}^{T} \square B_{it}^{+}(W_{it}^{-} + \Delta W_{it}\, y_{it}) - \sum_{i=1}^{m}\sum_{j=1}^{n}\sum_{k=1}^{r} \mathrm{P\,C}_{tk}^{+} D_{itk}{}^{+}) \\
s.t q_{th}^{-} &\geq \sum_{i=1}^{m}(W_{it}^{-} + \Delta W_{it} y_{itop} - D_{itk}{}^{+}) \\
&\forall h, k = 1,2 \ldots k \;, t = 1,2, \ldots T \\
W_{imax}^{\mp} &\geq W_{it}^{-} + \Delta W_{it} y_{2t} \geq D_{it}^{+} > 0 \forall i, t, k \\
D_{itk}{}^{\pm} &\geq D_{itkOP}{}^{-}, t = 1,2, \ldots T
\end{aligned} \tag{12}
$$

the value F_{opt}^{-} , $D_{itk}^{\pm}$ is obtained from the model (12). Using the solutions of the models (11 and 12), the following equations are obtained:

$$
\begin{aligned}
F_{opt}{}^{\pm} &= [F_{opt}{}^{-}, F_{opt}{}^{+}] \\
D_{itkOP}{}^{\pm} &= [D_{itkOP}{}^{-}, D_{itkOP}{}^{+}]
\end{aligned} \tag{13}
$$

as a result, the optimal water allocation for the planned period is calculated as follows:

$$
A_{itkOP}{}^{\pm} = w_{itkOP}{}^{\pm} - D_{itkOP}{}^{\pm} \forall i, t, k \tag{14}
$$

Data

Technical factors used in the study corresponded to the farming years 2013-2014. Research was conducted through simple random sampling of 250 regional farmers. Moreover, in order to calculate the possibility of water flow rate (low, medium or high) based on rainfall data gathered from 3 decades ago to the present and using the standardized precipitation index, the percentages of dry, wet and normal years were obtained. These percentages were used to determine the frequency possibility of low, medium and high water flows (Sabouhi *et al.*, 2008). The Standard Precipitation Index is defined as follows:

SPI= (Pi – S)/ P (15)

where SPI is the standard precipitation index, Pi is desired annual rainfall, P is long-term average rainfall, and S is the standard deviation of long-term rainfall. If the index is greater and/or less than 1, it means that wet and drought conditions exist, respectively.

Rates between 1 and -1 represent a year with normal and average rainfall (Table 1). The results showed that the number of years with an amount of precipitation less than the average (i.e. drought) is higher than the number of years that are higher than the average amount of precipitation (wet years).

Over the past 30 years, 14 % of cases had wet states (high), 47 % had drought (low) and 39% had normal (average) cases.

RESULTS AND DISCUSSION

The rate of target allocation of water for agricultural crops was calculated by the gross irrigation requirement. Likewise, its maximum and minimum rates were also considered in terms of the highest and lowest water use efficiency in the region. The variable of maximum allocation of water to different crops was calculated based on the most unfavorable efficiency of the irrigation area. Table 2 summarizes this information.

Since water is the most limiting factor for agriculture in the region, any change in this variable would be a good way to estimate the relationship between profits and losses for the water used. This information is presented respectively in Tables 3 and 4.

In addition, based on the information mentioned in Tables 1, 2, 3, and 4, the optimal allocation

Table1: Data on water provision (m m^3)

Flow level	Relevant probability %	Water supply
Low	47	(270,260)
Average	39	(2070,2055)
High	14	(2900,2880)

Table 2: Allocation of water needed for crops during the most critical growth stages during the planning horizons (m^3)

Required water	Planning horizon
Wheat	(3780,3520)
Barley	(2980,2920)
Melon	(13600,13721)
Watermelon	(15640,15980)
Ruby grapes	(2211,2310)

Table 3: Average profit of crops per consumption of an excess water unit released in the most sensitive growth stages

Profit of models	Planning horizon
Wheat	(2010,1763)
Barley	(2445,1895)
Melon	(1793,1985)
Watermelon	(1862,1998)
Ruby grapes	(3895,5425)

Table 4: Average loss of crops per consumption of an excess water unit released in the most sensitive growth stages

loss of Profit of models	Planning horizon
Wheat	(1984,1563)
Barley	(1485,1360)
Melon	(8750,8940)
watermelon	(8820,8980)
Ruby grapes	(1254,1345)

of water under the drought scenario for wheat, barley, melon, watermelon and Ruby grapes are presented in Table 5.

The solution of the target function $f^{\pm}$ was positive at the final value of the expected net profit of the crop under a normal scenario.The results of the optimal allocation of water in drought conditions showed that using ruby grapes would lead to the lowest expected value. Based on the information mentioned in Tables 1, 2, 3, and 4, the optimal allocation of water under the normal scenario for wheat, barley, melon, watermelon and ruby grapes are presented in Table 6.

Solving the target function $f^{\pm}$ is within the final expected value of net profit of the crops and in accordance with the positive normal scenario. The results of the optimal allocation of water under normal conditions also showed that cultivation of ruby grapes could lead to the highest expected profit.

Table 5: The results of the model under drought scenario during the most sensitive time of irrigation

Row	Crop	Percent of occurrence	Target water demand (m^3)	Allocated water (m^3)	Expected value
1	Wheat	47	3780	2018	(-1.86,-3.75)
2	Barley	47	2980	1938	(-1.84,-3.60)
3	Melon	47	13721	7031	(-1.88,-3.65)
4	Watermelon	47	15980	7555	(-1.89,-4.41)
5	Ruby rapes	47	2310	1182	(-1.82,-3.53)

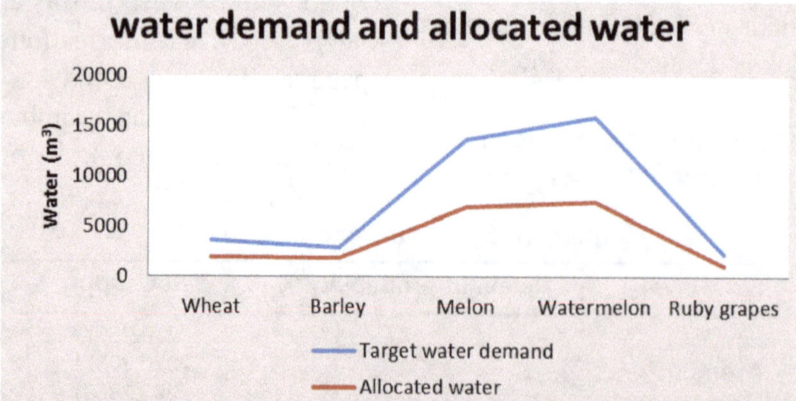

Figure 1: Target water demand and allocated water in drought scenario

Table 6: The solution of the model under a normal scenario during the most sensitive time of irrigation

Row	Crop	Percent of occurrence	Target water demand (m^3)	Allocated water (m^3)	Expected value
1	Wheat	39	3610	1576	(3514,243)
2	Barley	39	2960	1007	(2564,324)
3	Melon	39	13690	6640	(8249,831)
4	Watermelon	39	15420	7850	(8117,361)
5	Ruby rapes	39	2298	1102	(9119,491)

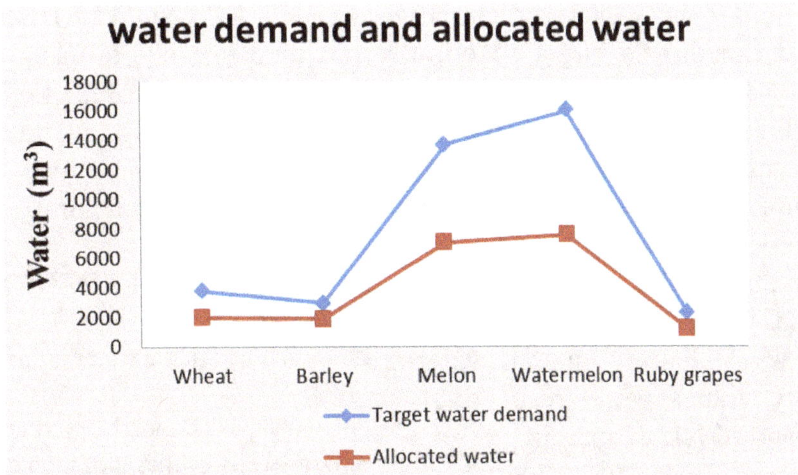

Figure 2: Target water demand and allocated water in normal scenario

Table 7: The solution of the model under a wet scenario during the most sensitive time of irrigation

Row	Crop	Percent of occurrence	Target water demand (m^3)	Allocated water (m^3)	Expected value
1	Wheat	14	3520	1495	(451,558)
2	Barley	14	2920	978	(462,467)
3	Melon	14	13600	6554	(452,471)
4	Watermelon	14	15590	8031	(352,355)
5	Ruby rapes	14	2211	1024	(462,667)

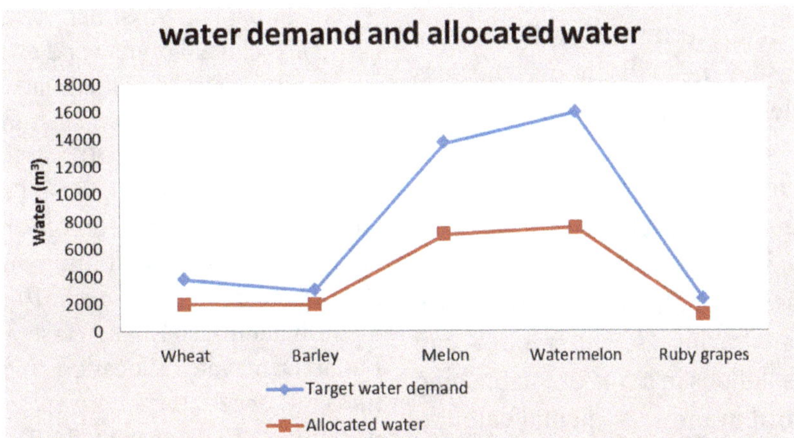

Figure 3: Target water demand and allocated water in wet scenario

Ultimately, based on the information in Tables 1, 2, 3, and 4, the optimal allocation of water under the wet scenario for wheat, barley, melon, watermelon and Ruby grapes are presented in Table 7.

The solution of the target function $f^{\pm}$ was positive at the final value of the expected net profit of the crop under normal scenario. The results of optimal allocation of water in wet conditions showed that cultivation of ruby grapes could provide the highest expected profits.

CONCLUSIONS

In this study, management of water allocated to agricultural sector was analyzed using randomly dynamic programming in the context of uncertainty. Technical factors used in the study referenced the farming years 2013-2014. It was conducted through a simple random sampling of 250 farmers in the region for crops wheat, barley, melon, watermelon, and ruby grapes.

By inserting the amounts of water scarcity and water allocation in the target function, the profit earned from optimal allocation of water was obtained. The results revealed that 14% of these cases occurred during the past 3 decades had a normal year (high), 47% experienced drought (low) and 39% had a wet (average) year. The results also showed that the rates of final water allocation in drought conditions for wheat, barley, melon, watermelon and ruby grapes were, respectively 2018, 1938, 7555,7031,1182 m3, in wet conditions, they were 2034, 1953, 7050.7570, and 1196m^3, respectively, and in normal conditions, they were 2025, 1942, 7046, 7559, and 1189m^3.

The results also showed that ruby grapes were the best crop with the highest expected profit in all conditions. In general, whenever the expected net income value of some crops is positive, the government will act optimistically and the high levels of water required are promised to the farmers. In turn, they would prefer to be more conservative and would promise the least amount of water provided to the farmers. Hence, if the promised water to the farmer is not wasted, he will choose the loss incurred from a lesser harvest.

RECOMMENDATIONS

It has been suggested to choose the type of crops based on the irrigated conditions. Moreover, if the farmers have enough freedom to choose and use different variables, the model can provide practical solutions in terms of establishing the amount of profit in farmers' mental calculations. Since in this study, the expected profit was obtained in drought, wet and normal scenarios under the most sensitive water requirement conditions, it is wise to consider several measures so that sustainable water could be provided to the farmers to grow crops on time and earn the minimum rate of household income. It would reduce the migration of Sistani villagers to cities and neighboring provinces. It is significant to note that if the east of the country becomes haunted it will endanger the security of the area and the whole country.

ACKNOWLEDGEMENT

I would first like to thank my thesis advisor Dr Javad Shahraki of the Economics Faculty at Sistan and Baluchistan University. Further, we would like to thank the reviewers of this manuscript for their very helpful advice.

REFERENCES

1- Abbott, M., & Cohen, B. (2009). Productivity and efficiency in the water industry. *Utilities Policy*, 17(3), 233-244.

2- Divakar, L., Babel, M. S., Perret, S. R., & Gupta, A. D. (2011). Optimal allocation of bulk water supplies to competing use sectors based on economic criterion–An application to the Chao Phraya River Basin, Thailand. *Journal of Hydrology*, 401(1), 22-35.

3- Dorfman, R. (1969). An economic interpretation of optimal control theory. *The American Economic Review,* 59(5), 817-831.

4- Ebrahimzade, E., & Lashkaripoor, G. (2001). Drought crisis in Sistan and adjust its strategies. *Political and Economic*, 167(15), 226-231.

5- Ghaffari Moghadam, Z., Keikhah , A., & Sabouhi, M. (2012). Optimum water resources allocation using game theory. *Water Resources Research*, 8(2), 12-23.

6- Ghaffour, N., Missimer, T. M., & Amy, G. L. (2013). Technical review and evaluation of the economics of water desalination: current and future challenges for better water supply sustainability. *Desalination*, 309, 197-207.

7- Ghasemi, A., Ghasemi, M., & Pessarakli, M. (2012). Yield and yield components of various grain sorghum cultivars grown in an arid region. *Journal of Food, Agriculture and Environment,* 10(1), 455-458.

8- Homauonifar, R., & Rastgaripour, F. (2011). Latian Dam water allocation among crops in conditions of uncertainty. *Journal of Economics And Agriculture Development*, 24(2), 259-267.

9- Karim, M., Ghasemi, M., Naroui, M.R., & Koohkan, S. (2012). Effects of Yaghoti grape cultivation on economy of rural families in Sistan and Baluchistan province-Iran. *International Journal of Agriculture and Crop Sciences,* 4(7), 386-389.

10- Kessler, J. J., & Van Dorp, M. (1998). Structural adjustment and the environment: the need for an analytical methodology. *Ecological Economics*, 27(3), 267-281.

11- Lynch, H. J., Campbell Grant, E. H., Muneepeerakul, R., Arunachalam, M., RodriguezIturbe, I., & Fagan, W. F. (2011). How restructuring river connectivity changes freshwater fish biodiversity and

biogeography. *Water Resources Research*, 10.1029/2010WR010330.
12- Mahan, R. C. (1997). *Efficient allocation of surface water resources in southern Alberta*. University of Calgary.
13- Mahan, R.C., Horbulykb, T.M., & Rowse, J.G. (2002). Market mechanisms and the efficient allocation of surface water resources in southern Alberta. *Socio-Economic Planning Sciences,* 36(1), 25-49.
14- Mihankhah, N., Chizari, A.H., & Khalilian, S. (2013). Optimum operational management from surface water resources with an application of dynamic programming. *Journal of Economics and Agriculture Development,* 26(4), 244-251.
15- Mohamadghasemi, M., Kouhkan, S.H.A., Akbari Moghadam, H., Rostami, H., & Goli Mahmoudi, H. (2008). Cost- benefit analysis of agricultural products in Sistan and Baluchestan. *Journal of Rural Dvelopment,* 11(4), 71-88.
16- Sabouhi, M., Rastegaripou, F., & Kahkha, A. (2008). Optimal allocation of water through the dam between urban and agricultural uses two-stage stochastic programming using fuzzy interval parameter in uncertainty. *Agricultural Economics,* 3(1), 33-55.
17- Salimifard, K.H., & Mostafaee Dowlatabad, K.H. (2013). Applying stochastic goal programming to water resource management. *Journal of Water and Soil*, 27(2), 282-291.
18- Sharaki, J., & Mohamadghasemi, M. (2014). *Economic allocation of water resources in the region of Sistan using dynamic optimization.* Paper presented at the Intrnational Conference on sustainable development, stratege and challengs, Tbriz-Iran .
19- Thompson, D., & Powell, R. (1998). Exceptional circumstances provisions in Australia is there too much emphasis on drought? *Agricultural Systems,* 57(3), 469-488.
20- Tu, M. Y., Hsu, N. S., & Yeh, W. W. G. (2003). Optimization of reservoir management and operation with hedging rules. *Journal of Water Resources Planning and Management*, 129(2), 86-97.
21- Zeng, X., Hu, T., Guo, X., & Li, X. (2014). Water transfer triggering mechanism for multi-reservoir operation in inter-basin water transfer-supply project. *Water Resources Management*, 28(5), 1293-1308.

16

Assessment Improving and Organizing Mechanism of Water Users Cooperatives Using AHP Method in Esfarayen County, Iran

Zahra Hosseinpur [1], *Mohammad Hossein Menhaj* [2*] and *Mohammad Kavoosi-Kalashami* [2]

Abstract

Institutionalization of Water Users' Cooperatives (WUAS) is one of the strategies for cooperative irrigation management, which lays the foundation for the participation of farmers in tackling water shortage challenge to reach the objectives of sustainable management of agricultural water in an efficient manner. The main purpose of the present study was to identify the grounds for improving WUAs and prioritizing the effective factors for developing and strengthening these cooperatives. The statistical population was comprised of experts and members of the board of directors as well as elite farmers who were WUAs members in 2015 (n=40). The interviewees were selected through consensus sampling. The analytic hierarchy process was used to prioritize the mechanism of improvement and development of WUAs. Based on the final weights of the criteria, the mechanisms received the following scores: supportive mechanism (0.266), agricultural irrigation management (0.261), economic mechanism (0.183), education and extension (0.123), and legal mechanism (0.090) and WUA's members' attitude (0.076). Accordingly, the supportive, agricultural irrigation management and economic mechanisms were the most important mechanisms to improve WUAs. Hence, it is recommended that low-interest financing should be provided by the Agriculture Jihad Organization in collaboration with the Agricultural Bank for transforming from traditional irrigation to localized irrigation.

Keywords:
Organizing mechanism, Development, Water Users' Cooperative (WUAS), Analytical hierarchical process

[1] *M.Sc. Student in Rural Development, Faculty of Agricultural Sciences, University of Guilan, Guilan, Iran*
[2] *Assistant Professor, Department of Agricultural Economics, Faculty of Agricultural Sciences, University of Guilan, Guilan, Iran*
* *Corresponding author's email: mejnhaj.mh@gmail.com*

INTRODUCTION

One of the major crises threatening the worlds in the future is a water shortage catastrophe (Mirzaii *et al.,* 2011). Natural resources are regarded as commonwealth; not only for the present generation, but also as a valuable legacy they must be sustained for future generations. In order to maintain a prosperous, flourishing, independent and self-sufficient society, we have to protect our water, soil and plants (plains and jungles), «environment»; otherwise, the future world will be set by poverty, hunger and pollution. Water as a natural resource is a substance for the survival of human beings. Agricultural activities are heavily depending on utilization of water for production of foods, which is also necessary for survival of living organisms. Todays, agriculture activities in Iran are faced with a shortage of water as the result of dry and semi-dry regions. Meanwhile, the earth is getting overpopulated and demands for water are increasing. Even though, demand and supply of water is in a critical condition, evidence shows that both surface and groundwater resources are used inappropriate manner (Saeedi, 2003). All the mentioned points make the users and agencies in charge of water focus more on appropriate management of supplies of water resources (Njatpuor, 2008). Effective management of water resources requires appropriate information across many areas, including the use of the water resources, the role of water markets in determining or regulating allocation decisions, and environmental necessity (Horne, 2015). Water as a singular unit, involves both public and private institutions. Therefore, it can be considered as a common-pool resource, which may take the form of communal, private or state property (Suarez Bosa, 2015). For the past two decades, irrigation agencies around the world have been involved in transferring some of their management responsibilities to farmers. This process has been highly varied, but the overall trend has been quite consistent. The process of the actual transfer of specific management responsibilities from government bodies to organized groups of farmers has become known in the literature as an irrigation management transfer. The trend of promoting the involvement of farmers in management, either through formal management transfer or through other less formal mechanisms, has become known as participatory irrigation management (Groenfeldt, 2003).

The presence and participation of farmers as the final users of water seems more essential at present than at any other time for making decisions, designing and evaluating programs related to water resources. Humankind as a social creature lives in a social way, obtaining his needs through cooperation and with the help of others (Njatpoor, 2008). Increasing demand for agricultural water, forces farmers to cooperate for consistent and comprehensive plan for managing and improving water consumption for the purpose of food security, to alleviate water shortage crises. Hence, a water user organization should be one of the principal strategies (Heydariyan *et al.,* 2008). Increasing productivity in agriculture are the main factors behind the establishment of efficient and independent water user organizations and serve as a source of strong motivation for participation (Salehi Tahbaz *et al.*, 2010). However, the idea of participation in such associations has remained at a very idealistic level and suffers from lack of sufficient analytical tools, scientific method and theoretical framework (Amini and Khayati, 2006). In Iran, apart from the approved rules for establishing governmental organizations in agricultural management, which are not, generally, successful, the most important instructions in WUAs are regulations of efficient consumption of agricultural water. In the past few years, a limited number of WUAs have been established in some plain irrigation networks of Iran. However, they are faced with many challenges due to their imperfect structure and lack of experiences. Sometimes, these challenges were so serious that they ended up in suspension or failure (Heydariyan *et al.*, 2008). Reformation, establishment and expansion of new organizations in agricultural irrigation are the most practical actions that will help to increase productivity, which, will bring the stable development of the agricultural sector and hence, balanced development of rural communities (Shahroudi and Chizari, 2008).

In varying agricultural conditions and rural environment, different factors are effective information and success of WUAs, Hence, render the success of these organizations remain uncertain. Therefore, feasibility study of the WUAs development program should be a good device for identifying these organizations potential and capabilities in each region. In order to conduct a research on this ground, it is necessary to consider the obligations of water resource management, regarding the recurring drought and move towards the capabilities of local communities, also consider the requirements which are contained in the third, fourth and fifth Iranian Development Programs, including the needs for transferring the management of irrigation networks to farmers and in improving the productivity (Ataei and Izadi, 2014). The stability of WUAs is a key to the success. This stability depends on increasing knowledge, changing attitudes and farmers, skills improvement (Peter, 2004). In the past two decades, as a result of the worldwide problem of water scarcity and the inability of the irrigation projects to reach the predetermined economic goals, special attention has been made to the formation of WUAs for changing the behavior of the farmers and encouraging them in participating in the management of irrigation water (Howarth and Lal, 2002). It has been mentioned in the ultimate declaration of the third World Summit that "most countries would suffer more from crises of water management (participation and cooperation) than from water shortage"; therefore, finding ideal management of water resources and an appropriate structure for these resources, seems necessary (Mirzaii *et al.*, 2011). According to the most experts, low efficient irrigation in developing countries is due to the lack of farmers' real participation in the decision-making process, performance, management and maintenance of irrigation networks. Given the fact that the exploitation of water resources can be systematized without appealing to native knowledge and the participation of local communities, a foundation called water-users cooperatives (WUAs) has been recognized (Heydariyan, 2003). Irrigation stability needs the overall water management system to be designed in such way that all eager farmers could cooperate in order to enhance water resources more productive. Hence, in the past two decades, due to a water shortage problem all over the world and increasing challenges among water users, special attention has been made to the formation of WUAs to involve the farmers in the management of agricultural water. WUA, as an important local association, plays a pivotal role in achieving optimal management of agricultural water usage, which is possible through the optimal participation of beneficiaries in decision-making, planning and implementation of irrigation programs (Heyd and Neef, 2004). In fact, the presence of WUAs is to develop the management of cooperative irrigation and increase the efficiency of water consumption through beneficiaries' interferences as much as possible (Hassabou and El-Gafy, 2007). Based on the local conditions, Iranian farmers suffer from insufficient rainfall and continuous droughts, which can be controlled only by conservation, storage and proper distribution of the available water, in the best way, which may be possible through the establishment of a participatory management (RazaviZand, 2008). Therefore, by establishing a new system like WUA, it is possible to provide a basis for planning and management of water resources. However, the number of WUAs established by government is not sufficient to achieve the ultimate goal (Mirzaii *et al.*, 2011). Direct participation of farmers in irrigation management is an effective device to improve the knowledge and utilization of irrigation water. As the result, increasing productivity and profitability in agricultural activities is one of the factors listed as the main elements of creating efficient and self-sufficient WUAs, and as a strong economic motivation, it can facilitate the process of irrigation management transfer (SalehiTahbaz *et al.*, 2010). Reformation and expansion of WUAs are the most principal movement toward higher productivity and more policy that is comprehensive that may be instrumental in stable improvement of agriculture organization and balanced rural communities (Shahroudi and Chizari, 2008). Since 1980 due to water shortages around the world and increase conflicts among water users, much attention have been made to the issue of water users associations to participate in the Agricultural Water Management (Heyd and Neef, 2004). In fact, the lack of monitoring the quality and excessive water exploitation, exacerbate the water shortage, which is a serious threat (Salmani *et al.*, 2010). In general, the term 'WUA' refers to a group of farmers who are formally in charge of administration and management of some parts of irrigation networks. They are usually a group of people located in the irrigation units, departments or regions. These groups have different names, such as; water users group, water user organization, irrigation union, etc. (Vuren *et al.*, 2004). The associations have four main tasks related to the exploitation (measuring the volume of water need and water supply, transport and water distribution networks, irrigation, water distribution during periods of water shortage and the reuse or transfer of the excessive water);conservation (according to the experts, conservation should be performed based on exploitation requirements and necessities; therefore, a system performing exploitation should also accept conservation management); financial support and ultimately the legal authority to resolve disputes between water users; consequently, each association will be successful and sustained providing to achieve these goals, and if the members try to cope with these four obligations effectively (Najafi *et al.*, 2013).

In a study on sustainable management of optimal consumption of irrigating water, Shahroudi and Chizari (2006) demonstrate that, increasing farmers' knowledge, attitude and skills about how to manage agricultural water through WUA may be one of the main ways for the farmers' cooperation to overcome water scarcity and reach the goals of agricultural water management. Azizi *et al.* (2009) investigate the effective structures of farmer's cooperation in irrigation management using an analytical approach in the irrigation network of Doruodzan Dam in Fars Province. They demonstrated that, among considering variables, farmers, attitude towards coordination in irrigation management, and understanding the current difficulties, had the most direct effect on the rate of farmers' co-operation. The rate of being social and the farmers, positive attitude towards agricultural services and extension centers, had the most indirect effect of farmer's co-operation. Based on the results of this investigation, Executive agencies should invite farmers to cooperate in irrigation management, such as preparing positive outlook toward participation and administrative system.

Mirzaii *et al.* (2010) studied the role of cooperation in irrigation management and indicates that, increasing overpopulation pressure, improving living standards, and increasing demands for preserving the environment, all caused governments to provide better ways for water management. In a study based on an optimal method in the management of the water resources in Qazvin Plain, Mortezanejad *et al.* (2012) showed that, the most important and effective solution is to pass powerful laws to prevent excessive digging of water wells, providing technical and extensional information in order to prevent unauthorized drilling. Their results identified four factors: reformation in pricing and education, development of agricultural cooperation and reducing bureaucracy, knowledge and skill improvement, informing and supervision. These elements, overall, explained about more than 60% of all differences, and the most significant of them were reformed in pricing and education. Maqabl *et al.* (2014) investigates the mechanisms of development and reinforcement in WUAs in Aras River region using the AHP method. They reported that, the final weight criteria: supportive mechanisms, education, extension, policymaking, communication and law, were the main criteria. The sub-criteria, included changes in traditional irrigation to drip and rainfall irrigation, holding educational and promotional classes to increase villagers' knowledge and awareness of how to play a role in the establishment and administration of WUAs through the help of Organization of Agricultural Department and Cooperative Office and District Water Office. Narayan (1995), in his study on 121 WUAs of Asia, Africa and Latin America found that, stakeholders' skills and knowledge increase, with regard to their participation, have a positive effect on successful water management and how making it happen, had a positive effect on the successful formation and continuation of cooperative activities. Moreover, in many cases, strengthening local organizations have facilitated the foundation of cooperatives. Another point is that the comparison of the places in some, where the socioeconomic status of farmers has received attention prior to the establishment of WUAs, and places where government and private institutions have ignored attitudes, interests and needs of the farmer's shows that the establishment of WUAs has been unsuccessful in almost all areas using the second mode of operation. Akbari *et al.* (2012); looked at the effective factors in the Iranian WUA success in the agricultural regions of Tajan, Moqan and Varamin. They found that, in these areas, the problems of WUAs are members' dissatisfaction with management of cooperative irrigation, network inefficiency and unfair water distribution, lack of confidence in project managers, lack of governmental support and lack of social solidarity. On the other hand, the most important success factors are: removing technical and managerial issues, renewing channels, educating farmers, collecting water and preserving the farmers' rights.

Given the fact that, the AHP can be expanded to decision-making problems in agricultural irrigation, this study conducted a research investigates and prioritize the major factors, which improve and organize the cooperatives to water users. Therefore, the purpose of this study was to evaluate the mechanism of improving and organizing WUAs in Esfarayen County of Iran.

MATERIALS AND METHODS

The study area

North Khorasan Province is located between the latitudes of 36°34' and 38°17' N., and longitudes of 55°52' and 58°20' E. It has an area of 24434.11 km^2, which is the fifteenth biggest province of Iran. The North Khorasan Province is limited to Turkmenistan from the north and eastern north with a borderline of 301 km, to the Razavi Khorasan Province from south and east, to Semnan Province of western south and to Golestan from west (Statistical Annual Year-Book of North Khorasan Province, Iran's Statistical Center, 2012).

Khorasan Province is divided into two parts based on unevenness: a) mountainous areas b) pots and smoothly splits. The highest point is Shah Jahan Peak with the altitude of 3051 m, which is located in the Aladaq mountain ranges and the lowest point is a TazeYab Village with the altitude of 4000 m lower than sea level located in the external part of Atrak River. The average height from sea level is 1326 meters. Northern Khorasan, which has a different environment than its neighboring areas has a special climate, and is adjacent from north and south to the great deserts. However, the climate is varied in this province because of its especial topographic conditions and the western-eastern location of mountains of Kope Daq in the north and Aladaq in the south. This creates relatively optimal conditions. The environmental conditions are different inside the province due to the mountains, forests, expanded plains and abundant water resources. However, the climate of the province is, in general, moderate and mountainous because the KopeDaq Mountains in the northern part and the Aladaq and Shah Jahan mountains in the south have surrounded central plains and created optimal conditions for agriculture, animal husbandry and fruit production. The most important water resources are Atrak River, Cheshmeh Saran, and four large dams named Bydvaz in Esfarayen, Shork and Barzou Shirvan and Shirin valley and Manet and Smlqan. Besides, the underground waters are significant in providing water resources (Statistical Annual Year-Book of North Khorasan Province, Iran's Statistical Center, 2012).

Esfarayen Town is located 60 km from the Bojnoord County in the Northern Khorasan. It has two types of climate: mountainous cold and moderate, and warm and dry. This town has two parts (Central, Bam and Safi Abad), two cities (Esfarayen and Safi Abad), seven rural districts, and 175 villages that all have settlers.

The average rainfall in this town is 241 mm and the average temperature varies between 8 to 13°C (North Khorasan Governor Office, Esfarayen City Development Document, 2010).

The county water is provided from 264 wells, 142 aqueducts (including 38.5% provincial aqueducts), 83 springs watering 274.46 million m3, and permanent and seasonal rivers with 112.4 million m3. The main permanent rivers are Bidvaz and Roien, the main seasonal river is KalShoor. In 2005, underground water volume was 1533.2 million m3, and it included 27% of all aquifers of the province (North Khorasan Governorate, Esfarayen City Development Document, 2010).

Considering the current drought, 79,215 hectares of fertile lands suitable for agronomy and horticulture in this region reduced to 27,731 hectares due to water shortage (Organization of Agricultural Jihad, 2014). Although there are some capabilities like Bidvaz Dam, irrigation networks, drainage and proper lands for agriculture, and the area is suitable for establishing WUAs, as there is no cooperation with farmers, they could not use their potential capability, and only two WUAs are registered there with the Corporation of Regional Water in the town. To date, 2440 members have registered in the WUAs (Regional Water Company Esfarayen City, 2014). Furthermore, about 2700 hectares are covered by the Nazargah WUA (founded in 2006), and 4850 hectares by Bidvaz WUA (founded in 2004). In order to facilitate and accelerate the increasing numbers of WUAs in this area and provide these associations with proper frameworks, this study tried to investigate the main effective factors in developing and organizing the WUAs and rank inclusive solutions, behavioral factors and effective structures, and adjust them for the planners of economic water in the agriculture sector of Iran. Therefore, the goal of this study was to investigate experts' viewpoints about improving activities and organizing regional WUAs (Figure 1, 2, 3).

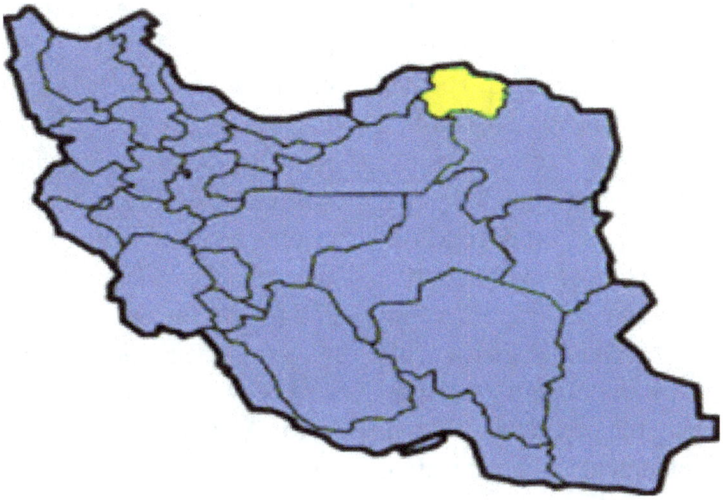

Figure 1: North Khorasan Province in Iran

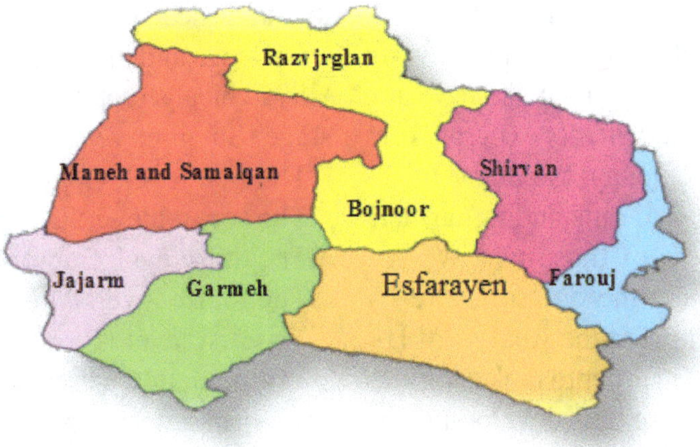

Figure 1: North Khorasan Province in Iran

Figure 3: Landscape of irrigation network of Esfarayen County (Bydvaz dam)

Each research topic has its own unique features. However, regardless of the subject of study, all researches have same administrative process, which is the objective of the research. If a proper method is not used, the scientific goals or scientific knowledge is not achieved. Todays, the reliability of scientific achievements gained and influenced by research methods, only (Kalantari, 2009). Selection of a research method is the main part of any study, and it depends on the research subject, expansion of administrative facilities, goals, appropriate assumptions and moral considerations (Ahmadinejad, 2011). This study was carried out by documents and field methods. After reviewing the previous research, the mechanisms of development and organization of WUAs were determined and the criteria related to each mechanism were categorized. Then, according to studies related to the drafts, questionnaires were set up in order to compare activities two-by-two, using the collected data. The population consisted of 40 experts from the Regional Water Company, the board of directors of two WUAs, and outstanding farmers who were the members of those two WUAs. Finally, after completing the questionnaire by the 40 members, we utilized the Analytical Hierarchy Process (AHP) in order to rank activities instrumental in improving and developing the WUAs. The data were analyzed by Expert Choice software package.

Alphonce (1997) suggested that Analytic Hierarchy Process (AHP) has some potential for resolving certain decision problems in agriculture. In these regards the AHP applied by using the set of approaches to decision-making in agricultural irrigation (Srdjevic, 1997). Compared with five different models for the estimation weight coefficient, AHP was found to produce the most credible results (Schoemaker and Waid, 1982).

The Analytic Hierarchy Process (AHP) is a method of measurement for formulating and analyzing decisions (Berrittella *et al.*, 2007). Saaty (1980) provided a theoretical foundation for the AHP that is a decision support tool which can be used to solve complex decision problems, taking into account tangible and intangible aspects. Therefore, it supports decision makers to make decisions involving their experience, knowledge and intuition (Berrittella *et al.*, 2007). Hierarchical structure consisting of the following stage (Sungguh *et al.*, 2013):

The first level: Goal

The second level: Criteria

The third level: Sub-criteria

The fourth level: Alternatives (Figure 4).

The first level is a goal to be achieved, in this study, namely the creation of a model of organization based on water resource utilization in the basin. The second level, the criteria, indicates

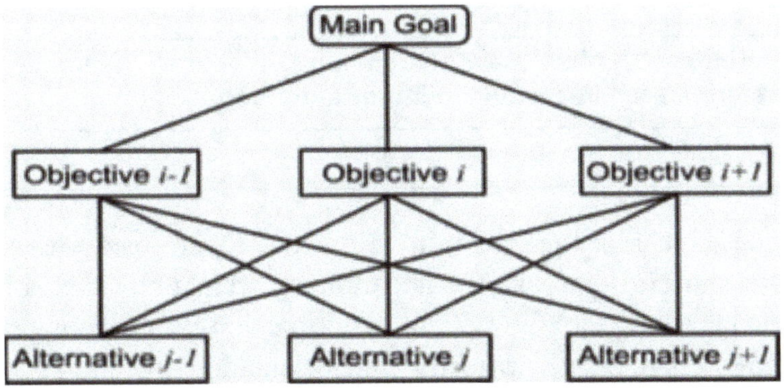

Figure 4: AHP hierarchy of goals, objectives and alternatives

the variables used in the analytic hierarchy process. The third levels, the sub-criteria, are indicators of the criteria. The fourth level is an alternative model of organization that will be created. The most important stage of the analytic hierarchy process is an assessment of the comparison pair. Unlike the conventional methods, AHP uses pairwise comparisons, which allow verbal judgments and enhances the precision of the results. The pairwise comparisons are used to derive an accurate ratio and scale priorities (Dalalah *et al.,* 2010). Developed by Saaty (1980), AHP provides a proven, effective means to deal with complex decision-making and can assist in identifying and weighing criteria, analyzing the data collected and expediting the decision-making process.

According to six criteria that can be seen in Figure 5, sub-criteria specific to each of them were defined, some of which are as follows:

Sub-criteria of education and extension mechanisms, including meeting with promoter's agricultural service center, participation in education and promotion classes, promotional screenings, visiting the successful cooperatives, etc.

Sub-criteria of economic mechanisms, including increasing agricultural production, creating competition in production, better use of water resources, etc.

Sub-criteria of protection mechanisms, including cover part of the cost lateral, regulatory services and facilitation of water organization, allocation of a certain percentage of prices of water cooperative, etc.

Sub-criteria of legal mechanisms, including stability of laws and regulations, management of water resources, change and transformation in the structure of agricultural water management, water user's cooperative's management pyramid codification of the law etc.

Sub-criteria of agricultural water management mechanisms including amount of land ownership, personal property on the land, area under cultivation, the price of water (right of water), etc.

Sub-criteria of WUA member's attitude improvement, are: establishing cooperatives to increase the farmers' income, create cooperative of water used, so that to distribute water in a fair way among farmers, with the membership of farmers in the cooperation, the management of irrigation water may be fulfilled in the appropriate way, etc. (Table 1) (Figure 5).

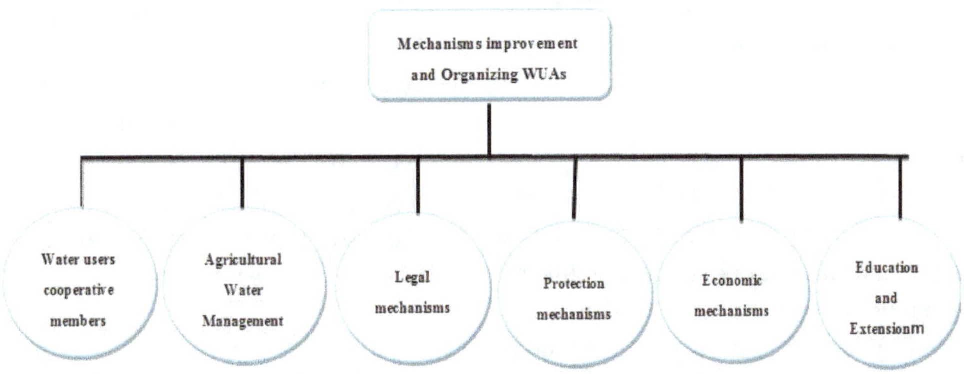

Figure 5: The hierarchy process tree to determine the criteria, sub and alternatives

Table 1: Paired comparison scale

Comparison situation	Middle preferences	Absolutely important	Much more important	More important	Equally important to a little more	Same
Weight or value	2,4,6,8	9	7	5	3	1

When this technique is used for the selection or prioritization, in general, an expert is finding a replacement because he knows which alternative should be selected as a priority (Maghabl *et al.*, 2014).

Data collection

Four processes are essential in collecting data: questionnaire design, data collection, population determination and paired comparative table design. In the questionnaires, the number of questions is proportional to the number of criteria, and they are designed as paired comparisons of the criteria. Samples were selected through censuses. In this study, 40 people, including board members, professionals in the Regional Water Authority Company and farmers who are members of water users associations were interviewed.

Paired comparative tables

Paired comparisons are the foundation of AHP. AHP is capable of evaluating quality issues, as there are no units for measuring them. These comparative forms the matrix, in which, elements are achieved from comparison of each element in rows with the elements in columns. After arranging the hierarchy tree and identifying goals and criteria (Figure 4), the paired comparative tables, in which there were questionnaires and collected data, are analyzed. As Table 1 shows, the table must have 9 degrees, and each comparison case must be placed in front of a degree. Measuring the project data are a complicated and lengthy operation for which Express Choice software package was used.

Calculation operations

The activities for development and improvement of WUAs based on their priorities were arranged considering six criteria: supportive, agricultural irrigation management, economics, extension – education, WUAs attitude and legal mechanism. The paired comparison was studied by Expert Choice software package. After identifying the ultimate weights by this software, the ultimate prioritization was carrying out. The purpose of the calculation operation is the evaluation of the final weights in order to prioritize in AHP method.

Average geometrical calculation: The following instructions are used in this calculation (Maghabl *et al.*, 2014).

$$(a_{ij}^{(k)} \pi_{k=1}^{n})^{\frac{1}{n}} = Aij \quad (1)$$

A_{ij}: geometric average criteria

a: sub-criteria compared with the options

ij: names of two alternatives compared with each other

k: sample number

n: all the people interviewed about a sub-criteria, And π: is the multiplication sign

The weights must be normalized after evaluation. There are different methods for normalization, but in AHP, the following rules are used (Ghodsi pour, 2000):

$$r_{ij} = \frac{a_{ij}}{\sum_{i=1}^{m} a_{ij}} \quad (2)$$

a: the name of sub-criteria

r_{ij}: the normalized sub-criteria

ij: two alternatives that must be compared

The normalization operation will be done based on the geometrical average of answers to alternatives and sub-criteria.

After the normalization of numbers, the balanced average of normalized numbers is extracted. For this purpose, the normalized numbers of each row of alternative calculation, and, finally, its average would be extracted. Instruction of the calculation of the weighted mean of following command was used (Maghabl *et al.*, 2014):

$$W = \frac{1}{N}\left[\sum_{j=1}^{N} r_{ij}\right] \quad (3)$$

W: weighted mean
N: the number of compared alternatives
r_{ij}: the normalized amount of each row
Instructions to calculate the weighted mean for the calculation of the mean of each row of the alternatives is as follows (Maghabl *et al.*, 2014).

$$W = \sum_{i=1}^{n} w_{ai}\, w_{ci} \quad (4)$$

W: the final weighted mean first row alternative
W_{ai}: alternative weighted mean i
W_{ci}: weighted mean sub-criteria j
n: whole criteria and alternatives in first row.
The whole criteria and alternatives in the first row, and criteria and sub-criteria were measured according to this instruction.

RESULTS AND DISCUSSION

Considering the results achieved from data analysis, supportive mechanisms as the first priority (weighted mean of 0.266) were ranked as the most important mechanisms for the improvement and organization of WUAS in the studied region (Esfarayen). Moreover, Arabi and Mohebbi (2008), Knowler and Bradshaw (2007), Mortezanejad *et al.* (2012), Maghabl *et al.* (2014), Narayan (1995), Peter (2004) and Wijayaratna (2002) all showed that supportive mechanisms are one of the most effective mechanisms in developing WUAs. The agricultural irrigation management mechanisms with the relative weight of 0.261were found to be the second priority and were the most important mechanisms for the improvement and organization of WUAs in the studied area (Esfarayen County), which is in agreement with Mohammadi *et al.* (2010).The third priority was found to be economic mechanisms with a relative weight of 0.183, and was confirmed by Ahmadvand and Sharifzadeh (2009), Maghsodi and MoemenQharib (2013) Narayan (1995). The extension and education mechanisms were the fourth priority with a relative weight of 0.123, and this achievement was Similar with the results of Ehsaniand Bustos *et al.* (2001), Farshi (2004), Heidari *et al.* (2006), Khaledi (2003), Restrpo (2005), and Zahtabyan (2005). The next was legal mechanisms with a relative weight of 0.090, which was confirmed with Restrpo (2005), and the last was WUAs member's attitude with a relative weight of 0.076which is similar whit Ahmadvand and Sharifzade (2009) (Figure 6).

CONCLUSIONS

The results of this research reveal that for improving and organizing WUAs in the region, the support mechanism, should be first, in the priority list of policymakers agenda in Iran. The current supportive plan, including political, administrative, financial and technical are insufficient and inappropriate. Hence, it is essential and necessary to review these supportive plans. Therefore, it is suggested to set up some committees to study and arrange the appropriate supportive plan for each WUAs based on the local conditions, and to provide facilities and financial credits for farmers, so that they can change traditional irrigation to modern one.

Since this study recognizes irrigation water management mechanism second in the row,

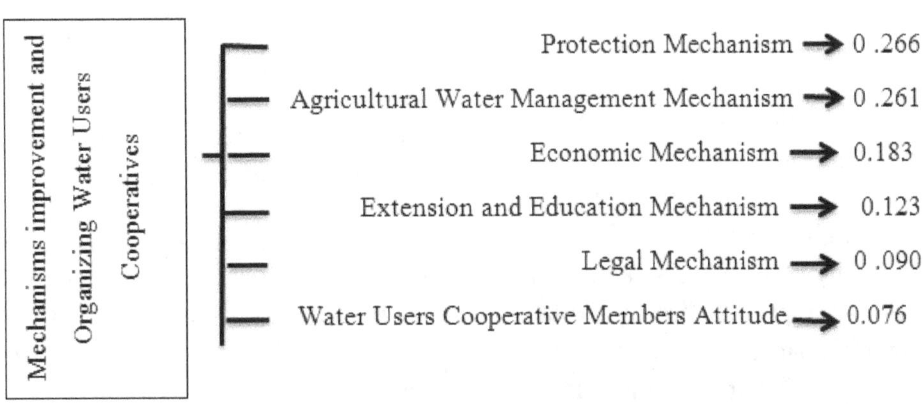

Figure 6: Ranking of the mechanisms for improvement and organization of Water Users Cooperatives with regard to the weighted average

therefor, the mechanism should be listed in the government program for the national water management plans. In order to induce positive attitude in farmers toward different water management methods, it is necessary to provide an appropriate approach for encouraging them towards water resource management, particularly for underground water.

Considering that economic mechanism third in the row, which implies that economic grounds for development of WUAs do not exist in the region? Therefore, more attention should be paid to economic organizations before their establishment, such as local markets for products, the profitability of associations, the quantity of production in the region, etc. Accordingly, economic and feasibility studies are necessary before the establishment of WUAs.

With regard to the fourth priority, i.e. extension and education mechanisms, it is suggested to provide technical trainings to develop and protect irrigation structures, especially measurement devices, and volumetric flow meter. Furthermore, education and legal assistance should be provided to prevent farmers' trespass on each other's territory and impermissible harvest. In addition, proper opportunities must be provided to enhance the knowledge and awareness of users, representatives, and boards of director members by holding training courses, visiting successful cooperatives, holding seminars and conferences.

Ranked in the next priority, are legal and attitude of cooperative water user's mechanisms. Given the importance of these mechanisms, sub-criteria are needed to be planned. As a result, considering the importance of the WUAs establishment and the fact that the country should inevitably follow the trend of water transfer from the public sector to the users' associations, it is essential that native experts of each region to conduct comprehensive studies in this regard, and by identifying and eliminating the impediments, attempt to successfully establish organizations which participate farmers in order to lay the foundation for the development and improvement of WUAs.

ACKNOWLEDGEMENT

The Authors would like to express their appreciation to all those who help us for gathering data. Furthermore, authors would also like to acknowledge with appreciation the role of anonymous reviewers and IJAMAD director in charge for improving our manuscript through their useful comments.

REFERENCES

1- Ahmadinejad, P. (2011). *Rankings sustainable agriculture co-operatives in wheat cultivation, production in Hamadan province,* Unpublished thesis, Faculty of Agricultural Sciences, Rural Development, University of Guilan.

2- Ahmadvand, M., & SharifZadeh, M. (2009). Feasibility study establishing WUA, Case Study: KovarDasht in Fars province. *Journal of Agricultural Extension and Education*, 5(2), 141-154.

3- Akbari, M., Omid, M. H., Zarafshani, K., Eskandari, Gh. H., & Fami, H. Sh. (2012). Factors Influencing the Success of Water User Associations in Iran: A Case of Moqan, Tajan, and Varamin. *Journal of Agricultural Science and Technology,* 14(1), 27-36

4- Alphonce, C. B. (1997). Application of the analytic hierarchy process in agriculture in developing countries. *Agricultural Systems*, 53(1), 97–112.

5- Amini, A., &Khayati, M. (2006). Factors affecting the failure of the cooperative formed by water users, (Using Regression fuzzy). *Agricultural Economics and Development,* 53(14), 69-91.

6- Arabic, A., & Mohebbi, M. (2008). The establishment of water users and related challenges in the Dasht of irrigation and drainage networks Abbas: The fifth technical workshop on the participation of water users in the management of irrigation and drainage networks. *Iranian National Committee on Irrigation and Drainage*, 74-53.

7- Ataei, P., & Izadi, N. (2014). Compare features acceptor farmers and incompliant WUAs. *Journal of Water Research in Agriculture,* 28(1), 251-261.

8- Atmanti, HD. (2008). Analytical Hierarchy Process sebagai Model yang Luwes. Prosiding INSAHP5, *Teknik Industri Universitas Diponegoro Report, Semarang, Uzbekistan.*

9- Berrittella, M, Certa, A., Enea, M., & Zito, P. (2007). An analytic hierarchy process for the evaluation of transport policies to reduce climate change impacts. *Fondazione Eni Enrico Mattei Center Report, Milan, Italy.*

10- Bosa, M. S. (2015). Water Institutions and Management in Cape Verde. *Water*, 7(6), 2641-2655.

11- Bustos, R., Marre, M., &Chmbouleyron, J. (2001). Performance of water user association in the Lower Tunuyan area, Argentina. *Irrigation and Drainage Systems,* 15(3), 235-246.

12- Dalalah, D., AL-Oqla, F., & Hayajneh, M.(2010). Application of the Analytic Hierarchy Process (AHP) in Multi-Criteria Analysis of the Selection of Cranes. *Jordan Journal of Mechanical and Industrial Engineering*, 4(5), 567 – 578.

13- Dhymavy, A., Mdhj, D., & Chenari, V. (2010). Evaluation of existing structures in irrigation management transfer to WUAs in practice companies exploiters in Khuzestan province, Third National Conference of irrigation and drainage networks 1-3 March, Ahvaz, Iran 185-175.

14- Ehsani, M., & Khaledi, H. (2003). Cognition and promotion the efficiency of agricultural water in order to water and food securing country, Proceedings of the Eleventh Conference of the Iranian National Committee on Irrigation and Drainage, 674-657.

15- Farshi, A. A. (2004). Management of agricultural water consumption, the first conference on problems of irrigation systems, drainage and management of agricultural water consumption, Tehran, 123-112.

16- Ghodsipour, S.H. (2000). *Analytical Hierarchy Process (AHP),* Publication CenterAmirkabir University of Technology (Tehran Polytechnic).

17- Groenfeldt, D. (2003). Participatory Irrigation Management, *Water-Culture Institute Report,* Available at www.waterculture.org.

18- Hassabou, A. H. M., & El-Gafy, I. K. (2007). Assessment indicators for water users' associations in Egypt. *E-WAter, Official publication of the European water association (EWA)*, 2, 1-14.

19- Hedy, H., & Neef, A. (2004). Participation of local people in water management: evidence from the Mae Sa watershed, Northern Thailand.

20- Heydari, N., Islami, A., ghadamiFiroozabadi, A., Criononi, A., Asadi, M., &KhajehAbdollahi, M.h. (2006). Determine the water use efficiency of crops in different regions of the country (Kerman, Hamedan, Moghan, Golestan, Khuzestan), The first National Conference of irrigation and drainage networks, Ahvaz, Faculty of Science Engineering of water. 238-230.

21- Heydariyan, S. A. (2003). Transfer management, Methods, barriers and strategies, the third technical workshop participation water users associations in the management of irrigation systems, Iranian National Committee on Irrigation and Drainage, Tehran, 81-25.

22- Heydariyan, S. A. (2008). Irrigation management transfer, Global efforts and results, Iranian National Committee on Irrigation and Drainage, Tehran.

23- Horne, J. (2015). Water Information as a Tool to Enhance Sustainable Water Management _ The Australian Experience. *Water*, 7(5), 2161-2183,

24- Howarth, S. E., & Lal., N. K. (2002). Irrigation and participation: Rehabilitation of the Rajapur Project in Nepal. *Irrigation and Drainage Systems*, 16(2), 111-138.

25- Kalantari,Kh. (2009). *Structural equation modeling in Social Research* – Economic, Saba Culture Publication, Tehran.

26- KhalkhylyAzizi, T., & Zamani, Gh. H. (2009). Structures effective participation of farmers in irrigation management, using path analysis. *Journal of Agricultural Economics and Development*, 24 (1), 83-90.

27- Knowler, D., & Bradshaw, B. (2007). Farmers, adoption of conservation agriculture: a review and synthesis of recent research. *Food Policy* 32(1), 25-48.

28- Maghabl, R.A., Pak Nia, F., Naderi Mehdi, K., & Nasiri, M. (2014). Survey mechanisms development and strengthen cooperatives water users: A Case Study of the Aras River Basin, application metod AHP. *Journal of Agricultural Economics and Development,* 2(1), 63-55.

29- Maghsodi, T., & MoemenGharib, M. (2013). Experts to examine the feasibility of WUAs. Case Study: Ramshir irrigation networks. *Research Journal of Agricultural Extension and Education*, 7(1),7-20.

30- Mirzaii, A., Mirdamadi S.M., & Hosseini, M. (2010). Cooperative water users the new approach of participatory irrigation management. *Journal of Agriculture and Natural Resources*, 7 (27), 18-21.

31- Mirzaii, A., Mirdamadi, S.M., Hosseini, M., & Sajjad, A. (2011). The impact of promotional tools and sources of information on the participation farmers in cooperatives Golestan water users, *Cooperation,* 23(5), 128-143.

32- Mohammadi, I., ShaabanaliFami, H., & Asadi, A. (2010). Analysis of factors influencing the agricultural water resource management in the city of ZarrinDasht from the perspective of farmers. *Journal of Agricultural Sciences and Natural Resources,* 16, Special Issue 2, 1-11.

33- Mortezanejad, M. Yaghobi, J., Setuodeh Nya, A., & Daghestani, M. (2012). Strategies to optimize the management of water resources in the irrigation network of WUA the perspective of, case study: the Qazvin network. *Journal Water Resources Engi-*

neering, 5, 69-77.
34- Najafi, N., Khosravipour, B., Ghanyan, M., Baradaran, M., & Dhymavy, A. (2013). Identify factors deterrent the success of WUAs from the perspective of Exploiters Karkhe areas of northern and southern Khuzestan province. *Rural Research*, 4(1), 188-165.
35- Narayan, D. (1995). The Contribution of People`s Participation: Evidence from 121 rural water supply projects, Environmentally Suistainable Development Occasional Paper Series, International Bank for Reconstruction and development, Washington, DC, (1), 1-122.
36- Nejatpour, H., Sharifi, H., & Moradi, R. (2008). How can beneficiaries participation in the exploitation and maintenance of water resources optimize be institutionalized?, First conference of the new approaches to public participation in the study, Construction, operation and maintenance of irrigation and drainage network, Shiraz, 7 - 8 February.
37- North Khorasan Governorate, Esfarayen City Development Document. (2010).
38- Peter, R.J. (2004). Participatory Irrigation Management. International Network on Participatory Irrigation Management Center. Washington DC., USA.
39- RazaviZand, S., Mlekmyan, L., Kamran, F., Ziaee, K., & AkhlasPour, R. (2009). Evaluation of barriers to effective participation of organized functionality case study WUAs Bardsir city. *Journal of Social Research*, 10(1), 100-119.
40- Restrpo, C.G. (2005). Irrigation management devolution in Mexico. International conference irrigation management transfer, 1-27.
41- Saaty, T. L. (1980). *The analytic hierarchy process: planning, priority setting and resource allocation*. McGraw-Hill, New York.
42- Saeedi, M. (2003). *Introduction to the participation of non-governmental organizations*, Samt Publications, Tehran.
43- Salmani, M., Tourany, A., & Khorasani, M. A. (2010). Based on the risk level of rural drinking water reservoirs and distribution network, case study: the central city Minoodasht. *Journal of Village*, 1(4), 177-155.
44- Shahroudi, A. A., & Chizari, M. (2006). Water user's cooperative strategy in achieving sustainable management of agricultural water optimized consumption. *Jihad Journal*, (247), 92-109.
45- Shahroudi, A. A., & Chizari, M. (2008). Behavioral analysis of area farmers Management of agricultural water consumption in Khorasan Razavi province, Compare participants and non-participants in the cooperative water users. *Agricultural Extension and Education Sciences,* 4(2), 81-98.
46- Srdjevic, B. (1997). On the use of systems analysis in horticultural crops irrigation. *Acta Horticultural,* 449(1), 245–250.
47- Statistical Yearbook North Khorasan Province, Statistical Center of Iran. (2012).
48- Sungguh, H.M., Bisri, M., Kusuma, Z., & Prijono, S. (2013). Application of the Analytic Hierarchy Process (AHP) in the Selection of River Basin Organizations Models in Toba Asahan River Basin – Indonesia. *Information and Knowledge Management*, 3(8), 2224-5758.
49- TahbazSalehi, N., Kuhpahi, M., & Nazari, M. (2010). Evaluation of the Participatory Irrigation Management in Iran, Case Study: Cooperative water users associations Tajan. *Agricultural Economics and Development*, 2(24), 205-216.
50- Vuren, G.V., Papin, C., & Haouari, N. (2004). Participatory Irrigation Management: Comparing Theory with Practice a Case Study of the Beni Amir irrigation scheme in Morocco, Projet INCO-WADEMED, Acts du Seminaries Modernization de l'AgricultureIrrigu´ee Rabat.
51- Wijayaratna, C.M. (2002). Requisites of organizational change for improved participatory irrigation management Report of the Apo Seminar of Organizational Change for Participatory Irrigation Management. Philippines, 23-27 October 2000 (SEM-32-00).
52- Zahtabyan, Gh.R. (2005). The low efficiency of irrigation in Varamin region. Seventh Seminar of the National Committee on Irrigation and Drainage, 264-240.

17

Factors Affecting Land Allocation to Saffron and its Expansion in Marand County, Iran

Kolsoum Azizi Mizab [1] *and Azadeh Falsafian* [2*]

Abstract

Currently, water, as the most limiting factor in production, determines the priority of planting in different areas of Iran. Saffron is one of the good candidates for drought conditions since it has high economic value and low water requirements that can help with sustainable development. By identifying the factors influencing the decision on saffron cultivation and its expansion, appropriate policies can be implemented to improve the planting of this crop. Marand, located in East Azarbaijan Province, Iran, is one of the areas where farmers have started to grow saffron in recent years. The allocation of 68 hectares of agricultural land to this crop has turned the county into the hub of saffron production in the northwest of the country. This study investigated the factors affecting the decision on saffron cultivation and its development in Marand. To this end, a total of 140 farmers from two groups of saffron growers and non-saffron growers were chosen, and the Heckman's two-step procedure was then employed. The results of estimating the first step of the Heckman procedure showed that age, familiarity with saffron growing, attending saffron training courses, the number of extension courses, marketing status, and profit status of saffron all had a positive effect on the decision on growing saffron. Moreover, the results of estimating the linear pattern of the second phase corroborated the view that the farmer's education level, the total area under agricultural and horticultural cultivation, as well as features of agricultural land had a positive impact, and access to water resources had a negative effect on the cultivation area of saffron.

Keywords:
Heckman's two-step procedure, land allocation, Marand, saffron

---

[1] *MSc. in Agricultural Management, Tabriz Branch, Islamic Azad University, Tabriz, Iran*
[2] *Assistant Professor, Department of Agricultural Management, Extension and Education, Tabriz Branch, Islamic Azad University, Tabriz, Iran*
* *Corresponding author's email: falsafian@iaut.ac.ir*

INTRODUCTION

Due to the climatic conditions of Iran where water is a major factor limiting agricultural development, it seems necessary to change the planting pattern toward the crops that have low water requirements. It is more helpful to achieve sustainable development in Iran. Among crops, saffron is a plant with little expectation of high economic efficiency, because it needs water when the other crops do not demand water or, at least, are not facing the water shortage. Marand is one of the counties of East Azerbaijan that has the potential to grow saffron due to its moderate-to-cool temperate climate. The county started planting this crop in 1996, and every year, the area under the cultivation of this commercialized valuable product that can bring foreign income so that it reached 68 hectares in Marand in 2014 (Marand Agriculture Department, 2014). Due to good soil and mild climate, saffron bulbs of Marand are of high quality, and the county is one of the important centers of saffron bulbs in Northwester Iran. Given saffron's potential to bring exchange as well as its features such as high medicinal and spice value, ease of maintenance and transportation, and most importantly, the talent of the Marand region to produce high-quality saffron, it is imperative to consider its planting and development.

In this regard, we should be addressed from two different aspects: one is increasing the number of saffron farmers and examining the factors affecting the farmers' determination to cultivate saffron, and the other one is increasing the saffron cultivation area and studying the factors affecting the development of the area on the farmers' side. This issue is important, because the factors affecting farmers' decisions to produce a certain crop do not necessarily match the factors that determine the cultivation area of the crop, but can be two different sets of variables. In other words, the decision to allocate land for saffron cultivation is done in two stages by producers: In the first step, the question is whether or not to allocate the land to the cultivation of saffron and the second step addresses the decision on the size of the area under cultivation. Salami and Einollahi (2001) argue that equating these two groups of factors and perceiving the farmers' decision as a one-step procedure can lead to failure to identify the factors leading some farmers to abandon growing this crop and may also result in the lack of access to the actual behavior of producers and erroneous conclusions. In addition, they may finally lead to incorrect policy recommendations. In such circumstances, one should use a model to evaluate and identify important factors that could make this distinction and offer reliable results. Given this, in this study, Heckman's two-step procedure was used in which the model was broken into Logit and regression models for determination and differentiation of factors in each stage of the decision to cultivate saffron and decision to increase the area under saffron cultivation in Marand. The results of this research can be useful in identifying saffron producers' real behavior and increasing the efficiency of policies taken to develop saffron cultivation.

Many studies have been carried out to identify factors affecting the acceptance and development of the cultivation area crops and the use of the models with limited dependent variables in a variety of topics. This section reviews the methodology and results of some of these studies.

Wekesa et al. (2003) first investigated factors affecting the adoption of cultivation of the improved varieties of maize, including access to cash, access to credit, and attending training classes using the Logit model. Then, Tobit model was estimated to review the adoption of the chemical fertilizers where the results showed that, in addition to the factors mentioned above, variables such as the farm size, off-farm income, age, and education level of the farmer were also effective in the adoption of the agricultural fertilizers. Bergtold et al. (2012) studied management and demographic factors affecting the profitability of winter cover crops (clover) in America using the Logit model. The results indicated that the high loan rate, high sales, conservation tillage, and the use of nitrogen for cover crops had positive impacts, and the years of agriculture, farm size, and the high cost of the cover crop had negative effects on the profitability of the cover crop. Keil et al. (2009) examined the factors influencing the allocation of land to corn and concluded that land's being mountain terrain or being under rice cultivation, literacy, wealth, gender and ethnicity of the farmer as well as the proximity to the road were some factors affecting households to allocate land to corn. In a study on identifying the factors underpinning the development of sugar beet cultivation in Khorasan Province, Iran, Salami and Einollahi (2001) used the Tobit model and cross-sectional data to analyze various economic, social, and technological factors affecting the changes of the cultivation of sugar beet cropping pattern in Khorasan Province. According to the findings of this study, factors such as the grade of the product in the previous year, distance of the field to the sugar plant, crop insurance, agricultural machines, the ratio of the cultivated land to total area, the ratio of income from the cultivation of sugar beet to total income, and the interaction of education and the relative profitability of the product are among the most important determinants of sugar beet cultivation development in Khorasan Province. They showed the importance of the application of the econometric Tobit models and Heckman's two-step procedure to determine the factors affecting the sugar beet cultivation. They also indicated the necessity of making difference between two groups of factors influencing the farmers' decision on crop and determinants of the crop acreage. The results revealed that the factors influencing the farmers' decision to grow sugar beet are not identical to the factors affecting the acreage of this product; therefore, the use of econometric models such as Heckman models that consider this distinction is essential.

Shafiee (2007) studied the factors underpinning the development of olive cultivation in Kerman Province, Iran using the Logit model. The results showed that overall education, number of children over 14 years old, the extension agent's relationship with the farmer, mutual trust with the farmer, the total area under cultivation, farmers' familiarity with olive, and the savings and income levels less than 10% affected the acceptance of olive planting by farmers. Moreover, the number of land pieces, the technological level, credits, and receiving loans did not have a significant impact on the adoption of olive growing. Hayati et al. (2010) utilized Heckman's two-step procedure to examine the factors affecting willingness to pay by visitors of Elgoli and Mashrouteh Parks in Tabriz, Iran. The results of the estimated Probit Model showed that monthly income, family size, gender, satisfaction with social security, and the number of visits to the park during the year were among the factors affecting willingness to pay, and in the second stage, age, education, duration of each visit, and education level affected willingness to pay. Bakhshi (2011) used the Tobit and Heckman's two-step procedures to identify the economic, individual, and social factors influencing the adoption and development of canola crops in Tabriz and Marand. The results showed that the level of education, the loan amount, the actual price of canola, the relative share of farm income in total income and the relative profit of canola had a positive and cost of machinery, farm to road distance and the number of land pieces had a negative effect on the cultivation area of this crop. Furthermore, to distinguish the factors affecting the cultivation of canola and the cultivated area, she used Heckman's two-step procedure. The results of the Probit model showed that machinery ownership, participation in extension classes, the share of farm income, education level, as well as canola planting experience had positive effects, and age and number of land pieces had negative effects on the probability of accepting to plant canola. The results of the estimation of the second step confirm that the amount of allocated loan, the relative benefits of canola, and the number of family labor influenced the cultivation of canola positively, and land to road distance and machinery costs per hectare influenced it negatively.

MATERIALS AND METHODS

As mentioned in the introduction section, the decision to allocate a land to the cultivation of a crop like saffron by producers is made in two stages. In such cases, the Heckman's two-step procedure is an appropriate option to separate these two stages, which are described in further detail below.

In Heckman method, the model is broken into two separate models to determine the effective factors in each of the two above-mentioned variables. In the first model pertaining to the first stage of the decision to grow crops on the farmer's part, the dependent variable has only two values. This means the dependent variable appears as a binary choice that only takes 0 or 1. In the literature, while the dependent variable is a dummy variable with two 0 and 1 values, the Logit, Probit, and Tobit models are used as variables that can be estimated in such an environment. In this study, the factors influencing the decision to grow saffron were estimated by the Logit model. In the second model pertaining to the second step to decide about the area under cultivation, the dependent variable is the area under the cultivation of the crop, and therefore, the linear regression model is used. In sum, in the two-step method of Heckman employed in this study, factors that can influence the farmers' decision to grow saffron were included in the Logit model as the independent variables and the factors that can be effective on the area under the cultivation of saffron were included in the linear regression model as the independent variables. It is noteworthy that, in the second model, a new variable called Inverse Mills Ratio was added, which was created using the estimated parameters of the first model. With this new variable, the second model was related to the independent variables in the first phase.

Logit Model

The Logit cumulative probability distribution is as follows (Green, 2000).

$$P(Y=1)=1/(1+exp(-\beta x) \quad (1)$$

where $P(Y=1)$ shows the decision to cultivate saffron, X is the explanatory variables, and β shows the characteristics of the model.

The alternative form of the above equation is as follows:

$$P_i=1/(1+e^{-(\alpha+\beta x_i)}) \quad (2)$$

where X_i is the ith independent variable, e is the base of the natural logarithm, p_i is the likelihood to grow saffron, β is the parameter of the model in the logistic function that must be estimated, and α is the intercept.

Linear Regression Model

This relationship is only estimated for observations where saffron cultivation is used. As was noted, at this stage, the Inverse Mills Ratio variable λ_i was added to the set of independent variables in the regression model. The presence of the variable inverse Mills ratio in the above linear regression model resolves heteroskedasticity of the basic model where the use of the OLS is permitted.

$$Y_i=\beta x_i+\sigma\lambda_i+e_i \qquad i=1,2,3,\ldots,n \quad (3)$$

where β and σ are the parameters of the model, ei is the error term in the above model, and λ_i is the inverse Mills ratio obtained as follows:

$$\lambda_i=\varnothing(\beta' x_i \mid \sigma)/\phi(\beta' x_i \mid \sigma) \quad (4)$$

where λ_i is obtained using the estimated parameters of the Logit model for all observations in which saffron has been cultivated. In the linear model of the second stage, the coefficients represent the ratio of absolute changes to the dependent variable in the explanatory variables. In other words, it states how much the dependent variable varies for every one percent change in the explanatory variable. To investigate the existence or lack of heteroskedasticity in the Logit and Probit models, usual methods such as the Breusch-Pagan, White, and Goldfeld-Quandt tests cannot be used. Davidson and MacKinnon (1984) offered a statistic called LM2 to test heteroskedasticity in the Logit and Probit models. This statistic is based on the LM method in which an artificial regression is used by employing the Logit or Probit model estimates, and this artificial regression is used to test heteroskedasticity between the error terms (Greene, 1993). Due to the importance of elasticity in economic issues, the estimated coefficients were converted to partial elasticities of the factors (Abrishami, 2011).

The Empirical Model

According to the materials presented, the empirical model for the two-stage Heckman model to saffron is as follows:

The First Stage:

$$Y_i = \alpha_0 + \alpha_1 X_{1i} + \alpha_2 X_{2i} + \alpha_3 X_{3i} + \alpha_4 X_{4i} + \alpha_5 X_{5i} + \alpha_6 X_{6i}$$

where Y_i shows the possibility of decision to cultivate saffron that is equal to unity for farmers cultivating it and zero for those not cultivating it, αi are the model parameters that must be estimated, X_1 is age (years), X_2 is familiarity with growing saffron estimated as on a five-point Likert scale, X_3 is attending saffron training course defined as a dummy variable zero and one, respectively, for failing to take the course and attending it, X_4 is the number of extension courses attended, X_5 is the relative profitability of saffron measured as on a five-point Likert scale, and X_6 is the saffron marketing status that is composed of several items in the post-harvest stages including transportation, warehousing, packaging and marketing of saffron, the role of intermediaries, as well as the current price of saffron.

The Second Stage:

$$A_i = \beta_1 Z_{1i} + \beta_2 Z_{2i} + \beta_3 Z_{3i} + \beta_4 Z_{4i} + \beta_5 Z_{5i}$$

where A_i is the area under saffron cultivation (ha), Z_1 is the availability of water resources measured on a five-point Likert scale, Z_2 is the education level defined as a rating variable, Z_3 is the total area under cultivation including cropping and horticulture (ha), Z_4 is the features of arable land as measured on a five-point Likert scale showing favorability of land for saffron cultivation, and λ is the inverse Mills ratio.

In this study, the data were collected from two groups of saffron producing and non-producing farmers for the 2013-2014 growing season. Saffron in Marand is mostly planted in three areas: *BonabJadid, Javash*, and *Osdaghi.* The data set was collected by the help of a questionnaire distributed to farmers in these three regions in the form of appropriate sampling. The study population was all local farmers including 380 people in which 316 farmers are non-producers of saffron and 64 were producers of saffron. In this study, a total of 140 questionnaires were completed, of which 64 questionnaires were filled by farmers in the three districts of *Marand* in consensus model and 76 questionnaires were collected from non-producers of saffron.

RESULTS AND DISCUSSION

Table 1 shows the results of examining the factors affecting the decision to cultivate saffron in the Logit model.

As can be seen in Table 1, the variables of *familiarity with saffron planting, attending saffron training courses,* and *the number of extension courses attended* are significant at the one percent probability level. The variables of age and relative profitability of saffron are significant at the five and 10 percent levels, respectively. In addition, it can be concluded from this table that all variables have a positive effect on the probability of farmers' willingness to accept saffron planting. It means that the higher the level of familiarity with the process of planting saffron, attending training courses, the number of courses attended, promotion courses attended, marketing status, and the profitability of the saffron is, the higher the tendency to grow saffron is likely to be. Results of estimation of the linear regression model to estimate the factors affecting the area under saffron cultivation are given in Table 2. The variables included in the model are *access to water resources, education level,* and *the total cultivation area ranging from agriculture to horticulture, features of arable land, and inverse Mills ratio.*

To test the variance heteroskedasticity, the Breusch-Pagan test was used, and its value was calculated as 9.48, which is significant; accordingly, the model has variance homoscedasticity. After resolving model variance heteroskedasticity, it is observed that all the variables are significant and the value of adjusted R^2 is equal to 0.847 that shows a good fitness of the model. The variable IMR is significant at the 5% level showing a significant difference among the variables affecting the probability of acceptance of saffron planting and the ones affecting the cultivation area of saffron. The variable *access to water resources* is significant at the 5% level and has a negative effect on the area under the cultivation of saffron. It means that under constant conditions, if access to water resources is decreased by one unit, saffron cultivation will increase by 0.132 units. In fact, when farmers' access to water resources declines, they reduce the area under the cultivation of other crops and increase the area under the cultivation

Table 1
The Results of Estimating the Logit Model to Investigate the Factors Influencing the Decision to Cultivate Saffron

Variables	coefficients	t-statistic values	Elasticity	Marginal effects
Intercept	-12.267***	-4.43	-	-
Farmer age	.024**	1.94	1.361	.009
Familiarity with saffron planting	.707***	3.26	3.264	.25
Attending saffron cultivation training course	1.250***	3.33	-	.393
The number of extension courses attended	.561***	2.9	.739	.199
The relative profitability of saffron	.567*	1.81	3.04	.201
Marketing status	.968***	2.68	4.06	.343
Percentage of right predictions= 84.29		Cragg-Uhler R^2= 0.66		
Likelihood ratio test= 95.23		Mcfadden R^2= 0.493		
Mckelvey – Zavoina R^2= 0.753		Efron R^2= 0.597		

*p<0.1, **p<0.05, ***p<0.01

Table 2
The Results of the Linear Regression Model to Examine Factors Affecting the Area under Saffron Cultivation

Variables	Coefficients	t-statistic values
Access to water resources	-0.132**	-2.37
Level of education	0.068***	2.78
The total area under cultivation	0.261***	5.21
Features of arable land	0.189***	3.62
Inverse Mills ratio (IMR)	-0.127**	-1.98
$\overline{R^2}$=0.847		

**p<0.05, ***p<0.01

of saffron. Other variables are significant at the 1% level, and all have a positive impact on the rate of saffron cultivation. In other words, saffron cultivation will increase as the education level of farmers, the total acreage under the cultivation ranging from agriculture to horticulture, and the features of the arable land are increased. It should be noted that a farmer's high area of agricultural and horticultural land is the evidence of his/her more prosperity and income diversity that can be interpreted in this way: The farmers who have more land and income diversification may have more risk-taking capacity, and therefore, may devote more area to the cultivation of saffron.

CONCLUSION AND RECOMMENDATIONS

In this study, assuming that the cultivation of saffron in Marand has a relative advantage, the factors influencing the decision to plant and develop the local area under its cultivation were analyzed. The Heckman's two-step method was used for this purpose. The results of the first stage of Heckman showed that age, familiarity with the saffron cultivation, attending saffron training courses, the number of extension courses attended, saffron relative profitability, and marketing status all have a positive impact on the probability of acceptance of saffron cultivation. The results of the linear model estimation of the second phase showed that farmer education, the total acreage of agricultural and horticultural land, and property in arable land have a positive effect, and on the other hand, access to water resources has a negative effect on the area under saffron cultivation. Finally, according to the results, the following suggestions are presented:

The results showed that holding saffron producing training courses is among the factors affecting the land allocation to saffron cultivation; therefore, it is recommended that the Agriculture Jihad Organization should use educational programs and extension programs to develop the farmers' skills and provide the ground for farmers to take advantage of these courses. According to the results, communication with Agriculture Jihad through attending extension courses has had a positive impact on the land allocation to saffron cultivation. This is partly a sign of positive activity of agriculture experts of Marand. Thus, consolidation of this process and its improvement by Agriculture Jihad Organization is recommended to motivate farmers to grow saffron.

Improving the transportation, warehousing, packaging, marketing and singling out profit-seeking intermediaries and increasing the government support for the supply of inputs, and the price of saffron are all in the form of marketing variable and have a positive impact on the adoption of saffron cultivation. Overall, the improvement of marketing is one of the most important factors affecting the local farmers' interest in saffron cultivation. Therefore, the government should also provide the necessary support in this regard. One of the results of this research holds that, according to saffron farmers, cultivation of saffron will increase by reducing access to water resources because of saffron's low water requirements. Due to the severe limitations of water resources in the country, on the one hand, and the potential to produce saffron in the country, on the other hand, especially in the current situation, it seems that any support by the government to create more incentives for farmers to enhance saffron production can help the current status.

REFERENCES

1- Abrishami, H. (2011). *Principles of Econometrics (Vol. II)*. Tehran University Press, Iran.

2- Bakhshi, N. (2011). *Identifying factors affecting the development of canola in the counties of Tabriz and Marand*. Unpublished Thesis, Tabriz University, Faculty of Agriculture, Tabriz, Iran.

3- Bergtold, J. S., Duffy, P. A., Hite, D., & Raper, R.L. (2012). Demographic and management factors affecting the adoption and perceived yield benefit of winter cover crops in the southeast. *Journal of Agricultural and Applied Economics, 44*(1), 99-116.

4- Davidson, R., & Mackinnon J.G. (1984). Convenient specification tests for logit and probit models. *Journal of Economics, 25*(3), 241-262.

5- Greene, W. (1993). *Econometric Analysis*. 2nd ed., Macmillan Publishing Company Inc., New York, USA.

6- Hayati, B., Ehsani, M., Ghahramanzadeh, M., Raheli, H., & Taghizadeh, D. (2010). Factors affecting willingness to pay to Elgoli and Mashrouteh parks of Tabriz: Heckman twostep method. *Journal of Agricultural Economics and Development, 24*(1), 91 -98.

7- Keil, A., Saint-Macary, C., & Zeller, M. (2009). *Maize boom in the uplands of Northern Vietnam: Economic importance and environmental implications*. Discussion Paper No. 4/2008, Department of Agricultural Economics and Social, Universitat Hohenheim, Germany.

8- Marand Agriculture Department (2013). *Agricultural statistics*. Unpublished report, IT center, Marand. Salami, H., & Einollahi, M. (2001). The use of econometric model Tobit and Heckman twostep procedure to determine the factors affecting sugar beet cultivation in the province of Khorasan. *Iran Journal of Agricultural Science, 32*(2), 433-445.

9- Shafiee, L. (2007). Identifying factors affecting the development of olive cultivation in the province. *Journal of Agricultural Economics and Development, 58*, 1-22.

10- Wekesa, E., Mwangi, W., Vekuijl, H., Danda, K., & De Groote, H. (2003). Adoption of maize production technologies in the coastal low lands of Kenya. Kenya Agricultural Research Institute (KARI) & International Maize & Wheat Improvement Center (CIMMYT). 1 - 34.

18

Using the PILOTE Model to Improve Water Productivity for Rice in Rasht, North of Iran

Maryam Shadpour-Rashti[1], Mohammadreza Khaledian[2*], Mohammadhassan Biglouei[3] and Mojtaba Rezaei[4]

Abstract

In order to calibrate and validate the PILOTE model for rice in a humid climate, this study was carried in a randomized complete block design with three replications on a popular local variety, Hashemi, during 2001, 2002, 2005, 2006 and 2007 crop seasons. This research was done at the Rice Research Institute of Rasht, Iran. Evaluation of simulated and measured grain yield and dry matter values was done using Nash-Sutcliffe efficiency (EF), Root Mean Square Errors (RMSE) and normalized root mean square errors (NRMSE) indices. The results revealed that RMSE for validation and calibration were 0.69 and 0.72 $Mgha^{-1}$, respectively. NRMSE for calibration was 9.5 % and for validation was 14.1 %. NRMSE for grain yield and dry matter were 8.74 and 13.37 %, respectively. EF values were between 0.84 and 0.98. The results showed that the PILOTE model can be used to manage properly rice irrigation in different regimes. Scenario analysis showed that the best irrigation regime was intermittent irrigation with 8-day interval.

Keywords:
Hashemi variety, irrigation management, Rasht, simulation

[1] MSc. Student, Irrigation and Drainage, Water Engineering Department, Faculty of Agricultural Sciences, University of Guilan, Rasht, Iran

[2] Water Engineering Department, Faculty of Agricultural Sciences, University of Guilan, Rasht, Iran, and and Department of Water Engineering and Environment, Caspian Sea Basin Research Center.

[3] Water Engineering Department, Faculty of Agricultural Sciences, University of Guilan, Rasht, Iran

[4] Rice Research Institute of Iran, Agricultural Research, Education and Extension Organization (AREEO), Rasht, Iran

* Corresponding author's email: Khaledian@guilan.ac.ir

INTRODUCTION

Water is an important factor for agricultural production in Iran. However Guilan province has an annual rainfall equal to 1200 mm (1960-2010) that regularly takes place in fall and winter that are out of rice cultivation seasons; accordingly, the rice can not be irrigated by rainfall. In addition, Sefidroud River, which is the main source of irrigation water of Guilan province, is taking water from upstream provinces, and because of dams built on the upstreams of this river, water allocation has reduced on Sefidroud irrigation network. The water statistics of Guilan Regional Water Authority showed that river discharge have reduced to half in the past decade than the three decades before it. The solution to this problem is to increase the irrigation water productivity in agricultural section. Rice is the main product in this region that is cultivated in vast areas and needs a lot of water. As such, improving water productivity is the main priority in this region. Alternatively, sustainability of agriculture will be in danger. Alternative irrigation can be a suitable substitute for flooding irrigation in this region;however, before this change, it is crucial to test different option. Similar research studies conducted so far are of two major types: field experiments and simulation models. Field experiments are expensive and time consuming. If these models were calibrated and validated satisfactorily, the problem would be solved. After this stage, different management scenarios can be tested without any cost, and the best choice can be selected at the lowest possible time. As lack of water is a pretending problem in this area, water-crop models are suitable to test water saving managements. Many models of this kind have been developed by some researchers. Some of these models are ORYZA 2000 (Bouman et al., 2001), WOFOST (Bouman & Tuong, 2001), SWAP (Van Dam, 1977) and CropSyst (Stokle et al., 1992). Amiri et al. (2011a) have evaluated rice yield in the different irrigation management using ORYZA 2000. Grain yield was simulated by RMSE equal to 150-182 kg ha^{-1} and NRMSE equal to 6 %. Saadati et al. (2011) evaluated WOFOST plant growth model for irrigation management of rice with two varieties i.e. Binam and Hassani varieties. NRMSEs of yield simulation were 7.94 and 13.71 % for Binam and Hassani varieties, respectively.

Saadati (2011) evaluated rice yield with two varieties called Binam and Hassani in a humid climate using CropSyst model. The yield NRMSE of Binam variety was between 0.81 to 12.58 % in calibration year. The NRMSE of Hassani variety was between 2.4 to 19.42 %. The NRMSE of Binam was between 0.83 and 16.4 % and the NRMSE of Hassani was between 2.82 and 21.27 %, in validation year.

The PILOTE model is a water-yield model that requires a few inputs and is based on LAI simulation. This model simulates the balance of water and yield at day time steps. This model was developed for sorghum and sunflower by Mailhol et al. (1997). Later, this model was completed for corn and durum wheat for tillage and no tillage systems by Khaledian et al. (2009). They used it to improve water productivity (Khaledian et al., 2011, 2013). The PILOTE model was used in combination with HYDRUS-2D model to improve subsurface drip irrigation performance and irrigation program for corn by Mailhol et al. (2011). This model, as all other models, should be calibrated and validated, and then it can be used to study different irrigation management scenarios to improve water productivity. Due to lack of water availability in rice cultivation season, irrigation specialists are looking for a way to increase irrigation water productivity. The PILOTE model is a suitable tool to study different irrigation management because of fewer input data compared with other water-yield models. The PILOTE model, thanks to its simple structure, can be easily calibrated for other crops, maintaining its robustness. The goal of this study was to calibrate and validate the PILOTE model to estimate Hashemi rice yield in Rasht under different irrigation regimes. Following validation, the model was used to find relevant irrigation regime in the region by the help of a scenario analysis during a long climatic series of 20 years.

MATERIALS AND METHODS

In this study, two experiments were carried out in the research field of the National Rice

Table 1
Physical and Chemical Properties of Soil (0-0.35 m) at Rice Research Institute of Iran (n=3)

Soil texture	Available potasium mg/kg	Extractable phosphoros mg/kg	Total nitrogen%	pH	Saturation percentage
Silty clay	242(±22.61)*	18.73(±6.00)	0.162(±0.01)	7.27(±0.12)	75(±0.0)

Research Institute, Rasht at 37°12′ North latitude, 49°38′East longitude, and 24.6 meters above the sea level. Climate data consist of rainfall, sunshine hours, wind speed, temperature, humidity, and radiation taken from the local climate station. In this study, evapotranspiration (ET_0) is calculated by ET_0 calculator software, which is based on Penman-Monteith equation (Allen et al, 1998).

Soil samples have been sent to the lab for chemical analysis: N, P, and K were determined according to Kjeldahl method for N, Olsen-sodium Bicarbonate Extractant Method for extractable P and Ammonium Acetate Extractant Method for available K (Page, 1982). The results of this analysis are summarized in Table 1. According to this table, the soil has the minimum essential nutrients for rice cultivation. Water chemical analysis results and meteorological data are shown in Tables 2 and 3, respectively. Water quality and climate conditions do not limit normal plant growth.

The soil samples were classified as a silty clay soil. In order to evaluate the PILOTE model performance on the Hashemi rice variety in Guilan province, two field experiments were done in a complete randomized block design with three replications: The first one, which was carried out in irrigation treatments including four levels, consisted of permanent flooding irrigation (I_1: usual irrigation regime in the region) and five (I_2), eight (I_3) and 11 (I_4) days irrigation interval in 2001 and 2002 crop seasons in the experimental plots of 3 meters ×5 meters. The second experiment was carried out under irrigation treatments including three levels (I_1, I_2 and I_3) in 2005, 2006 and 2007 crop seasons. In order to prevent losses of water and fertilizer from plot walls, plot boundaries were covered by nylon. Ten days after seedling, irrigation treatments were applied. This time was considered for better establishment of seedling. After physiological maturity, plant samples were taken from the center of each plot. The samples were dried in the oven and the yield values were estimated in each unit.

Table 2
Chemical Properties of Irrigation Water at the Rice Research Institute in Rasht, Iran

Na^+	Mg^{2+}	Ca^{2+}	SO_4^{2-}	Cl^-	HCO^{3-}	CO_3^{2-}	TDS	pH	EC	SAR
meq lit^{-1}							mg lit^{-1}	-	dS m^{-1}	-
3.1	1.2	3.8	2.2	2.6	3.2	0.8	600	8	0.8	1.96

Table 3
Weather Parameters During Rice Growth Season in Rasht (April-August), Iran

Weather parameters	Crop seasons				
	2001	2002	2005	2006	2007
Precipitation (mm)	144.7	248.2	160.9	135	129.9
Average of maximum daily temperature (°C)	32	31.2	23.6	23.7	23.6
Average of minimum daily temperature (°C)	11.8	13.0	18.9	18.5	18.8
Average of maximum relative humidity (%)	100	98.4	98.0	96.7	97.3
Average of minimum relative humidity (%)	30.8	32.2	60.0	55.8	58.8

Model description

The PILOTE is a soil water balance and crop yield model assuming that just the water is limiting crop growth. In the PILOTE model, the soil module consists of three water reservoirs from soil surface to the maximum root depth. The shallow reservoir R_1 with a depth of 0.10 m manages water balance, where soil evaporation is ruled by LAI which acts on the partitioning coefficient between soil evaporation and transpiration. The capacity of second reservoir R_2 enlarges with root growth. The next reservoir represents the left part before the maximum root depth is completely taken by R_2. Water first is captured from R_1 till transpiration and soil evaporation completely deplete the R_1 and following that water is obtained from R_2 only by transpiration. Water Stress Index (WSI) is calculated as the ratio of actual transpiration to the maximum transpiration (actual transpiration/ the maximum transpiration) using actual and Maximum Evapotranspiration (MET). The later is calculated as $MET=K_c \times ET_0$, where Kc is the crop coefficient as a function of LAI, and ET_0 is the reference evapotranspiration (Allen et al., 1998). Actual evapotranspiration diminishes linearly from MET with the depletion of R_2 under water stress condition. Therefore, WSI could be calculated and was then exported to the crop module. Crop module is based on the LAI simulation and its response to WSI. The simulation involves the temperature sum corresponding to the maximum LAI and two shape parameters. Biomass yield is calculated according to Beer´s law (Mailhol, 2005). Grain yield is the product of biomass yield and harvest index. The later is set to a potential value if average LAI from grain filling stage to pasty grain stage is greater than a threshold value, if not it linearly declines (Khaledian et al., 2009; Mailhol et al., 1997).

In order to evaluate model statistically, the simulated and observed values of the grain and biomass yields were determined. Three statistical indices including Root Means Square Error (RMSE), Nash-Sutcliffe efficiency (EF), as well as Normalized Root Mean Square Error (NRMSE) were calculated as below:

$$RMSE = \left(\frac{\sum (P_i - O_i)^2}{n} \right)^{0.5} \quad (1)$$

$$NRMSE = \frac{100 \left(\frac{\sum (P_i - O_i)^2}{n} \right)^{0.5}}{O_{mean}} \quad (2)$$

$$EF = 1 - \sum (P_i - O_i)^2 / \sum (P_i - O_{mean})^2 \quad (3)$$

In these formulas, P_i is the simulated value, Oi is the observed value, n is the number of estimated values and O_{mean} is the mean of the observed values.

The best value for RMSE and EF are zero and one, respectively. Indeed, NRMSE showed relative difference (percentage of mean value) between simulated and observed values. If the evaluated result (NRMSE) is less than 10 %, it will show an excellent relative difference. If NRMSE is more than 10 % and less than 20 %, it will show good relative difference. If NRMSE is more than 20 % and less than 30 %, it will show the fair relative difference. If NRMSE is more than 30 %, it will show poor relative difference (Bannayan & Hoogenboom, 2009; Jamieson et al., 1991).

The PILOTE model has four input parts: The first part is the soil physical properties such as field capacity, permanent wilting point, soil moisture preliminary reserve, and so on, which is shown in Table 4.

The second part consists of meteorological data that include daily mean temperature, radiation, rainfall, and evapotranspiration. The third part is about plant properties that involve root depth, root growth rate, maximum LAI, plant density, required degree-day for different growth stages, and the fourth part includes an irrigation schedule where date and depth of irrigation are introduced to the model. Required soil information was measured by soil sampling. Meteorological data were provided by National Rice Research Institute of Iran. Some plant information was taken from references (Amiri et al., 2011a; Rezaei & Nahvi, 2007; Saadati et al., 2011), some others were resulted by model calibration, and finally, some were provided by direct field measurements. Irrigation schedule during the study were determined by field taken and recorded irrigation data. Irrigation water depth was determined by a volumetric contour.

Sensitivity analysis

Sensitivity analysis is a useful technique for determining the parameters that have a significant effect on the model outputs, and doing sensitivity analysis is necessary before calibration to recognize sensitive parameters. These parameters were selected from measured or calibrated parameters. Sensitivity is conceptualized as the rate of change in the output resulting from a change of that input parameter (Wohling, 2005). The sensitivity index, SI, proposed by Ng and Loomis (1984) is defined as:

$$sl = \frac{(\frac{100}{N})\sum_{i=1}^{N}(Xin - Xci)Xci}{\Delta} \quad (4)$$

where X_{in} is the new value of the ith data with a changed value of the input parameter, X_{ci}: the value of output for the ith point in the control simulation run, N: the number of point and Δ: is the absolute change in the input parameter. SI in the given form is a measure of the percentage change in the output from that in the control simulation resulting from one percent change in the value of the input parameter. According to the results of sensitivity analysis, model calibration was started from the most sensitive parameter having a higher SI. Primary input parameters were introduced to the model by five percent changing process. In order to find relevant input parameters, model results versus measured data were compared in each calibration stage according to statistical indices. This procedure was applied for next sensitive parameters. Crop season of 2006 was used for model calibration and crop seasons of 2001, 2002, 2005 and 2007 for model validation.

A calibrated and validated model can be used as a reliable tool to analyze different irrigation scenario on a long climatic dataset to consider all dry and wet years in a region and to find the best irrigation regime. The main goal in an irrigation program can be maximizing water productivity (WP) while maintaining crop production

Table 4
Soil Properties Used in the PILOTE Model

Parameter	value
field capacity (FC)	40 %
permanent wilting point (PWP)	23 %
initial soil water reserve (0-0.35 m)	140 mm

Scenario analysis to improve water productivity

level.

According to Viets (1962) WP is defined as:

WP= yield/water use (5)

To discriminate the role of irrigation in crop production, WP relation can be modified as (Rodrigues & Pereira, 2009):

WP= grain yield/irrigation water depth (6)

where WP is in kg ha^{-1} mm^{-1}, grain yield is in kg ha^{-1} and irrigation water depth is in mm. The PILOTE model was employed to compare the irrigation depth and yield production in different irrigation treatments, that is, permanent flooding irrigation as traditional irrigation method in the region and intermittent irrigation method such as 5, 8, and 11 days intervals. The simulation was performed on a long climatic series, which was available in the region (20 years in 1982-2012 period). In some years climatic data were not recorded completely and five years were considered for model calibration and validation.

SPSS 15 was used to do statistical analysis including analysis of variance and Duncan multiple range test for mean comparison.

RESULTS AND DISCUSSION

Experimental results

Results showed that there was no significant difference between permanent flooding irrigation and intermittent irrigation (Table 5) in all the years studied. Furthermore, the table of mean comparison (Table 6) showed that applied irrigation treatments had no statistically significant effects on the measured yield except the intermittent irrigation with 11-day interval as compared to flooding irrigation in 2001. Reviewing Table 6 showed that performing intermittent irrigation

Table 5
Analysis of Variance

S.O.V.	df	Year		df	Year		
		2001	2002		2005	2006	2007
replications	2	154000	616000	2	162136	779959	177700
irrigation	3	758000[ns]	60000[ns]	2	21186[ns]	207332[ns]	33369[ns]
error	6	197000	58000	4	70592	77993	278562
CV (%)		13.6	7.8		7	9	9

S.O.V.: source of variation, ns: represent no statistically significant differences, the numbers in the table except of CV represent mean squares

Table 6
Analysis of Mean Comparison of Rice Grain Yield under Different Irrigation Regimes (kg ha^{-1})

Irrigation Treatments	Crop season				
	2001	2002	2006	2006	2007
Flooding	3845.8[a]	3148[a]	4043[abc]	5290[a]	4575[abc]
5-day interval	2955.6[ab]	2942[a]	3892[abc]	4871[abc]	4637[abc]
8-day interval	3508.3[ab]	3271[a]	3720[c]	4863[abc]	4755[ab]
11-day interval	2750.0[b]	3038[a]	-	-	-

Same letters means no difference at 95% by Duncan multiple range test

sometimes caused a minor yield increase or decrease, a finding that has been approved by other researchers in the region (Rezaei & Nahvi, 2007; Rezaei et al., 2013). Belder et al. (2004) reported that intermittent irrigation did not decrease rice yield as compared to permanent flooding irrigation. Our findings highlight the potential of intermittent irrigation as compared with flooding irrigation to reduce water consumption, while maintaining crop production, or in other words, this technique can improve WP. Yet, in all types of regions, climate can change over years. accordingly, it would be difficult to generalize the results to all years unless this method tests for all types of year in the region. To do that, a simple water-yield model can be helpful to test this scenario on a long climatic series including all types of year, that is, from wet to dry years. Yet, before using a model, it should be calibrated and validated. That is why, the PILOTE model was calibrated and validated in this study.

Model calibration and validation

Sensitivity analysis of the PILOTE model was determined by 25 % reduction and increase in input parameters and controlling the changed in model outputs. This difference for field parameters was determined by10% difference from the reference. Initial SWR was a sensitive factor in the PILOTE model (Mailhol et al., 1997).

The results of sensitivity analysis showed that simulation of grain yield is highly sensitive to the field capacity, maximum plant efficiency, maturity degree-day, Harvest Index (HI), and radiation use efficiency; accordingly, these values should be determined carefully. Confalonieri et al. (2009) evaluated WARM model in comparison to CropSyst and WOFOST models in Italy. As shown by the findings of the present study, WARM model was sensitive to 30 % of input model parameters against less than 10 % and 20 % with CropSyst and WOFOST models, respectively. The PILOTE model is sensitive to 16% of input model parameters, and the WARM and CropSyst models are more and less sensitive, respectively.

Table 7 shows input parameters and their corresponding values in the PILOTE model. They derived from literature, field measurements, and model calibration. Table 8 shows the result of calibration and validation of the PILOTE model. In this table the simulated values of the grain and biomass yields are compared to the observed values in 2006. Two data sets were used by val-

Table 7
Plant Parameters Used in the PILOTE Model

Parameter	Value	Unit	Identification method
Maximum root depth	0.35	meter	field observation
Root growth rate	0.01	m day^{-1}	field observation
Installation period of root	10	day	field observation
Plant coefficient	1.05	-	field observation
Evaporation coefficient	0.3	-	field observation
Sensitive coefficient to water stress	1.75	-	calibration
Real planting density	25	plant m^{-2}	field observation
Optimum planting density	25	plant m^{-2}	Amiri et al (2011a)
Grain humidity	14	percentage	Measurement
Radiation coefficient change to biomass	0.9	-	calibration
Harvest index, HI	0.5	-	field observation
Water stress threshold	3	m^2 m^{-2}	calibration
Reduction coefficient of harvest index	0.15		Rezaei and Nahvi (2007)
Floration degree-day	1200	degree-day	Rezaei and Nahvi (2007)
Maturity degree-day	1500	degree-day	Rezaei and Nahvi (2007)
Germination degree-day	0	degree-day	Rezaei and Nahvi (2007)
Base temperature	10	^{0}C	Saadati et al (2011)
Maximum LAI	3		field observation
α	6		calibration
β	1.5		calibration
γ	4		calibration

idation years. RMSEs for calibration of biomass and grain yield were 0.94 and 0.18 Mg ha-1 and for validation of biomass and grain yield were 0.96 and 0.36 Mg ha^{-1}. NRMSEs for calibration of biomass and grain yield were 9.9 and 3.5% and for validation of biomass and grain yield were 14.4 and 10.1%.

According to Table 7, the RMSE of calibration (for both biomass and grain yields) was equal to 0.69 Mg ha^{-1} and in this model the RMSE of validation (for both biomass and grain yields) was equal to 0.72 Mg ha^{-1}. The RMSE value of biomass was equal to 0.96 Mg ha^{-1} and grain yield RMSE value was equal to 0.33 Mg ha^{-1} indicating that model has simulated yield satisfactorily. Amiri et al. (2013) evaluated ORYZA 2000 and the RMSE of grain yield was equal to 0.15 – 0.182 Mg ha^{-1}. In addition, they tested WOFOST, and the RMSE of biomass yield was equal to 0.389-0.553 Mg ha^{-1}.

NRMSE value is 14.12 % in the validation years. According to Jamieson et al. (1991), relative difference was categorized in the good class as being between 10-20%. This value is 9.51 % in the calibration year showing an excellent performance (according to Jamieson et al. (1991)). Moreover, biomass yield was categorized in the good class and NRMSE is equal to 13.37 % showing a good relative difference. Saadti et al. (2011) evaluated WOFOST model to simulate yield of Binam and Hassani rice varieties and NRMSE was 8 and 14% being in the good to excellent classes, showing that WOFOST model simulated the yield of Binnam similar to the PILOTE model, but with Hashemi variety. In addition, the results of the study by Amiri et al. (2011a) indicated that WOFOST model could estimate biomass similar to the PILOTE model, and the NRMSE (relative difference) was in the good class. Amiri et al. (2013) evaluated the CERES-Rice model under irrigation and N application managements in Iran. RMSE of this research for grain yield was equal to 0.297 $Mgha^{-1}$ and RMSE of biomass was equal to 0.862 $Mgha^{-1}$; Furthermore NRMSE of grain yield was equal to 8 %; therefore, according to Jamieson et al. (1991), it was in the excellent class and NRMSE of biomass was equal to 10 %, being in the good class. The results show that CERES-Rice model and the PILOTE model are in the same class.

Comparing WOFOST model to the PILOTE model showed a negligible difference. According to Table 8, EF for grain and biomass yields,

Table 8
Model Statistical Evaluation with Root Mean Square Error (RMSE, Mg.ha^{-1}), Normalized Root Mean Square Error (NRMSE, %) and Nash-Sutcliffe Efficiency (EF)

Parameter	Validation	Calibration	Biomass yield	Grain yield
RMSE	0.72	0.69	0.96	0.33
NRMSE	14.12	9.51	13.37	8.74
EF	0.87	0.98	0.84	0.98

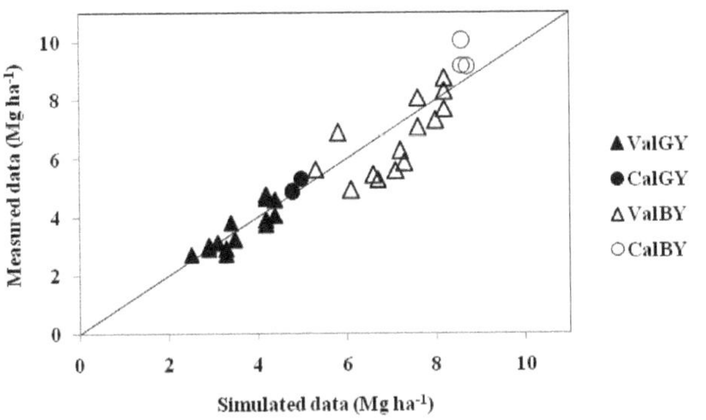

Figure 1. The 1:1 line for comparing field records and model outputs. Biomass yield and grain yield (Mg ha^{-1}) of rice simulated by the PILOTE model versus biomass yield and grain yield in different treatments of irrigation at Rice Reasearch Institute (Val: validation, Cal: calibration, GY: grain yield, BY: biomass yield, EF=0.91, NRMSE=13% and RMSE=0.71 Mgha^{-1}).

calibration and validation years were 0.98, 0.84, 0.98 and 0.87, respectively. In the research of Confalonieri et al. (2009) on three models i.e. WARM, CropSyst and WOFOST in Italy, modelling efficiency (EF) was similar to the PILOTE in the calibration year, but a little bit (0.03) higher in validation years.

The other method to compare observed and simulated values is to use a graph with a 45° slope line. Figure 1 shows the graph of simulated and observed values of grain and biomass yields in the selected field. If observed and simulated grain yields were compared to each other, it would be observed that the model was calibrated well in 2006. Confalonieri and Boechi (2005) evaluated CropSyst model and NRMSE was equal to 11-29 % in the calibration year for rice in the North of Italy. These values are categorized in good and fair classes. Yet, the result of the PILOTE model and NRMSE was different in this study; NRMSE was 9.51 %, categorized in the excellent class. Because of the calibration test stage proved to have been successful, the model was evaluated with two data sets (2001-2002) in addition to the last data set (2005, 2006 and 2007). It showed that the model have simulated grain yield better than biomass yield. This problem could be because of clear cutting, unsuitable drying, and leaf senescence that caused the field data to be weak. The biomass yield was in the inferior; accordingly, the model is able to simulate the economic yield very well. As such, the difference in simulated yield and observed yield is not considerable. Hsiao et al. (2009) evaluated AquaCrop model. The result was similar to the PILOTE model. Singh et al. (2013) worked on CropSyst model, the RMSE of biomass, and grain yields were 0.7 and 0.33 Mg ha^{-1}, respectively which is similar to the PILOTE model. On the other hand, CropSyst reaction with a lot of input parameters was the same as the PILOTE model with a fewer input parameters. Xiong et al. (2008) evaluated CERES-Rice model in China. Results showed that CERES-Rice model was able to simulate rice yield with a RMSE equal to 991 kg ha^{-1} and NRMSE equal to 14.9 %.

PILOTE model application to evaluation of water productivity

According to calibration and validation results, the PILOTE simulates satisfactorily the yield with NRMSE of 13 % and EF of 0.91. therefore, the model can be used as a proper tool for evaluation of different irrigation scenarios on a long climatic series regarding water productivity. Figure 2 demonstrates the results of the PILOTE simulations on the climatic series. WP was found to increase from 3.2 with flooding irrigation to 7.2 kg ha^{-1} mm^{-1} with 11-day interval irrigation in this region. Irrigation interval of 8 days results in a WP of 7.7 kg mm^{-1}. It can be said that in average a water application depth of 600 mm was saved with 8-day compared with flooding irrigation.

Table 9
Comparing Mean of Water Productivity in Different Irrigation Treatments During Climatic Series of 20 Years Using t-test

Irrigation treatments		p-value	Statistical test result
flooding	5-day interval	0.00	**
flooding	8-day interval	0.00	**
flooding	11-day interval	0.00	**
5-day interval	8-day interval	0.00	**
5-day interval	11-day interval	0.00	**
8-day interval	11-day interval	0.037	*

$*p<0.05$, $**p<0.01$

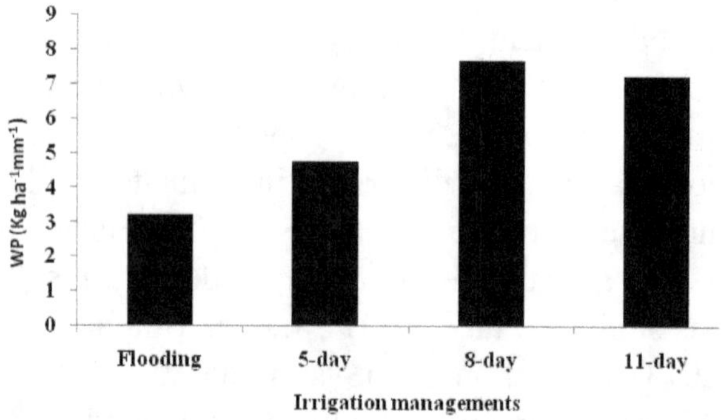

Figure 2. Mean water productivity (kg ha^{-1} mm^{-1}) as affected by different irrigation managements

Regarding statistical analysis (see Table 9), the best irrigation treatment was intermittent irrigation with 8-day time interval. The grain yield under this treatment condition was not different from that of others (see Table 6); accordingly, this treatment can be recommended because of both the highest water productivity and equal crop production. This irrigation is very interesting regarding water deficiency in the region.

CONCLUSION

According to the result of statistical evaluation, the PILOTE model was shown to provide an acceptable level of accuracy in the simulation of rice yield. Despite the fact that the PILOTE model has less input parameters than other models and is easy to use, it can simulate yield with high accuracy. After calibration and validation of the model, it was used to assess different irrigation management scenarios. Using intermittent irrigation, instead of flooding irrigation, could produce equivalent grain yield with a higher level of water productivity, which is very attractive in a region with limited water resources for rice production. As is the case with the region, the problem of water scarcity is highlighted; therefore, using this model for managing irrigation properly, can address this issue.

ACKNOWLEDGEMENT

This research was partially supported by the Rice Research Institute of Iran, Agricultural Research, Education and Extension Organization (AREEO), Rasht, Iran. We thank our colleagues from that institute who provided insight and expertise that greatly assisted the research.

REFERENCES

1- Allen, R.G., Periera, L.S., Raes, D., & Smith, M. (1998). *Crop evapotranspiration guidelines for computing crop water requirement*. The 56th report on the irrigation and drainage (FAO), Rome.

2- Amiri, A., Razavipoor, T., & Banayan, M. (2011a). Evaluation of rice water productivity in different irrigation condition and cultivation distance by using ORYZA 2000 model. *Electronical Field Crop Production, 4*(3), 1-19.

3- Amiri, A., Rezaei, M., Motamed, K., & Emami, S. (2011b). Evaluation of WOFOST plant growth model in condition of irrigation management.*Agronomy:Research and Building, 90*, 9-17.

4- Amiri, E., Rezaei, M., Bannayan, M., & Soufizadeh, S. (2013). Calibration and evaluation of CERESRice model under different N and water management options in semi-Mediterranean climate condition, *Communication in Soil Science and Plant Analysis, 44*(12), 1814-1830.

5- Belder, P., Bouman, B.A.M., Cabangon, R., Guoan, L., Quilang, E.J.P., Li, Y., Spiertz, J.H.J., & Tuong T.P. (2004). Effect of water saving irrigation on rice yield and water use in typical lowland conditions in Asia. *Agricultural Water Management, 65*, 193–210.

6- Bannayan, M., & Hoogenboom, C. (2009). Using pattern recognition for estimating cultivar coefficients of a crop simulation model. *Field Crops Research, 111*(3), 290-302.

7- Bouman, B.A.M., Kropff, M.J., Tuong, T.P., Wopereis, M.C.S., Ten-Berge, H.F.M., & Vanlaar, H.H. (2001). ORYZA2000: modeling low land rice. Los Banos (Philippines): International Rice Research Institute, and Wageningen: Wageningen University and Research Center, pp. 235.

8- Bouman, B.A.M., & Tuong, T.P. (2001). Field water management to save water and increase its productivity in irrigated lowland rice. *Agricultural Water Management, 49*(1), 11-30.

9- Confalonieri, R., Acutis, M., Bellocci, G., & Donatelli, M. (2009). Multi- metric evaluation of the model WARM, CropSyst, and WOFOST for Rice. *Ecological Modelling, 220*(11), 1395-1410.

10- Confalonieri, R., & Bocchi, S. (2005). Evaluation of Crop Syst for simulating the yield of flooded rice in northern Italy. *European Journal of Agronomy, 23*, 315-326.

11- Hsiao, T.C., Heng, L., Steduto, P., Rojas-Larra, B., Raes, D., & Fereres, E. (2009). Aqua Crop: the FAO crop model to simulate yield response to water: parameterization and testing for maize. *Agronomy Journal, 101*, 448-459.

12- Jamieson, P.D., Porter, J.R., & Wilson, D.R. (1991). A test of the computer simulation model ARC-WHEAT on wheat crops grown in New Zealand. *Field Crops Research, 27*(4), 337-350.

13- Khaledian, M.R., Mailhol, J.C., Ruelle, P., & Dejean, C. (2013). Effect of cropping strategies on the irrigation water productivity of durum wheat. *Plant, Soil and Environment, 59*(1), 29-36.

14- Khaledian, M.R., Mailhol, J.C., Ruelle, P., Mubarak, I., & Maraux, F. (2011). Nitrogen balance and irrigation water productivity for corn, sorghum and durum wheat under direct seeding into mulch when compared with conventional tillage in the southeastern France. *Irrigation Science, 29*(5), 413-422.

15- Khaledian, M.R., Mailhol, J.C., Ruelle, P., Rosique, P. (2009). Adapting PILOTE model for water and yield management under seeding system: The case of corn and durum wheat in a Mediterranean context. *Agricultural Water Management, 96*(6), 757-770.

16- Mailhol, J.C. (2005). *Contribution to the control of irrigation and its impacts*. Unpublished dissertation Irrigation Deptment, Faculty of Agricultural Sciences, University of Montpellier 2. France.

17- Mailhol, C., Olofayo, A.A., & Ruelle, P. (1997). Sorgum and sunflower evapotranspiration and yield from simulated leaf area index. *Agricultural Water Management, 35*(1-2), 167-182.

18- Mailhol, J.C., Ruelle, P., Walser, S., Schutze, N., & Dejean, D. (2011). Analysis of AET and yield predictions under surface and buried drip irrigation systems using the crop model PILOTE and Hydrus-2D. *Agricultural Water Management, 98*, 1033-1044.

19- Ng, N., & Loomis, R.S. (1984). *Simulation of growth and yield for the potato crop*. Simulation Monographs. Agricultural Publishing and Documentation Center, Wageningen, Pp.147.

20- Page, A.L. (1982). *Methods of soil analysis*. American Society of Agronomy, Inc. and Soil Science Society of America, Inc. Madison, Wisconsin, USA.

21- Rezaei, M., Davatgar, N., Khaledian, M.R., & Pirmoradian, N. (2013). Using intermittent irrigation to use saline water in rice production. *Acta Agriculturae Slovenica, 101*(1), 49-57.

22- Rezaei, M., & Nahvi, M. (2007). Effect of different irrigation management in clay soils on water efficiency and some features of local rice cultivars at Guilan. *Agricultural Sciences Research, 9*, 15-25.

23- Rodrigues, G.C., & Pereira, L.S. (2009). Assessing economic impacts of deficit irrigation as related to water productivity and water costs. *Biosystem Engineering, 103*(4), 536–551.

24- Saadati, Z. (2011). *Assessment of aquacrop, cropsyst and WOFOST models in yield simulating of two rice varieties, Binam and Hasani, under different irrigation regimes*. Unpublished thesis Water Engineering Deptment, Faculty of Agricultural Sciences, University of Guilan, Iran.

25- Saadati, Z., Pirmoradian, N., Amiri, A., & Rezaei, M. (2011). Evaluation WOFOST model in simulating two yields varieties rice under different irrigation regimes. *Water Research in Agriculture, 3*, 324-337.

26- Singh, A.K., Goyal, V., Mishra, A.K., & Parihar, S.S. (2013). Validation of CropSyst simulation model for direct seeded rice-wheat cropping system. *Current Science, 104*(10), 1324-1331.

27- Stokle, C.O., Williams, J.R., Rosenberg, N.J., & Jones, C.A. (1992). A method for estimating the direct and climatic effects of rising atmospheric carbon dioxide on growth and yield of crops: Part I. modification of the EPIC model for climate change analysis. *Agricultural System, 38(3)*, 225-238.

28- Van Dam, J.C. (1977). *Field-scale water flow and solute transport. SWAP model concepts, parameter estimation, and case studies.* Unpublished dissertation Netherlands: Faculty of Agricultural Sciences, University of Wageningen.

29- Viets, F.G.Jr. (1962). Fertilizers and the efficient use of water. *Advances in Agronomy,14*, 223–264.

30- Wohling, T. (2005). *Physically based modeling of furrow irrigation system during a growing season.* Unpublished dissertation Dresden: Faculty of Agricultural Sciences, University of Dresden. Pp. 209.

31- Xiong, W., Ian, H., Declan, C., & Erdalin, Y. (2008). A crop model cross calibration for use in regional climate impact studies. Ecological *Modeling, 213*, 365-380.

19

Economic Analysis of Water Demand in Greenhouses of Khash Township

*Javad Shahraki [1] and Neda AliAhmadi [2]**

Abstract

The main goal of this study was to investigate the way of affecting of water in production of cucumber. From aspect of econometrics, mutual relationships of production function and expenses have been analysed in which Translog cost function has been used. This function has been estimated using the conditional input demand functions, the shefard theorem, chemical fertilizer, animal fertilizer, labour force, seed, and pesticide in the framework of a system of equations by taking advantage of the irrelevant regression method. Applied inputs is for 151 cucumber Beneficiaries in Khash in farming year of 2011-12 that insider and crossing tendency of input demand for this product has been investigated through collecting questionnaires and using obtained coefficients. The results show that the translog cost model is a good fit compared to the data of research. According to the reviews, demand for water has a minor Insider-succession tendency more than one which shows possible stretch of demand function rather than the price of the inputs. Hence, appropriate pricing policies can be used to take a positive step in preventing the extreme usage of this input and directing beneficiaries for optimum use of these inputs.

Keywords:
Irrigation water demand, Price elasticity of demand, Succession elasticity of demand, Translog cost function

[1] *Assistant Professor of Agricultural Economics, University of Sistan and Baluchestan.*
[2] *Master of Agricultural Economics in MA great of the University of Sistan and Baluchestan.*
* *Corresponding author's email: ahmadi_15877@yahoo.com*

INTRODUCTION

Water is the Nowadays water is one of the main scarce input of natural resource and life on Earth. More than 70% of the Earth surface is covered with water (nearly 360 million out of 510 million cube meter). Despite of this enormous volume of water, only 2% of the Earth's water is fresh and drinkable and the remaining in unusable due to a variety of soluble salts, especially uneatable ones. From this two per cent of the fresh water, more than 90 per cent is in the form of frozen in two poles and is located away from the human access (Iran hydrology). Management and optimal decision making about water is one of the most important human concerns today due to increasing population and limited water resources.

Undoubtedly, providing Sufficient and guaranteed irrigation water for plants is the main axis of human challenge because one of the main consumers of world water resources is agricultural sector. (Kouhpayee, 2004). In water management we did not create balance just control, motivate and limit partners at both supply and demand sides. One of the management practices of increasing water efficiency in demand section, that specify the water resource value and has non-organic structures, is establishing actual price of water in both supply and demand sides because determination of actual price of water will considerably help optimal allocation of water among different products and also to logical and proper consumption of water which will finally increase application and efficiency of water. Continuing process of unreal and non-economical determining of price of water will escalate extreme consumption and indiscriminate waste. In other words, determine the actual water price will cause water usage considering allowed capacity from underground and surface water of aquifers and quotas. (Ghaffari Shirvan, 1998). On the other hand Managing water demand involves better and more effective harnessing of water which is possible through legislation of laws, planning rules and regulations, using economic tools, programming and supervising as well as contributing of beneficiaries. So the main objective of water demand management is controlling the amount of demand and efficient application of water meaning development in depth. Considering the limitation of to limitation of water supply in the Iran, it is clearly seen that emphasizing on demand management is a necessary issue to direct water sources to sustainable usage. Creating new perspectives consistent with sustainable development objectives of resources among beneficiaries and controlling the amount of water taken through existing tools can be highlights of water -demand management.

Including national and international studies have been done include:

Torkamani and Kalani (2001) have used multi-product Translog cost function to estimate cost functions and demand of inputs at the same time in agriculture of Fars province. Results show that Translog cost model is a good fit of the investigated data and chemical fertilizer is a complementary for consumed seeds and workforce, machineries, and chemical fertilizer have replacement relationships. Also, all prices stretching of demand for inputs is less than one indicating no stretching of demand for them.

Hajir Keyani and HajiAhmad (2002) have investigated the estimation of demand functions for production inputs and supply of water and dry-farming wheat in Iran agriculture. In this research Translog flexible-profit function as well as profit equations, supply and demand of production inputs using equation system seemingly duplicated unrelatedly were used. And results indicate that all stretching obtained from estimation of coefficients for both water and DEIM wheat have signs corresponding theoretical expectations. And all self-stretching for input demand indicate negative inverse relationship between price and amount of production inputs.

Abedi (2005) used multi-product Translog cost function to estimate cost functions and demand of inputs at the same time in agriculture of Zayande Roud and results show that Translog cost model is a good fit of the investigated data and according to investigations from the pattern, minor self-replacement and crossing stretching is less than one however it was near to one for water. Additionally, self-price and crossing stretching of demand for all inputs is less than one.

Bostani and Mohammadi (2006) investigated productivity and demand function of water in production of sugar cane in Gelid district. Results show that Cobb Douglas production function of consumed water inputs, poison, machineries, planted area and funds have meaningful effects on production in which consumed water and cost of machine operations is positive. Average of positive productivity and final water uses are 0.89 and 0.304, respectively. According to R^2 estimated production function is able to determine 85 per cent of changes in production. Also price stretching of demand is more than one which illustrates stretchable demand function rather than input price.

Shajari *et al.* (2008) investigated management of water demand using policy of water -price determination in Jahrom date farms and case study of Shahani date. Results show that there is a positive and meaningful relationship between number of productive date trees in the garden, number of used workforce and frequency of irrigation and amount of date produced. Final productivity and final production value of water in drip and GHARGHABI irrigation is 0.194 kg and 204.06 kg, respectively and was calculated 67.23 Rials. In final price of water, price stretching for water for drip and water irrigation was obtained -3.035 and -2.093 respectively which shows that demand for water in both methods is stretchable.

Al.Qunaibet and Johnston (1985) estimated demand function of water using time- series data in Kuwait. According to results, price stretching of water comparing to the price was calculated -0.9 .So they mentioned that increasing the price of water can lead reduction of water consumption. Diagnosis coefficient was also more than 80 per cent in this estimation. Glas *et al.* (1989) investigated agricultural structure of Northern Ireland using multi-product Translog cost function. This model contains two product of livestock and its derivations as well as plants and four inputs including fund, workforce, chemical fertilizer and related index of sum of livestock food and imported livestock to the region. They, using the mentioned model, estimated replacement stretching among inputs, self and crossing stretching of demand of inputs, parameters taken from scale and rate of Hiks technology change for each year. In this study, they concluded that agricultural structure of Northern Ireland is not homothetic and inputs of fund is stretchable while workforce is not stretchable and input includes sum of livestock food, seeds and livestock import is also not stretchable .

Ali and Parikh (1992) used estimation of Translog cost function as well as extraction of functions of cost shares using Shephards lemma in order to investigate the relationships among inputs .Equations of cost share along with total cost function were estimated using limitations of symmetry and unity by SUR method and the authenticity of these limitations were investigated by Coefficient of langraz.

Following goals are taken in to consideration: The main purposes of this research are:

1-Estimating Translog cost function of greenhouse cucumber of Khash

2-Estimation of Cucumber water demand function in greenhouses of Khash

3-Determining insider minor elasticity, Allen crossing and insider price stretching and crossing of demands of water inputs through Translog cost function

MATERIALS AND METHODS

To estimate water – demand function and determine price elasticity of water demand for greenhouse cucumber ,first ,production function of cucumber is estimated like Translog production function.

$$Y = \alpha_0 \prod_{i=1}^{n} X_i^{\alpha_i} e^{\frac{1}{2}\sum_{i=1}^{n}\sum_{j=1}^{m} \gamma_{ij} LnX_i LnX_j} \quad (1)$$

Where Y is Output, a_0 is efficiency, X_i and X_j are i, j, a_i and γ_{ij} of unknown parameters. Logarithmic form of this function is as follows.

$$LnY = Ln\alpha_0 + \sum_{i=1}^{n}(\alpha_i Lnx_i) + \frac{1}{2}\sum_{i=1}^{n}\sum_{j=1}^{n}(b_{ij} Lnx_i Lnx_j) \quad (2)$$

Each production function has one cost function. So, form of Translog cost function is as follows:

$$\ln c = \ln a_0 + \Sigma a_i \ln p_i + a_Q \ln Q + 1/2 \Sigma\Sigma \gamma_{ij} \ln p_i \ln p_j + \Sigma \gamma_{iQ} \ln Q \ln P_i + 1/2\, \gamma\, (\ln Q)^2 \quad (3)$$

Where I,j=1,……,N

In which in above equation, Pi and Pj are price of inputs and C is the total cost.

Based on the characteristics of neoclassic production theory which says that cost share is addable, the following limitations are considered for demand function.

a) unity assumption $\Sigma a_i=1, \Sigma \gamma_{iQ}=\Sigma \gamma_{ij}=\Sigma \gamma_{iQ}=0$

b) symmetric assumption $\Sigma \gamma_{ij}=\Sigma \gamma_{ji}$

On the other hand , because the total of cost shares is one ($\Sigma s_i=1$) ,using these conditions on cost function, we need n-1 share and the prices in main function will change in to relative. To avoid becoming zero for koarians Variance matrix, one of equations of cost share is omitted and it is normally omitted in experimental tasks which has the minimum contribution to production costs (Mcgeehan,1993).

To estimate the cost function, ISUR has been applied. The reason to use this technique is existence of unity between impaired sentences in equations of cost share. Since total of cost share is one, if the left side of demand equations are summed ,this sum should be one. Thus, the sum of impaired sentences of equations is zero. On the other hand, independent variables (price of inputs and amount of production) are the same in all demand equations. As a result, all variables which have not been entered will appear in impaired sentences and will unity among impaired sentences. The common way to estimate functions is that one of equations of cost share is taken ways from the simultaneous –function category, and parameters of other functions are estimated and then relevant parameters of taken-away equation are calculated based on other parameters of equations. Since sum of equations of cost share is one, omitting any equation is optional (we deleted poison equation here). Regressions seemingly duplicated unrelatedly tend to maximum method of being true and are independent from omitted equation.

Elasticity of substitution and price elasticity of demand

One common index to measure level of reaction to price changes of production input is Allen AES –Evzaza V_{ij} between to inputs of I and j which are obtained from first and second derivation of cost function to price of I and j inputs.

$V_{ij} = c\ c_{ij}/c_i c_j$

For logarithm cost function, Allen AES can be written as follows:

$V_{ij} = b_i + s^2_i - s_i / s^2_i$

And self-replacement stretching is as follows:

$V_{ij} = b_{ij} + s_i s_j / s_i s_j$

There is a positive replacement between production inputs meaning the replacement of these inputs. While, negative replacement stretching determines that inputs are completing each other. One main gaol of this study is estimating replacement stretching among production factors to investigate the level of replacement of complementary of each pair of production inputs in agricultural sector. Investigation of this issue has interesting applications in policy making in investment.

One factor that can illustrate the technology in one production unit in summary is price stretching of demand for production inputs (Eij) which is defined as follows:

$E_{ij} = \partial \log X_i / \partial \log P_i$

Where production amount (Xi) and the price of all production inputs (Pi) are fixed.

Price self-stretching of demand (Eij) of a production input shows relative changes of input demanded of that input resulting from relative changes of that input. Price crossing stretching of demand (Eij), i ≠j measures relative changes in amount of an I input being demanded because of relative changes of price of j input.

We can show that Allen AES is related with price stretching of demand for production factors. Price crossing stretching of demand for production inputs can be written as follows:

$E_{ij} = S_j V_{ij}$

And price self-stretching of demand is written as follows:

$E_{ij} = S_i V_{ii}$

Where I, j=L, H, SH, W, S, B (Sherafat,1996)

In this research price of inputs and products is determined according to market price. Share of costs has been calculated through dividing paid costs for that factor to total of varying costs.

RESULTS AND DISCUSSION

Estimation of function of water-input demand for cucumber using Translog cost function Estimation results of Translog cost function coefficient using regression seemingly unrelated are illustrated in table number 1. Poison cost has been omitted in estimation of this model.

Table content illustrates that most coefficient

Table 1: Estimation of total cost of production equation for greenhouse cucumber in Khahsh.

Parameter	Coefficient	t-value	Parameter	Coefficient	t-value
C(1)	101.61	8.13	C(15)	-0.14	-2.22
C(2)	-16.4	-8.04	C(16)	-2.79	-1.72
C(3)	-3.74	-1.72	C(17)	4.51	3.50
C(4)	10.05	2.23	C(18)	-0.81	-3.71
C(5)	17.79	4.55	C(19)	-2.59	-1.78
C(6)	-2.01	-2.32	C(20)	0.42	1.70
C(7)	0.31	1.64	C(21)	0.057	0.56
C(8)	0.25	2.47	C(22)	-1.67	-1.04
C(9)	-1.38	-4.47	C(23)	0.43	6.58
C(10)	-1.87	-6.77	C(24)	0.066	0.49
C(11)	0.14	2.16	C(25)	0.21	0.62
C(12)	0.15	1.08	C(26)	-0.65	-2.25
C(13)	-0.48	-1.61	C(27)	-0.13	-1.54
C(14)	-0.2764	-0.99	C(28)	-0.09	-0.65

R^2=0.82 D.W=2.02

in a=0.05 are meaningful and thus the issue that most equation coefficients are meaningful is a suitable result. According to the fact that determination coefficient is 0.82 per cent and shows a good fit of equation, it means that independent variables have been able to explain the reason of total cost change. And Durbin- Vatson data shows that Autocorrelation phenomenon does not exist. (c(1) Labor, c (3) livestock fertilizer, c (4) water, c (5)seeds, c (6) chemical fertilizer).

Based on research goal, share of water cost which is too similar to demand function has been used as follows in order to analyse more.

$$S_w = 10.05\text{-}1.38\log (P_L/P_s) - 0.48\log (P_H/P_S)\text{-} 2.79\log(P_W/P_S)+4.5\log(P_B/P_S)\text{-}0.81\log (P_{SH}/P_S) +0.21\log Q$$

t= (4.4) (0.30) (0.30) (1.62) (1.28) (0.21) (0.34)

As it can be seen, all coefficient of cost-share equation is meaningful based on amount of t. Investigating above coefficients show that cost-share equation of water has a positive relationship with amount of production and price of seeds and negative relationship with price of workforce, livestock fertilizer, water and chemical fertilizer. In other words, increasing the price of seeds as well as increasing production, water cost out of total cost increases and increasing the price of workforce, livestock fertilizer, water, and chemical fertilizer decreases relatively.

Calculating stretching

In this section, minor self and crossing stretching of Allen and self and crossing price stretching of demand using average cost of inputs are being calculated and investigated.

The following results are taken from table 2 and 3:

1- The self-stretching of water has a logical sign(negative) which is compatible with economic theories and show that there is an inverse relationship between prices and amount of inputs and that illustrates the inverse relationship

Table 2: Estimation of replacement stretching of inputs for production of cucumber.

	Labor	Animal manure	water	Seed	Chemical fertilizer
Labor	-0.9	0.3	0.3	0.8	0.2
Animal manure	0.3	-1.6	6.7	19.5	5.4
water	0.3	6.7	-1.6	16.4	4.1
Seed	0.8	19.5	16.4	-3	19.3
Chemical fertilizer	0.2	5.4	4.1	19.3	-0.3

Table 2: Estimation of Demand price stretching of inputs for production of cucumber.

	Labor	Animal manure	water	Seed	Chemical fertilizer
Labor	-0.36	0.12	0.12	0.12	0.03
Animal manure		-0.58	2.27	2.72	0.75
water			-1.69	2.81	0.57
Seed				-2.49	2.69
Chemical fertilizer					0.03

between demand for water and its prices.

2- Absoulute amount of self-stretching for inputs including workforce and chemical fertilizer is less than one which shows none stretching of demand for them. However absoulute amount self-stretching of inputs including water, seeds, livestock fertilizer is more than one which shows stretching of demand inputs.

3- Crossing stretching of demands for water with other inputs shows that it has a replacement stretching in comparison with work force, livestock fertilizer, seeds and chemical fertilizer.

4- Maximum level of replacement of water stretching among inputs is for seeds. In other words, increasing the price of water will increase applying the seeds. And the sensitivity of replacement between these two inputs is less than one and that shows that if the price water goes up , seeds with better quality and modified ones will be used to save water in order to decrease water consumption.

CONCLUSIONS AND RECOMMENDATIONS

1- According to high self-stretching of price , derived demand of water shows that policy in determination of price can be a useful tool to control extreme usage of water in production of green houses.

2- Since amounts of crossing stretching of demand(except work-force input) is more than one, policies to change the price in these inputs will have a considerable effect on planting composition.

3- The amount of self-stretching for water demand is more than one. Thus, changing the price of this input (deep water) may create a pressure on consumption of other kinds of water(surface water).

4- Existence of replacement connection among water, workforce, livestock and chemical fertilizer, and seeds expresses higher usage of these inputs in planting cucumber to save water.

5- Small replacement stretching of workforce in comparison with water lead that relevant policies to factor change affecting in demand of an input not to have considerable effect on composition of other consuming inputs.

As results show demand for water is stretchable. According to being stretchable for water demand, it can be certainly said that price determination of water has an important role in applying this input.

According to obtained results from this study, following solutions are recommended:

1- Possibility to implement the policy of water-price determination with determined water price in this research should be gradually done during the time unless farmers with low efficiency will be omitted.

2- Of course this important issue needs to be taken un consideration that determining water price is not only enough and it is necessary to apply collecting water price and using it in order to improve water resources along with this policy.

3- Creating association to supervise water resources, collecting water cost and expending it to improve the water resources is necessary.

4- Increasing the efficiency of water usage in farms and greenhouses (due to growing up the price of water for irrigation) to farmers and investors is an essential subject.

5- Installing smart counters on agricultural well and controlling the amount of water being taken as well as controlling the irrigation of farms using micro irrigation in greenhouse planting.

REFERENCES

1-Abedi, A. (2005).Using multi-product Translog cost function in simultaneous estimation of supply and demand function of agricultural inputs .case study Isfahan Zayanderoud, fifth two-year conference of agricultural economy, Iran, Sistan and

INTRODUCTION

That requires policy solutions and recommendations to solve the problem of subsidies. Over the past years, water pricing below the actual prices has had an adverse effect on the economy. Rapid growth of water consumption, reduced performance, cost subsidy burden on the government budget, and the indiscriminate use of water are some consequences of this policy.

The problem of water shortages in Iran in years of drought and the intense competition for this resource are well known. However, water consumption by the productive sectors in the country appears not to be rational because the Iranian economy has an intensive water consumption production system. This phenomenon is due to many factors: the old water culture in the region, the system of prices and tariffs of the resource, the institutional system of concessions of water use and other aspects that frame the management system. It is impossible to analyze the impacts of all of them in a single paper. However, each component is important in building a complete picture of the role of water and water policy in the future growth and development of the Iranian economy.

The first studies about water as an economic factor took part in 1950s, but operational difficulties limited the scope of their analyses. Lofting and McCaughey (1968).Became able to fix the problems by regarding water as a production factor in a traditional input-output model; then to evaluate the water needs of the Californian economy. Later on we can find many works which analyze the relationship between water needs and the different productive sectors using input-output models (Sánchez-Chóliz, Bielsa and Arrojo, 1992; Bielsa 1998; Duarte, 1999; Duarte, Sánchez-Chóliz and Bielsa, 2002; Velázquez, 2006; Dietzenbacher and Velázquez, 2006).

Several different methodologies have been explored in the analysis of water pricing (see, for example, the excellent reviews of Johansson *et al.*, 2002, and Dinar and Subramanian, 1998). Many analysts have employed variants on linear programming approaches, such as those developed by Berbel and Gómez-Limón (2000) and Doppler *et al.* (2002) as well as input-output model applications such as the work of Sáenz de Miera (1998).

There is an extensive literature which has employed computable general equilibrium models, and many studies with a similar objective to the one that is the focus of the present paper. One of the pioneers was an analysis by Dixon (1990), in which he offered indications to the public authorities of Melbourne, Sydney and Perth on appropriate water prices. Horridge *et al.* (1993), and Thabet *et al.* (1999) to analyze the impact and efficiency of water prices. Nevertheless, the use of CGE to analyze the reallocation of water rights between users is less common. Seung *et al.* (1998) studied the welfare gains of transferring water from agricultural to recreational uses in the Walker River Basin (located in northwestern Nevada and in California). Diao and Roe (2000) provide a CGE model to analyze the consequences of a protectionist agricultural policy in Morocco and show how the liberalization of agricultural markets creates the necessary conditions for the implementation of efficient water pricing (particularly through the possibility of a market for water in the rural sector). By using an applied CGE, Goodman (2000) shows how temporary water exchanges provide a lower cost option than the building up of new dams or the enlargement of the existing water storage facilities. To sum up, the analysis of water allocation then requires a comprehensive view of the economy, and applied CGE methodology gives a potential framework to assess and compare policy options. For this purpose we use a CGE to analyze the implementation of a water market in the Balearic Islands. This CGE will also allow us to quantitatively and qualitatively compare the advantages of markets over other alternatives such as desalinization plants. Kumar and Young (1996) explained how a Social Accounting Matrix (SAM) can be extended to incorporate water resources and analyze the implications of water pricing policies. In a similar way, Susangkarn and Kumar (1997) used a general equilibrium model to incorporate water as a separate productive sector. Decaluwé *et al.* (1999) developed a general equilibrium model to compare different water price policies as well as to analyze water production according to the use of different technologies. Seung *et al.* (2000a) used a CGE model to evaluate the impacts of water reallocation; in a other study, Seung *et al.* (2000b) they used a dynamic model to analyze the temporal effects of water reallocation from the agriculture sector to recreational uses in rural areas of Nevada. In a similar fashion, Briand (2004) developed a static CGE model to estimate the effects of a water price policy on production and employment in Senegal. Using a slightly different CGE formulation, Hewings *et al.* (2005) evaluated the impact of water reallocation from agriculture to other productive sectors in a recursive fashion that fully captured the feedback effects. The major impact here was on agricultural employment; the reallocation of water to more productive sectors (in terms of value added) could not compensate for the enormous net loss in employment. Qin *et al.* (2012) evaluated the economic impact of water tax changes in china. The main conclusion drawn indicates the main conclusion drawn indicates that water price rise led to a reduction in total production, total exports, GDP and household welfare. Another important finding is that the tax imposed on the water has the greatest impact on the agricultural sector. Cardenete *et al.* (2011) analyzed the effects that an increase in the price of the water delivered to the agriculture sector to promote the conservation of this resource would have on the efficiency of the consumption of water and the possible reallocation of water to the remaining productive sectors. The main conclusion drawn indicates that, although the tax policy applied does not correspond to a significant water saving in the above-mentioned sector, a reallocation of this resource is achieved which seems to generate a more efficient and more rational behavior from a production point of view.

The paper has two objectives. First, we analyze the possible effects that a reducing in the various economic sectors water subsidy would have on the Iranian economy and on water conservation and Secondly, we evaluate the water reallocation to sectors of the economy generated by water price increases in various economic sectors.

General equilibrium models and water policy analysis

General equilibrium models, models that are based on a comprehensive macroeconomic structure optimized micro-economic principles to provide a review of economic policy (Kehoe *et al.*, 1983).

Walras general equilibrium models are based on general equilibrium theory by Arrow -is expanded. Walras general equilibrium model of the design so that the overall macroeconomic issues in the field of production and consumption are in balance. The model assumes that the rational behavior of individuals in the quantities supplied and demanded will apply. In fact, Walras general equilibrium system is based on the mathematical characteristics of interdependencies between production and consumption sectors of the economy can be expressed in the form of equations. How can we change the system so that each of the independent variables, the impact of different economic sectors. Walras above problem using a system of trial and error to be solved. This method of removing excess demand due to price changes and the practice continues to this surplus to zero.

Comprehensive social accounting matrixes, given the wider economic context, represent a nation or an economy is the interactions between the various inputs. SAM's total physical and financial flows in the economy at a particular point in time shows. (Lofgren *et al.*, 2002).

SAM structure and design methods are not standardized. A SAM requires only two conditions:

1- The matrix must be square.

2- All rows (total revenue) and column (total costs) for each account should be equal.

Although we can obtain some important insights from partial equilibrium analysis, this framework, when put into practice, is of a limited use for the analysis of the efficiency of water rights allocation. The main criticism comes from the fact that water is used in almost all production activities, being an essential input in many of them, and also from the fact that water value is highly dependent on time and location. Any change in the distribution of property rights over water will probably have consequences on the sectoral composition of the economic product, on employment, on costs and prices, and on the income distribution between the rural and urban sectors. A market of property rights will undoubtedly increase efficiency but, as partial analysis also leads to partial answers, in the case of water policy, when many effects are a matter of political concern, it is also necessary to provide a framework able to capture all the relevant economic effects of a changed structure of water property rights.

Basic model structure

The social accounting marix consists of activities (agriculture, industry and mining, water supply and services), commodities (agriculture, industry and mining, water supply and services), the factors of production (labor, capital, and water), inputs (households, enterprises, government and external world) are. Separate accounts, products and activities because it is a commodity may be produced by one or several activities and an activity can produce multiple products.

In order to analyze the effects of economic policies, fit the model considered in policy design and fit of the specified model can be used to reduce the effects of different scenarios of the water subsidy on macroeconomic variables in general equilibrium of the economic. The production technology uses a nested multilevel CES, as shown in Figure 1.CES (constant elasticity of substitution) functions are widely used in CGE modeling to represent both production and utility. They have the advantages of being well behaved, with a decent degree of flexibility and consistent with assumptions used in CGE models (linear homogeneity/homothecity). The standard two-variable CES production function may be written as:

$$Y = A[\beta K^{\rho} + (1-\beta)L^{\beta}]^{\frac{1}{\rho}}$$

where Y is the output, K and L the two production factors, A is a scale parameter, β and (1-β), respectively, represent the share of factor K and factor L in total factor payments and r is related to the elasticity of substitution

$$\sigma = \frac{1}{1-\rho}$$

One drawback of the CES function is that the elasticity of substitution between any pair of goods or factors is constant. To specify that this elasticity between members of one subset of goods or factors is different from the one between members of one subset and members of another, it is necessary to combine the CES with another kind of functions.

The bottom right side of Figure 1 shows the Leontief water extraction technology (RW) meaning that producing water for crops requires underground water and energy. The Leontief (fixed coefficients) function is a special case of the CES function when $\sigma \rightarrow 0$. This function is commonly used to model the use of intermediate (manufactured) inputs which are combined with the other factors of production to produce the final good. The CGE models available in the literature do not explicitly consider that water for agriculture is a produced input. Following Boyd and Newman (1991) and Decaluwe′ *et al.* [1999] we assume that capital and land are also CES aggregates (KT). First level aggregate inputs, raw water and the composite capital land are specific production factors of the agricultural sector. Similar to that of Goodman (2000), our model is more flexible than the alternatives

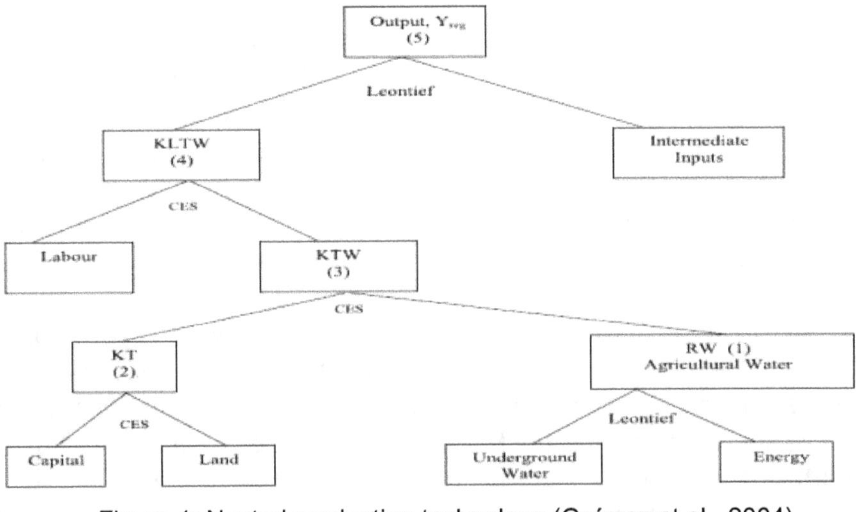

Figure 1: Nested production technology (Go′mez et al., 2004).

provided by, i.e., Seung *et al.* (1998), where land and water enter in the production function with fixed Leontief coefficients.

In a second level the composite land-capital and water are combined in the CES composite KTW, which in a third aggregation level is combined with labor (L) to obtain a CES composite (KLTW) of all capital, labor and nature production factors. Finally, the combination of all these factors with an aggregate of intermediate inputs is combined with a Leontief technology to obtain the final output (Go´mez *et al.,* 2004).

Nested production function equations

$$RW = \min\left[\frac{AS}{aa}, \frac{A_{en}}{ee}\right]$$

Where RW is the composite input water for crops, AS is the volume of underground water used, A_{en} are these of the Armington aggregate of the energy sector, and aa and ee are Leontief coefficients.

$$KT = A^{kt}\left[\beta^{kt}K^{\rho^{kt}} + (1-\beta^{kt})T^{\rho^{kt}}\right]^{\frac{1}{\rho^{kt}}}$$

Where KT is composite land and capital, K is capital; T is land, A^{kt} is the efficiency parameter, β^{kt} is the capital share parameter, r^{kt} is the substitution parameter and s^{kt} is the substitution elasticity.

$$KTW = A^{ka}\left[\beta^{ka}K^{\rho^{ka}} + (1-\beta^{ka})T^{\rho^{ka}}\right]^{\frac{1}{\rho^{ka}}}$$

Where KTW is composite KT and RW, A^{ka} is the efficiency parameter, β^{ka} is the share parameter, r^{ka} is the substitution parameter and is substitution elasticity.

$$KLTW = A^{la}\left[\beta^{la}K^{\rho^{la}} + (1-\beta^{la})T^{\rho^{la}}\right]^{\frac{1}{\rho^{la}}}$$

Where KLTW is composite KTW and labor, L A^{la} is labor, β^{la} is the efficiency parameter, is the share parameter, and r^{la} is the substitution parameter, and s^{la} is substitution elasticity.

$$Y = MIN\left\{\frac{KLTW}{vaa}, \frac{A_1}{iia_1}, \frac{A_2}{iia_2}, \ldots, \frac{A_g}{iia_g}, \frac{II_1}{iia_1}, \frac{II_2}{iia_2}, \ldots, \frac{II_i}{iia_i}\right\}$$

where g ϵBM is a set of traded production sectors, $i \epsilon BNM$ is a set of nontrade production sectors, Y is output of irrigated agriculture, II_i is use of the intermediate input of the sector i, A_g is use of the Armington(1969) aggregate of the sector g, and vaa, iia_g iiag, and iia_i, are technical fixed coefficients (Go´mez, 2004).

RESULTS

The goal of this research is analyzed the effects of reducing water subsidies on GDP. Reduce the impact of subsidies on the part of Water on GDP in each of the sectors were calculated and in table (1) is shown.

We have simulated changes in the reducing water subsidies on various economic sectors, with six different scenarios. As it can be observed, these simulations assume that the water price is increased significantly, with a consequent potentially heavy sacrifice on the side of the farmers. The reason for high impact in agricultural sector is that about 90 percent of water resource is used in this sector due to low water price. Thus, increasing water price led to increase in production price. This, in turn, causes the decrease in agricultural production. This is the point that government should support and help farmers in order to prevent production decrease.

According to table (1) the results of the impact of reducing water subsidies in the form of scenarios are given on GDP. Every six scenarios, reducing water subsidies (50, 100, 200, 300, 400 and 500), highest and lowest impact are related to agricultural and water supply sector, respectively. However, the impact of decreasing production in the industry and mining as well as services is negligible. And also the results show that economic growth in various scenarios to reduce water subsidies in agriculture, water supply, mining industry and services, respectively; from 0.57, 0.06, 0.16 and 0.35 percent in the first scenario to the 4.51, 0.66, 1.53 and 3.30 percent in scenario six is decrease.

It is very important to show that the presence or absence of a water market plays a crucial role in the assessment of the convenience of maintaining the existing facilities to increase the supply of raw water (or of building new facilities for the same purpose). In the case of our model, the existing water desalinization plant represents this kind of facilities. As water markets will result in a better allocation of water among various economic sectors

Table 1: effects on domestic production to reduce water subsidies in various economic sectors in 2001 (billion rials)

	Current Status	50% reduction in water subsidies	100% reduction in water subsidies	200% reduction in water subsidies	300% reduction in water subsidies	400% reduction in water subsidies	500% reduction in water subsidies
Agricultural sector	112630	112050	111443	110330	109331	108429	107612
Changes to the current situation (percent)		-.57	-1.11	-2.1	-2.98	-3.78	-4.51
Department of Water	3041.7	3039.7	3037.6	3033.6	3029.6	3025.6	3021.6
Changes to the current situation (percent)		-0.06	-0.13	-0.27	-0.4	-0.53	-0.66
Department of Mining and industry	426907	426207	425519	424174	422870	421606	420379
Changes to the current situation (percent)		-0.16	-0.32	-0.64	-0.94	-1.24	-1.53
Services	401349	399934	398539	395804	393141	390547	388019
Changes to the current situation (percent)		-0.35	-0.7	-1.38	-2.04	-2.69	-3.32

Source: Research findings

water production.

CONCLUSION

A complete model was determined in order to investigate water resources and then six scenarios for reducing water subsidies (50, 100, 200, 300, 400, and 500%) were applied. Finally, by comparing the findings achieved by simulated computable general equilibrium, the impact of price shock on GDP was analyzed.

The results show that in all six scenarios about reducing water subsidies (50, 100, 200, 300, 400 and 500%), highest and lowest impact are related to agricultural and water supply sector, respectively. However, the impact of decreasing production in the industry and mining as well as services is negligible. The reason for high impact in agricultural sector is that about 90 percent of water resource is used in this sector due to low water price. Thus, increasing water price led to increase in production price. This, in turn, causes the decrease in agricultural production. This is the point that government should support and help farmers in order to prevent production decrease.

SUGGESTIONS

1 – Government can prevent the decrease in production by redistributing incomes.

2- Low-interest loans and tax breaks for entrepreneurs is to motivate the implementation of water saving, water saving will result.

3 - The computable general equilibrium models, the parameters of the model and the stretch can be exogenous. And the results of the sensitivity analysis can be carried out, and compared the results with different scenarios for traction.

4 - Standards for Imports efficient technology in order to make optimum use of water resources.

REFERENCES

1-Armington, P.S. (1969). A theory of demand for products distinguished by place of production, International Monetary Fund, Staff Papers-International Monetary Fund, 16: 159-178.

2-Arrow, K.J. & Dubreu, G. (1954). Existence of an Equilibrium for a Competitive Economy, Journal of Econometrical, 22, 90- 265.

3-André, F.J., Cardenete, M.A & Velázquez, E. (2005). Performing an environmental tax reform in a regional economy, a computable general equilibrium approach. Annals of Regional Science, 39(2), 375-392.

4-Berbel J. & Gómez-Limón, J.A. (2000). The impact of water-pricing policy in Spain: an analysis of three irrigated areas. Agricultural Water Management, 43 (2), 219-238.

5-Bielsa J. (1998). Modelización de la gestión integrada Del agua en el territorio: magnitudes asociadas desde una perspectiva económica, Ph.D. Dissertation, Universidad de Zaragoza.

6-Boyd, R. & Newman, D.H. (1991). Tax reform and land-using sectors in the US economy: A general equilibrium analysis. American Journal of Agricultural Economics, 73(2), 398-409.

7- Briand, A. (2004). Comparative water pricing analysis: duality formal-informal in a CGE model for Senegal. Paper presented in the Conference Input-Output and General Equilibrium: Data, Modelling and Policy Analysis. Brussels, 2-4 September 2004.

8- Cardenete, M. A. & Hewings, G. (2011). Water price and water sectoral reallocation in Andalusia. A computable general equilibrium approach. Environmental Economics, 2, 17-27.

9- Cardenete, M.A. & Sancho, F. (2003). An applied general equilibrium model to assess the impact of national tax changes on a regional economy, Review of Urban Development Studies, 15 (1), 55-65.

10-Decaluwe, B., Patry, A. & Savard, L. (1999). When water is no longer heaven sent: comparative pricing analyzing in an AGE model,Working paper 9908. CRÉFA 99-05. University of Laval.

11- Dietzenbacher, E. & Velázquez, E. (2006). Analysing andalusian virtual water trade in an input-output framework. Working paper, series Econ I, Universidad Pablo de Olavide.

12- Diao, X. & Roe, T. (2000). The win-win effect of joint and trade reform on interest groups in irrigated agriculture in Morocco, in The Political Economy of Water Pricing Reforms, edited by A. Dinar, pp. 141–165, Oxford Univ. Press, New York.

13-Dinar, A., & Subramanian, A. (1998). Policy implications from water pricing experiences in various countries, Water Policy, 1, 239-250.

14- Dixon, P.B. (1990). A general equilibrium approach to public utility pricing: determining prices for a water authority, Journal of Policy Modelling, 12 (4), 745-767.

15- Doppler, W., Salman A.Z., Al-Karablieh, E.K. & Wolff, H-P. (2002). The impact of water price strategies on the allocation of irrigation water: the case of Jordan Valley, Agricultural Water Management, 55 (3), 171-182.

16- Duarte, R. (1999). Estructura productiva y contaminación hídrica en el valle Del Ebro. UN análisis input-output. Ph.D. Dissertation. Universidad de Zaragoza.

17- Duarte R., Sánchez-Chóliz, J. & Bielsa, J. (2002). Water use in Spanish economy: an input-output approach, Ecological Economics, 43, 71-85.

18- GAMS Development Corp. (2001). GAMS version 2.5, solver PATH, Software,Washington, D.C.

19- Gomez, C. M., Tirado, D. & Rey-Maquieira, J. (2004). Water exchanges versus water works: Insights from a computable general equilibrium model for the Balearic Islands. Water Resource, Res, 40.

20- Goodman, D.J. (2000). More reservoirs or transfers? A computable general equilibrium analysis of projected water shortages in the Arkansas River Basin, Journal of Agricultural and Resource Economics, 25(2), 698–713.

21-Hewings, G.J.D., Dridi, C. & Guilhoto, J.J.M. (2006) .Impacts of reallocation of resource constraints on the northeast economy of brazil, Discussion Paper 06-T-01, Regional Economics Applications Laboratory, University of Illinois, Urbana.www.uiuc.edu/unit/real.

22-Horridge, J.M., Dixon, P.B. & Rimmern, M.T. (1993). Water pricing and investment in Melborne: General equilibrium analysis with uncertain streamflow, Working Pap. IP-63, Cent. of Policy Stud. And the Impact Proj, Monash Univ., Melbourne, Victoria, Australia.

23-Johansson, R., Tsur, Y., Roe, T., Doukkali, R., Dinar, A. (2002). Pricing irrigation water: a review of theory and practice. Water Policy, 4 (2), 173-199.

24-Kehoe, T. J. & Jaime, S. P. (1983). A computational general equilibrium model with endogenous unemployment, Journal of Public Economics, 22, PP. 1-26.

25-Kumar R., Young C. (1996). Economic policies for sustainable water use in Thailand. International Institute for Environment and Development.

26-Lofting, E.M. & Mcgauhey, P.H. (1968). Economic valuation of water. An input-output analysis of California water requirements, Contribution nº116. Water Resources Center.

27- Lofgren, H., LeeHarris, R., Robinson. (2002). A standard computable general equilibrium (CGE) Model in GAMS, (Vol. 5). International Food Policy Research Institute, 2033 K Stteet, N.W., Washington, D.C., 20006-1002, U.S.A.

28- Qin, C., Jia, Y., Su, Z., Bressers, H.T., & Wang, H. (2012). The economic impact of water tax charges in China: a static computable general equilibrium analysis. Water International, 37(3), 279-292.

29-Saénz de Miera, G. (1998). Modelo input-output para el análisis de las relaciones entre la economía y el agua. Aplicación al caso de Andalucía. Ph. D. Dissertation. Universidad Autónoma de Madrid.
30-Sánchez-Chóliz, J., Bielsa, J. & Arrojo, P. (1992). Water values for Aragon, Environmental and Land Issues. Wissenschaftsverlag vank Kiel KG. Ed. Albisu, L.M. and Romero, C. EAAE, CIHEAM.
31-Seung, C., Harris, T., Englin, J. & Netusil, N. (2000a). Impacts of water reallocation: a combined computable general equilibrium and recreation demand model approach, Annals of Regional Science, 34, 473-487.
32- Seung, C., Harris, T., Englin, J. & Netusil, N. (2000b). Application of a Computable General Equilibrium (CGE) Model to evaluate surface water rellocation policies, The Review of Regional Studies, vol 29, nº2, 139-156.
33-Susangkarn, C., Kumar, R. (1997). A computable general equilibrium model for Thailand incorporating natural water use and forest resource accounting, Asian-Pacific Economic Literature, 12 (2), 196-209.
34-Thabet, C., Macgregor, B., & Surry, Y. (1999). Effets macro-économiques de la politique du prix de l'eau d'irrigation en Tunisie. Économie rurale, 254(1), 28-35.
35-Velázquez, E. (2006). An input-output model of water consumption, Analyzing Intersectoral Water Relationships in Andalusia. Ecological Economics, 56 (2), 226-24.

21

Application of Linear and Non-linear Programming Model to Assess the Sustainability of Water Resources in Agricultural Patterns

Seyed Abolghasem Mortazavi*, Reza Hezareh [1], Sina Ahmadi Kaliji [2] and Samira Shayan Mehr [3]

Abstract

Water resources sustainability is one of the major issues in the agricultural sustainability. In this study sustainability of water resources has been investigated by use of linear and non-linear models in six models based on optimal utilization of water resources in the north parts farms of Iran because of incorrect use of agricultural water resources, from 2011 to 2012. Also "gross margin per a unit of water consumption" and "employment per a unit of water consumption" are used as indicators for assessing the sustainability of cropping patterns. The results show that cropping pattern of fractional goal programming (FGP) model has been near to current situation and has shown realistic conditions according to expertise and advantage of this area in cultivation of certain crops. So the FGP model has desirability in each of indicators than other five models.

Keywords:
Consumers, Hedonic Pricing, Cowpeas, Nigeriawater resources sustainability, Fractional goal programming, Cropping pattern, The north parts farms of Iran

[1] *Faculty Member and MSc Student, Department of Agricultural Economics, University of Tarbiat Modares, Tehran, Iran.*
[2] *Former MSc student, Department of Agricultural economics, Gorgan University of Agricultural Sciences and Natural Resources, Gorgan, Iran.*
[3] *MSc Student in Agricultural Economics, University of Kordestan, Iran.*
* *Corresponding author's email: Samortazavi@modares.ac.ir*

INTRODUCTION

Sustainability of natural resources, especially water and soil that are considered as the main sources of production are one of the most fundamental challenges that all countries are faced them (Amini Faskhodi and Nori, 2010). Water is a basic need for the sustainability of life on earth (Amini Faskhodi and Nori, 2010). Sustainability is not a new concept and human always has tended to maintain its vitality resources (Oron and Gillerman 2008). In recent decades, attention to sustainability and management of water resources as vital sources of life have become the main and significant issue (Borimnejad and Sharifat, 2012). Low productivity leads to loss of water resources. Productivity of water use in the agricultural sector of Iran is very low (Bakhshi *et al.*, 2011). In the north parts farms of Iran, water transfer efficiency from the source to field is low that leading reduces of crop production. 33% of the water extracted from various sources of this area is wasted (Faskhodi and Nori, 2010). Actual efficiency in the north parts farms of Iran is 32 percent and 80 billion m3 of agricultural water has been wasted. This has led to that agriculture is not profitable in this area (Faskhodi and Nori, 2010). Based on, farmers' attention to the correct exploitation of these resources is essential. Ratios are considered a natural way to deal with related issues to the sustainability of agricultural systems. In this framework, the purpose is not just to maximize or minimize the objective function, rather optimization the inputs to the outputs is proposed. So in order to explain the stability, should be paid to the comparison of input and output. Accordingly, sustainable use of agricultural water resources is one of the most important issues in the sustainability issues that in this study are considered as input. On the other, appropriate levels of income are considered as an economic factor and creation of employment opportunities as needed social factors for sustainability of agricultural systems. To this aim, in this study sustainability of water resources has been investigated by use of linear and non-linear models in six models based on optimal utilization of water resources in the north parts farms of Iran because of incorrect use of agricultural water resources, from 2011 to 2012. Also "gross margin per a unit of water consumption" and "employment per a unit of water consumption" are used as indicators for assessing the sustainability of cropping patterns.

Studies that have been done in this case, Lara and Minasiam (1999) are presented Fractional Programming as a tool for studying the sustainability of agricultural systems. This study, the essentials of the technique in both the single and the multi-objective cases are outlined. Their results show, fractionalizing can also be a useful way of getting a reduced set of efficient solutions. So, FP has a theoretical as well as a practical usefulness. Borimnejad and Sharifat (2012) present a multi-objective model for efficient allocation of water resources from the perspective of sustainable development. The results gained from this study indicate that with the increase for water demand, the efficient allocation can be carried out in a way that different users of water resources will suffer the least and thus move towards sustainability. Amini Faskhodi *et al.*, (2010) in their paper introduces two ratio objective functions in order to assess the sustainability of a rural farming system by simultaneously optimizing these ratio indicators and subsequently determining the optimal cropping pattern. Aimed to this, the Fractional Programming (FP) procedure has been used. Results of this study show, the associated environmental impacts of farming activities can be reduced by more efficient use.

MATERIALS AND METHODS

Usually in agricultural economics studies for calculating water sustainability indicators, fractional and goal programming models are used that we will try to choose the best model to calculate the indicators by creation a relationship among the models. In this study, to determine the pattern and assessment of agricultural sustainability indicators in north parts farms of Iran, Linear Programming, Goal Programming, Fractional Programming and Fractional Goal Programming model are used with multiple objectives that are connected as rings. Accordingly, the considered model can be used the terms of three objectives of maximizing gross revenue, employment and minimizing of water consumption. In order to linearize multi-objective programming models, three methods of restricted, weighted and non-inferior set estimation methods are used (Sabouhi, 2012). We use weighted method for this study. In fact, the weight of each objective enters in the objective function due to their weight.

According to the proposed model, Simple linear programming model is minimized or maximized the objective function given the restrictions. But, in this case, the inputs used in production are assumed to be homogeneous. In other words, restrictions of related to sustainable agriculture are not in the model. However, Fractional programming model can enter restrictions of sustainable agriculture in the objective function (Zamani *et al.*, 2010). Also this model makes easy manage of answers (Borimnejad and Yazdani, 2006). The next model is Goal programming that is a branch of multi-objective optimization, which in turn is a branch of multi-criteria decision analysis (MCDA), also known as multiple-criteria decision making (MCDM). This is an optimization programmer. It can be thought of as an extension or generalization of linear programming to handle multiple, normally conflicting objective measures (Pati *et al.*, 2006). Also, in Fractional Goal Programming model, objectives are linear fractional functions (Caballero and Hernandez, 2010). Thus if these ratios had it verify certain target values, we would have a set of fractional goals (Gomez *et al.*, 2006).

Models of this study

The linear and goal programming model are defined as follows:

$$\begin{aligned}
&Max\left\{\sum g_i \times x_i, \sum L_i \times x_i\right\}\\
&st:\\
&1)\sum_i a_{ij} \times x_i \le b_j \quad \forall j\\
&2)\sum_i t_j \times x_i \le 0 \quad \forall_i\\
&3)\sum_i w_i \times x_i \le \lambda tw \quad \forall_i\\
&4)x_i \ge f_i \quad \forall_i\\
&x_i \ge 0
\end{aligned} \tag{1}$$

$$\begin{aligned}
&Minz = w_m \times n_m\\
&st:\\
&1)\sum g_i \times x_i, \sum_i L_i \times x_i + n_m - p_m = u_m \quad \forall_i\\
&2)\sum a_{ij} \times x_i \le b_j \quad \forall_j\\
&3)\sum_i t_i \times x_i \le 0 \quad \forall_i\\
&4)\sum_i w_i \times x_i \le \lambda_{tw} \quad \forall_i\\
&x_i \ge 0
\end{aligned} \tag{2}$$

The Fractional and Fractional Goal Programming model are defined as follows:

$$\begin{aligned}
&Max\left\{\frac{\sum g_i \times x_i}{\sum w_i \times x_i}, \frac{\sum L_i \times x_i}{\sum w_i \times x_i}\right\}\\
&st:\\
&1)\sum_i a_{ij} \times x_i \le b_j \quad \forall_j\\
&2)\sum_i t_i \times x_i \le 0 \quad \forall_i\\
&3)x_i \ge f_i \quad \forall_i\\
&x_i \ge 0
\end{aligned} \tag{3}$$

$$\begin{aligned}
&Minz = w_m \times n_m\\
&st:\\
&1)\frac{\sum g_i \times x_i}{\sum w_i \times x_i}, \frac{\sum L_i \times x_i}{\sum w_i \times x_i} + n_m - p_m = u_m \forall_m\\
&2)\sum_i a_{ij} \times x_i \le b_j \quad \forall_j\\
&3)\sum_i t_i \times x_i \le 0 \quad \forall_i\\
&4)x_i \ge f_i \quad \forall_i\\
&x_i \ge 0
\end{aligned} \tag{4}$$

In these models:

i is products of cropping pattern in the north parts farms of Iran (rice, wheat, barley, cucumber, tomato and watermelon); **j** is production inputs (land, fertilizer, pesticides, labor, machinery, nitrogen and potassium fertilizer); **n** is negative deviations; **p** is positive deviations; $\mathbf{g_i}$ is gross revenue of each **i** product; $\mathbf{L_i}$ is the amount of used labor in each **i** product; $\mathbf{w_i}$ is the amount of required water to produce an hectare of **i** product; $\mathbf{a_{ij}}$ is technical coefficients of **i** product of **j** input; $\mathbf{b_j}$ is available amount of input **j**; $\mathbf{t_j}$ is coefficients of the **i** product in period; **λ** is irrigation efficiency (this factor represents the amount of irrigation efficiency that use to calculate the total plant available water); $\mathbf{t_w}$ is the total available water resources; $\mathbf{f_i}$ is at least of **i** product cultivation levels; $\mathbf{w_m}$ is each of goals weight (typically two ways are used to calculate the objective weights, AHP method and relative weighting method, which in this study relative weighting method is used and due to unification of deviation variable units, this variables are divided on the right values into their equations and are multiplied at 100); nm is negative deviations of m goal and pm is positive deviations of m goal (as mentioned, for goal programming model, levels of objectives are obtained by solving two linear programming model with the goal of maximizing profits and minimizing employment).

Objective function is maximizing of employment and gross profit in equation (1). In this equation, restriction (1) is availability of production inputs, restriction (2) is crop rotation (in order to not culture of fall and spring crops, this limitation is used as a limitation rotation), restriction (3) is available water in the region and restriction (4) is the minimum strategic crops of this area (rice, wheat, tomatoes). Since these crops are the only strategic crops and farmers of this area have expertise in cultivation of these crops, to prevent of products removal, in proposed cropping pattern minimum cropping is considered for these products. After solving the model and obtaining the objective function values, these values are considered as a goal in the equation (2). Objective function is minimum achievement of goals level given the importance of each of these in equation (2). In this equation, restriction (1) is goal restriction (goal amount of solving the model (1) for each of the objectives). Other restrictions of this equation are such as the previous equation. Objective function is maximizing level of gross revenue and employment than water used in equation (3). Equation (4) is calculated after solving the model (3) and clarification each of goals. Objective function and restrictions of this model is such as equation (2).

Required data set is collected through random sampling and completing the questionnaires of 115 farmers at 2011-2012 crop years in the north parts farms of Iran that are statistical population of this study. Major activities of farmers production are includes of: $\mathbf{x_1}$: rice (ha), $\mathbf{x_2}$: irrigated wheat (ha), $\mathbf{x_3}$: irrigated barley (ha), $\mathbf{x_4}$: watermelon (ha), $\mathbf{x_5}$: cucumber (ha), $\mathbf{x_6}$: tomatoes (ha). In this study **TL** is total of workforce (person/day), **TGP** is total gross profit (million rials), **TW** is total water used (100 m^3), **LPGM** is linear programming model with the objective of maximize the gross margin, **LPL** is linear programming model with the objective of maximize the employment, **LPG** is linear goal programming model, **FGPM/TW** is fractional programming model with the objective of maximizing the gross margin to water used, **FPL/TW** is fractional programming model with the objective of maximizing the employment to water used and **FGP** is fractional goal programming model.

RESULTS AND DISCUSSION

In this study, two linear programming models were calculated with the objectives of maximizing employment (LPL) and gross margin (LPGM). As these results show, barley cultivation due to the low gross revenue and also cucumber cultivation due to the high cost of cultivation and having roughly equal income to other products have not included in the cropping pattern. So that it also happened in the LPL model. Rice cultivation because of having potential employment in this area and making income in both linear model patterns has contributed high cropping pattern. The results of solving linear and nonlinear models show in table (1).

The next step, results of LGM (2,346,894.399 million Rials) and LPL (2,702,174.435 person/day) model as goal for LPG models have used. By using results of the cropping pattern from each of the models have calculated amounts of TL, TGM and TW. Then, due to objectives of this study, division between the gross profit and employment on the function of water used, two fraction objective functions have been created that shown in model (5). Two models of

Table 1: Results of activity patterns in linear and nonlinear programming models

Models	Patterns					
	X1	X2	X3	X4	X5	X6
LPGM	23147	49.9697	0	13908	0	50.458
LPL	70.29383	1408	31.3464	40.25231	0	420
LPG	30.24906	1408	56.3210	20.37435	0	420
FGPM/TW	19800	1408	41.8559	10.13638	0	3805
FPL/TW	19800	1408	60.1624	40.17187	0	420
FGP	19800	1408	1624	111817	17.1845	420

Source: Research findings

Table 2: Results of water sustainability indicators in linear and nonlinear programming models

Models	Indicators				
	TL	TGM	TW	TL/TW	TGM/TW
LPGM	689.1995059	399.2346894	016.2258876	883.0	038.1
LPL	425.2702174	796.3090520	2826853	955.0	093.1
LPG	2352785	4.3193235	3.2367696	993.0	348.1
FGPM/TW	287.1908357	245.2379749	513.2046247	932.0	162.1
FPL/TW	453.2106586	453.2106586	9.1920405	953.0	096.1
FGP	4435297	6420362	5.4545515	975.0	412.1

Source: Research findings

FGPM/TW and FPL/TW with amount of 1.162 and 0.932 as goal for model (6) have determined for simultaneously maximizing of this two fraction function. As show in table (1), cropping of cucumbers and tomatoes due to the high water requirement compared with employment have small contribution in the FPL/TW model that this issue is true for cropping of cucumbers in FGPM/TW and cropping tomatoes in FGP model.

In the following, to compare the results of this activity patterns in relation to two indicators of TL/TW and TGM/TW have been explained as instrument for assessing sustainability of agricultural systems in the different models. As the table (2) shows, the amount of each indicator compared to other linear model has been developed in LPG model. Indeed, TGM/TW indicator in LPG model has grown 29% than LPGM model and 18% than LPL model. Also TL/TW indicator in LPG model has grown 12% than LPGM model, but this indicator has decreased 3% than LPG model.

On the other hand given the results of Table (2), results of LPG model have considerable growth in all indicators comparison with other nonlinear models. So that two indicators of TGM/TW and TL/TW in this model than FGPM/TM and FPL/TW models have increased respectively 21%, 28% and 4%, 2%. Indeed, as can be seen linear and nonlinear of goal programming model than single-objective fractional and linear models is created more favorable conditions in both indicators. The fractional programming models than linear programming models are created more favorable conditions in TGM/TW indicator than TL/TW indicator. So that FGP model than other models has shown favorable condition. Cropping pattern by FGP model will be created more favorable conditions in each of both indicators.

The shape (1) shows suitable condition in order to compare each of the indicators in different models than current condition. TGM/TW and TL/TW indicators in fractional goal programming model have been developed respectively 25% and 2% than current condition. This topic show high capacity of this region to develop sustainability indicators, specifically, the TGM/TW indicator.

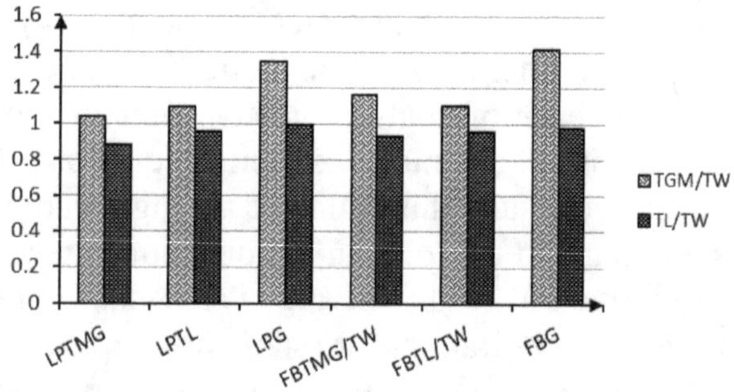

Figure 1: Condition of water sustainable indicators in linear and non-linear models and current condition.
Source: Research findings

CONCLUSION AND RECOMMENDATION

In this study in order to assess the sustainability of cropping patterns in the north parts farms of Iran, six models have been investigated. The results show that FGP model has desirability in each of indicators than other five models. On the other, cropping pattern of this model has been near to current situation and has shown realistic conditions according to expertise and advantage of this area in cultivation of certain crops. Results show that changes in cropping patterns according to the proposed goal programming models could be improved a significant portion of farmers lost profits that lead to decrease of the TGM/TW indicator in current situation.

According to the results of this study, the fractional goal programming model is appropriate tool for analysis on water resources suitability and policy makers could benefit of this instrument to achieve more accurate planning.

REFERENCES

1- Amini Faskhodi, A. & Nouri, S.H. (2010). Assessing sustainability and crop pattern of farming systems according to optimization of water and soil use, nonlinear mathematical programming models. Journal of Science and Technology of Agriculture and Natural Resources, 55(15), 99-109.

2- Bakhshi, A., Daneshvar kakhki, M. & Moghadasi, R. (2011). Application of positive mathematical programming model to analysis of water pricing alternative policies in Mashad plain. Economic and Agricultural Development, 3, 284-294.

3- Borimnejad, V. & Yazdani, S. (2006). Analysis of sustainability in water resource management in agricultural sector using fractional programming: case study of Kerman State. Research and Development in Agronomy And Horticulture, 63, 2-16. (In Farsi).

4- Borimnejad, V. & Sharifat, M. (2012). Water resources sustainable allocation: case study; Alborz province, Iran. Advances in Environmental Biology, 6(2), 912-915.

5- Caballero, R. & Hernandez, M. (2010). Resolution of the linear fractional goal programming problem. Revista Electronica de Comunicaciones y Trabajos de ASEPUMA, 11, 27-40.

6- Gomez, J. A. & Martinez, Y. (2006). Multi-criteria modeling of irrigation water market at basin level: A Spanish case study .European Journal of Operattional Research, 173. 313-336.

7- Lara, P. & Stancu-Minasiam, I. (1999). Fractional programming: a tool for the assessment of sustainability. Agricultural Systems, 62, 131-141.

8- Oron, G. & Gillerman, L. (2008). Membrane technology for advanced waste water reclamation for sustainable agriculture production. Desalination, 218, 170-180.

9- Pati, R., Vrat, P. & Kumar, P. (2006). A goal programming model for paper recycling system. Omega, (36), 405-417.

10- Sabouhi, M. (2012). Application of Mathematical Programming in Agricultural Economics. Publication of Zabol University.

11- Zamani, O., Sabouhi Saboni, M. & Nader, H. (2010). Determining cropping pattern towards sustainable agriculture by use of fazi fractional programming. Journal of Agricultural Science and Sustainable Production, 2(4), 101-112.

22

Estimation of Potential Evapotranspiration and Crop Coefficient of Maize at Rupandehi District of Nepal

Govinda Bhandari

Keywords:
Maize, Lysimeter, Potential evapotranspiration, Crop coefficient, Aridity index, Nepal

Abstract

This study was conducted to determine the potential evapotranspiration (PET) of maize, the crop coefficient (Kc) under full water requirement as well as the cause of decrease in maize yield. It was determined that the seasonal PET of maize is about 486.6 mm. The Kc under full water supply was found to be: 0.11, 0.35, 1.51 and 0.34 for initial, development, mid-season and the late season stages respectively. The study also revealed that maintenance of sufficient moisture need of maize has a significant effect on growth, development and fruiting of the crop.

Environment Professionals' Training and Research Institute Pvt. Ltd. Kathmandu, Nepal
** Corresponding author's email: govinda.eptri@gmail.com*

INTRODUCTION

Potential evapotranspiration (PET) is an important factor in the estimation of water requirement of crops. In the FAO Irrigation and Drainage paper No. 24 on Crop Water Requirements, four methods were given by Doorenobs and Pruit (1977) for the calculation of PET ; Blaney-Criddle, radiation, Penman (adjusted) and pan evaporation. Maize (*Zea mays* L.) is the second important crop in Nepal after rice in terms of area and production. It is grown in the sub tropical to cool temperate climates. For higher yields, crop water requirement is 500-600 mm depending upon the climate and duration of the crop, there should be adequate water during the crop establishment period. Water deficit during the grain filling period results in reduced grain weight. However, during the maturity and harvesting period, rainfall has negative impact on maintaining grain quality (Nayava and Gurung, 2010). Maize is grown during April- July in Terai. During this period, about 500 to 800 mm rainfall (R) occur in the Terai. Annual mean temperature (T) during the growing period of maize in Terai is 21^{0}C to 33^{0}C. In fact Terai region has only 9.1% of the total maize area and contributes 20.4% of the total maize production in the country. During the last 36 years period from 1970/71 to 2007/08, the production of maize in the Terai region increased from 230,700 MT to 383,141 MT. It is interesting to note that the maize yield in Terai was initially low. In Terai, area under maize increased by 60% and its yield increased only 71%. The average yield of maize from 1970/71 to 2007/08 in Rupandehi district is 1822 kg ha^{-1}. The yield of maize in Rupandehi district is observed lowest in 1996 and the highest in 1976 which is 934 kg ha^{-1} and 3241 kg ha^{-1} respectively (MoAC, 2007). Maize is planted in all three seasons (summer, winter and spring) in the Terai region. The planting of summer maize starts from April to May and harvests on August to September. The growing season of winter maize covers from November / December to March/ April. Similarly, the growing season of spring maize is in March/April and harvests on June to early July. For the summer and winter crops, farmers prefer to grow long duration varieties, where as for the spring maize short duration varieties are preferred. Winter and spring maize is planted, wherever the irrigation facilities are available. After the introduction of high yielding varieties, the yield increased in Terai comparatively more than those of the hills and mountain (Nayava and Gurung, 2010).

The maintenance of low plant population in maize fields is one of the major factors affecting low productivity of maize in Nepal. Since yield of a crop is the result of final plant population, the establishment of optimum plant population is essential to get maximum yield. The competition between plants may not occur, and resources are not efficiently used at very low plant population. Under low plant population, grain yield is limited by inadequate number of plants whereas at higher plant population, it declines mostly because of an increment in the number of aborted kernels and/or barren stalks. At higher plant population, the rate of yield reduction is in response to decreasing light, moisture nutrient and other environmental resources available to each plant. Nepalese farmers have been practicing broadcasting method of seed sowing, which can not maintain uniform plant population in the field. The maintenance of plant population of maize in farmer's field condition is lower than the recommended level; and hence the yield is reduced. This is one of the causes of lower maize yield in Nepal (Shrestha and Timsina, 2011).

Recently the effect of climate change to agriculture in Nepal was studied by different researchers. The Asia Pacific Network report shows, when the temperature rises up to 1^{0}C, there is a positive role in percentage change in maize yield in all the agro-ecological zones. When the temperature rises up to 2^{0}C and at the same time carbon dioxide is doubled, the yield will decrease in Terai, hills will not be much affected. On the contrary, mountain environment will have better yield if the carbon dioxide is doubled and temperature rises up to 4^{0}C, the maize yield in Terai will suffer 25% lesser than the present yield (Nayava and Gurung, 2010). Overall rainfall during summer monsoon of the year 2005-06 and 2006-07 was about 16% below the normal which reduced cultivation area of agricultural production in the country (MoE, 2011).

Scientific statements regarding changing climate of Nepal are pronouncedly focused on temperature rise at 0.06 °C per annum. Such a rise in average temperature is variable across the country (Gautam and Pokhrel 2010), higher in the mountains and Himalaya (0.08°C) as compared to low-lying Terai (0.04°C). As a result, series of speculation have been reported such as days and nights becoming warmer, retreating snow lines, increased evapotranspiration and peak runoff, decreasing regular water discharge in streams and recharge of natural water stores, changing water tables and changing pattern of precipitation in terms of form, season, duration and amount (MoE, 2010 a; Gautam and Pokhrel, 2010). Also reported are decreasing total rainy days and increasing number of drier days (evapotranspiration>precipitation) and days receiving over 100 mm rain (MoE, 2010 a; Practical Action, 2009). Precipitation, though not much varying in total amount, is mentioned being erratic and ill timed (Gurung, 2008). Weather variability associated with rising temperature and changing pattern of precipitation is expected to have utmost adverse impacts on various components of agricultural systems. The impacts, though expected to become higher in the mountains compared to low lying terai region, are detrimental to both the regions and ultimately to agricultural production, food security and people's economic sustenance.

The aim of this study is to estimate Kc; Aridity index, AI; and PET of maize using a Lysimeter and Blaney-Criddle method that could be used in Nepal.

MATERIALS AND METHODS

General description of the study area and a lysimeter.

The study area was located at National Wheat Research Program (NWRP), Bhairahawa in Rupandehi district as shown in Figure 1. It is situated between 27° 32' N latitude and 83° 25' E longitude. It is 105 m above the sea level.

In the terai, mostly sandy loam and loam types of soils are reported. Farmers reported black clay as the most fertile soil. However, this soil is not suitable for maize cultivation except in drier years, as it holds water longer than other soil types and tends to waterlog. Land preparation is difficult in clay soils, especially red clay. Brown and gray colored loams are the most suitable for maize cultivation (Paudyal *et al.*, 2001). Loam and clay loam soil textures are the dominant soils of the study area. Loamy soil is easy to plow but yield declines substantially if rainfall is low.

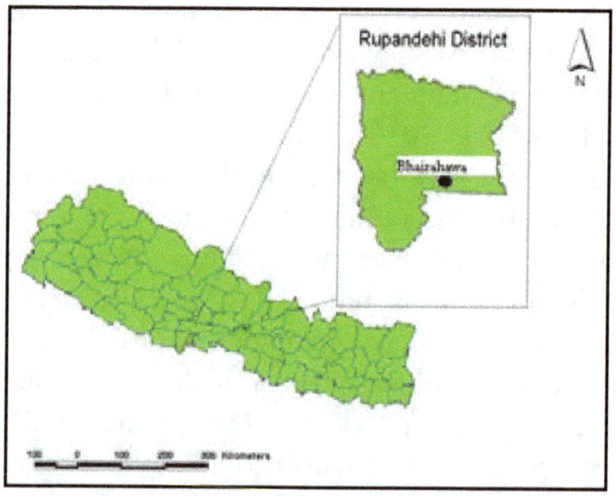

Figure 1: Location map of the study area in a Rupandehi district, Nepal

Experimental design and field layout

A lysimeter is made from readily available materials, plastic bucket with an area of 5811 square centimeter and 10 cm deep and is kept at one corner of the maize growing plot. The first bucket (B1) with tiny holes is completely filled with soil and leveled to the ground surface and holds maize seeds and the second bucket (B2) underground with a hole is connected with a pipe to pass out the infiltrated water to the receiving vessel which is kept deep underground as shown in figure a and b respectively. The receiver vessel kept underground and the B2 are connected with a water pipe so as to collect the percolated water. (Figure a and b)

Figure a: Water pipe connecting to B2 bucket

Figure b: B1 bucket with sowed maize seeds and a receiver vessel over B2 bucket which is connected with a receiver vessel

Planting

Rampur Composite, a major downy mildew disease resistance variety of maize is sown on 22 May, 2011 as in fig a. Screening of this variety against downy mildew disease was carried out at Suwan farm, Thailand in 1981/82 (Rajbhandari, 1982). (Figure a and b)

Figure a: Sowed maize seeds

Figure b: A completely installed lysimeter

Growth stages

Four growth stages are considered. They are initial stage, developmental stage, mid-season stage, and late season stage. The initial stage lasted for 10 days (May 22 – May 31, 2011). The developmental growth stage lasted for 30 days (June 1– June 30, 2011). The mid-season growth stage (flowering and fruiting) stage lasted for 31 days (July 1 – July 31, 2011) and the late season stage lasted for 14 days (August 1– August 14, 2011). (Figure a and b)

Figure a: Mid-season growth stage

Figure b: Late season stage

Irrigation regime

One liter of water is maintained daily except the rainy days so as to provide the sufficient moisture for the growth of maize in a lysimeter. Irrigation days amounted to 42 days and rainy days amounted to 43 days out of the 85 days of the total growing period. The percolated water is used for the irrigation to conserve the chemical composition of the soil.

Potential evapotranspiration

Evapotranspiration is one of the major contributors of the total irrigation water requirement. The amount of evapotranspiration is largely governed by weather parameters. Evapotranspiration might be potential or actual. Potential evapotranspiration assumes the continuous supply and no deficit of moisture in the air (WMO, 1994).

The measurement of PET includes the moisture evaporated to the atmosphere from plants and soil. As the soil and vegetation is confined within a small tank (the lysimeter) the measurements are made of the water input: Rainfall (R) and Additional water (A) and output: Percolated water (P) collected in the receiving vessel, PET can be estimated from the equation below:

$$PET=R+A-P \quad (1)$$

Crop coefficient

K_c is calculated by using the value of PET (as obtained in equation 1) in a Blaney-Criddle formula (as in equation 2)

$$PET = 2.54\ K_c\ F \quad (2)$$

and $F = \sum P_h\ T_f / 100$,

where,

K_c = an empirical coefficient, depends on the type of the crop

P_h = monthly percent of annual day time hours, depends on the latitude of the place

T_f = mean monthly temperature in °F

F= sum of monthly consumptive use factors for the period

Aridity Index

AI represents the severity of dryness of a region. Aridity is defined as the more or less repetitive climatic condition, which is characterized by a lack of water (Perry, 1986). It should be noted that aridity can be considered on seasonal or monthly basis (Coughlan, 2003). The AI ranges from 0.05 to 0.65 for the dry seasons. AI less than 0.65 correspond to Dry lands that, according to the United Nations Convention to Combat Desertification (UNCCD), may suffer desertification processes. So, AI should be greater than 0.65 during the maize growing seasons. In this study, UNEP aridity index (Hare, 1993) is used to estimate the AI which can be expressed as:

$$AI = P/PET \quad (3)$$

where, P=Precipitation in mm and PET=Potential Evapotranspiration in mm.

Other data collected

Five plants are selected randomly from the field and compared with the five plants of the lysimeter to compare the different parameters.

Height and width: Plant height and width of stalk is measured by a measuring tape.

Corn yield: Late April planted corn yields the highest and produces the most profit per hectare. Yield potential is lower when corn is planted late. Number of cobs and grains was determined by counting them respectively.

Grain and biomass weight: Grain weight and dry biomass was determined by using electronic analytical balance.

Statistical analysis

Data collected were subjected to analysis of mean and standard deviation by using MS Excel and SPSS. It is a non-restrictional design in the sense that there is no blocking and any treatment can be replicated any number of times. Five maize plants are selected randomly from the lysimeter and agricultural land as control group and experimental group respectively. Such random assignment of items to two groups is technically described as principle of randomization. Thus, this design yields two groups as representatives of the population. As the experimental units are homogenous, completely randomized design (CRD) has been used for the data analysis and interpretation of results.

Source of Inputs

Seed, fertilizers, and manure are the major inputs used for maize. The Agricultural Inputs Corporation (AIC), a public sector undertaking, was the only institution marketing fertilizers until a few years ago, when its monopoly ceased following changes in government policy. It was envisioned that the private sector would step into supply these inputs, but this has not come to pass. The private sector does supply inputs to Terai districts, but supplies limited quantities in the eastern and central midhills and almost none in the mid-western and far-western midhills and highhills. Some NGOs have been supplying seed, fertilizer, and plant protection chemicals in some areas for vegetable cultivation, but not for maize. Negligible amounts of pesticides are used in maize production in the midhills and highhills. However, pesticide use is common among farmers in the central Terai. All of the pesticide used, especially in the Terai, is purchased from agrovets in nearby markets. Agrovets supply a limited amount of hybrids and improved maize seeds in comparatively accessible areas. Their interest, however, remains on hybrid seed, which has higher profit per unit of seed sold.

Farmyard manure is one of the inputs that every household uses in maize fields. Though farmyard manure is also used for other crops, the largest part of the manure is used for maize production (Paudyal *et al.*, 2001). Soil application of 60:30:0 kg NPK/ha is used to get high yield (Dhital *et al.*, 1990). The grain yield increases significantly with the increasing dose of nitrogen (Tripathi and Pathak, 1984). The chemical fertilizer especially nitrogen fertilizer is universally accepted as a key component to higher corn yield and optimum economic return (Gehl, *et al.,* 2005). The loss of soil fertility and lower use of fertilizer input is also another important factor responsible for low yield of maize. The amount of nitrogen to be applied depends largely on the plant population/unit area of land. Optimum plant population can result in increased production only if there is proper supply of nutrients, particularly nitrogen. Srivastav *et al.*, (2002) found non-significant response of early and late varieties to three levels of N (60, 90 and 120 kg ha^{-1}) and plant densities (53000, 71000 and 95000 plants ha^{-1}) during summer but a significant response during winter season. Adhikari, *et al.*, (2004) reported that the highest grain yield of 9352 kg ha^{-1} was produced when the crop was fertilized with 120 kg N ha^{-1} on the crop planted under the plant population of 53,333 plants ha^{-1}; and they noted the lowest yield (6657 kg ha-1) with the crop supplied with 60 kg N ha^{-1} under plant population of 44,444 plants ha^{-1}.

Problems encountered

1- Southern leaf blight, a new disease of maize is a great threat in the Terai. This disease has reduced plant height, grain yield, biomass and grain weight.

2- Army worm, blister beetle, cut worm, moth, stem borer, termite, weevils, white grub etc are the agents causing damage to maize at different stages in Terai.

3- Erratic rainfall

4- Drought is prolonging year after year in Nepal. Development of drought tolerant varieties is the answer to the drought problem. Adjustment of planting season could help to escape the critical stage of crops to escape from the drought.

RESULTS AND DISCUSSIONS

The results obtained from this study shows that when the maize is given its full water requirement, 486.6 mm of water is required. The maximum amount of water i.e. 318 mm is required and utilized at the mid stage (flowering and fruiting). However, for higher yields, crop water requirement is 500-600mm depending upon the climate and duration of the crop, there should be adequate water during the crop establishment period (Nayava and Gurung, 2010).

In this study, K_C is 0.11, 0.35, 1.51 and 0.34 for initial, development, mid-season and the late season stages, respectively as indicated in Table 1. However, values of Kc depend on the month and locality. The range of monthly values is 0.50-0.80. Average value of Kc for the season for maize is 0.65 (Subramanya, 2007). Similarly, the crop coefficient value was found to be higher as the number of days increases. The

Table 1: PET, T, R, Kc and AI

Months/Stages	R (mm)	PET (mm)	Kc	T in ^{0}F	AI
May/Initial	44.00	24.00	0.11	85.35	1.83
June/Crop development	202.10	74.10	0.35	87.04	2.73
July/Mid	429.90	317.90	1.51	85.47	1.35
August/Late	106.60	70.60	0.34	86.43	1.51

Potential evapotranspiration, PET; Temperature, T; Rainfall, R; Crop coefficient, Kc and Aridity index, AI

PET increases as increase in the amount of rainfall and temperature and vice versa as shown in figure 2 and figure 3. However, the PET at mid season has been observed maximum due to the increase amount of rainfall and decrease in temperature. The AI increases with the increase of temperature and observed to be maximum at the crop development stage and minimum at mid stage as shown in table 1.

The crop coefficient (K_c) is affected by a number of factors, which include: the type of crop, stage of growth of the crop and the cropping pattern (Allen *et al.,* 1998). Doorenbos and Pruitt (2000) indicated that plant height and total growing season influence crop coefficient values. The higher the plant height and the longer the growing season the higher the crop coefficient values and vice versa.

Yield components

The relationships between crop yield and water use are complicated. Yield may depend on the timing of water application or on the amount applied. Information on optimal scheduling of limited amounts of water to maximize yields of high quality crops are essential if irrigation water is to be used most efficiently (Anac *et al.,* 1997). Timing, duration and the degree of water stress all affect crop yield.

In this study, height, number of grains, 1000 grains weight and biomass of maize grown in a lysimeter was found to be greater than that of field maize as shown in Table 2. The average yield of maize is estimated to be 1044 kg ha^{-1} when amount of water is provided enough. About 50% reduction in the yield (580 kg ha^{-1}) is observed when there is the scarcity of water. However, the number of grains of maize in a lysimeter and field was negatively affected by the emergence of southern leaf blight disease and other insects and could not reach maximum.

Table 2: Different parameters of Maize

Parameters	Field	Lysimeter
Average plant height (cm)	218	236.2
No. of cobs	4	3
Width of plant with cobs (cm)	6.5	5.75
Grains of a plant/cob	162	216
1000 grains weight (gm)	160	281
Biomass (gm)	649.9	716.99

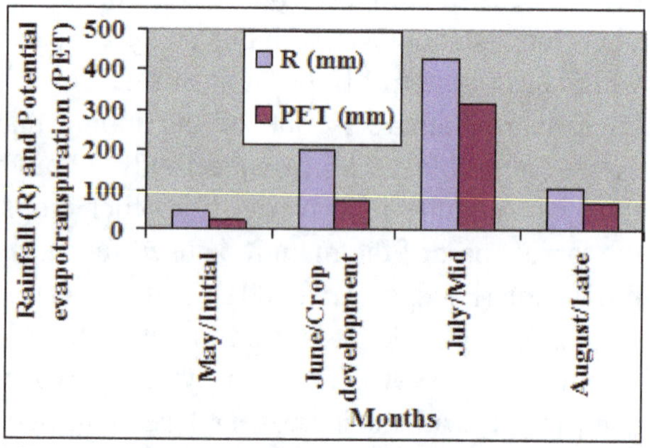

Figure 2: Rainfall and PET at different stages of maize growing period

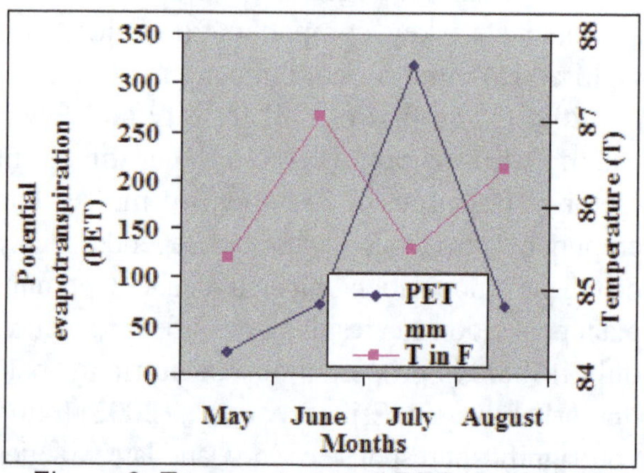

Figure 3: Temperature and PET at different stages of maize growing period

CONCLUSION

Crop coefficient is 0.11, 0.35, 1.51 and 0.34 for initial, development, mid-season and the late season stages respectively. So the range of monthly values of crop coefficient of maize is found to be 0.11-1.51. Potential evapotranspiration is 24 mm, 74.1 mm, 317.9 mm and 70.6 mm for initial, development, mid-season and the late season stages respectively. The total PET is 486.6 mm. The PET of maize is more dependent on rainfall and temperature besides other factors. The PET increases as the amount of rainfall increases. Maintenance of sufficient moisture in a lysimeter has resulted higher yield and therefore the yield of maize can be increased by providing sufficient moisture and manure, controlling disease and pests during initial, crop development and mid season.

ACKNOWLEDGEMENT

The author would like to thank to the director of NWRP, Bhairahawa who gave permission to carry out this research. The author wishes to thank Mr. Ramesh Bhandari, Mr. Dayanidhi Pokharel and his assistants for their cooperation during the whole research period. The facilities and technical assistance provided by the EPTRI, Kathmandu are highly acknowledged.

REFERENCES

1- Adhikari, B. H., Sherchan, D. P. & Neupane, D. D. (2004). Effects of nitrogen levels on the production of maize (*Zea mays* L.) planted at varying densities in Chitwan valley. In: D. P. Sherchan, K. Adhikari, B. K. Bista and D. Sharma (eds.) Proc. of the 24th National Summer Crops Research Workshop in Maize Research and Production in Nepal.

2- Allen R, Pereira L, Raes D, Smith M. (1998). Crop evapotranspiration: guidelines for computing crop water requirements. FAO Irrigation and Drainage Paper No. 56, FAO, Rome, Italy.

3- Anac M.S., Ali Ul M., Tuzal I.H., Anac D., Okur B. and Hakerlerler H. (1999). Optimum irrigation schedules for cotton under deficit irrigation conditions. In: C. Kirda, P. Moutonnet, C. Hera, D.R Nielsen, (Eds.). Crop Yield Response to Deficit Irrigation. Dordrecht, the Netherlands, Kluwer Academic Publishers. Pp. 196-212.

4- Coughlan, M.J. (2003). Defining drought: a meteorological viewpoint. Science for Drought. In: Proc. of the National Drought Forum. Brisbane, Australia, Pp. 24-27.

5- Dhital B.K., Sthapit B.R., Joshi K.D., Pradhanang P.M., Subedi K.D., Vaidya A., Gurung G and Kadayat K.B. (1990). A Report on Maize Research at Lumle Agricultural Center (LAC) and its Off-Station Research (OSR) Site (1989/90) Seminar paper No. 13, Lumle Agricultural Centre, Pokhara, Kaski.

6- Doorenbos J. and Pruitt W.O. (2000). The mechanism of regulation of 'Bartlett' pear fruit and vegetative growth by irrigation withholding and regulated deficit irrigation. Journal of American Society of Horticultural Science. 111: 904.

7- Doorenbos, J. and Pruitt, W.O. (1977). "Guidelines for predicating crop water requirements", FAO Irrigation and drainage paper No.24, FAO, Rome.

8- Gautam, A.K. and Pokhrel, S. (2010). Climate change effects on agricultural crops in Nepal and adaption measures. Presented in Thematic Working Group (agriculture and food security) meeting, Feb 23rd, 2010, Kathmandu, Nepal.

9- Gehl, R.J., Schnidt, J.P., Maddux, L.D. & Gordon, W. B. (2005). Corn yield response to nitrogen rate and timing in sandy irrigated soil. Agron. J., 97:1230-1238. June 28-30, 2004 at NARC, Khumaltar, Nepal, Pp: 216-219.

10- Gurung, G. (2008). Impacts of Climate Change: Some field observations.

11- Hare, F.K. (1993). Climate variation, drought and desertification. WMO, Geneva, Switzerland, Pp. 44.

12- MoAC (Ministry of Agriculture and Co-operatives). (2007). Statistical Information on Nepalese Agriculture (Time Series Information). MoAC, Singha Durbar, Kathmandu, Nepal, Pp. 380.

13- MoE (Ministry of Environment). (2010 a). National Adaption program of action (NAPA) to climate change (report). Ministry of Environment, Kathmandu, Nepal.

14- MoE (Ministry of Environment). (2011). Status of Climate change in Nepal. Ministry of Environment. Government of Nepal.

15- Nayava, J.L., and Gurung, D.B. (2010). "Impact of Climate Change on Production and productivity: A Case study of Maize research and development in Nepal", The journal of Agriculture and Environment Vol: 11, Ministry of Agriculture and Cooperatives, Government of Nepal.

16- Paudyal, D.C., Manandhar, G., Koirala, K.B. (2001). Maize in Nepal: Production Systems, Constraints, and Priorities for Research. Kathmandu: NARC and CIMMYT, Pp. 56.

17- Perry, A. H. (1986). Precipitation and Climate Change in Central Sudan. In: H.R.J. Davies (ed) Rural Development in the White Nile Province, Sudan. A Study of Interaction between Man and Natural Resources, The United Nations University, Tokyo, Pp. 33-42.

18- Practical Action (2009). Temporal and spatial variability of climate change over Nepal (1976-2005). Pratical Action, Kathmandu Office.

19- Rajbhandari, G.R. (1982). A review on various technological aspects of maize. Proceedings of the Ninth Summer Crops Workshop. Parwanipur, Bara, Nepal, Pp. 60-68

20- Shrestha, J., and Timsina, K.P. (2011). "Agronomic Evaluation and Economic Analysis of Winter Maize under Different Plant Population and Nitrogen Rates in Chitwan, Nepal" Nepalese Journal of Agricultural Sciences. Vol. 9, Pp: 5-13.

21- Subramanya, K. (2007). "Engineering Hydrology" Tata McGraw –Hill Publishing Company Limited, New Delhi, India, Pp 369.

22- Tripathi, B.P., and Pathak, L.R. (1984). Response of maize varities to different levels of nitrogen under upland rainfed conditions. Paper presented at the 12th Summer Crops Workshop, Rampur, Chitwan, Nepal.

23- WMO (1994). Guide to Hydrological Practices: Data acquisition and processing, analysis, forecasting and other applications. WMO-No. 168. World Meteorological Organization, Geneva, Pp. 735.

23

Effect of Socio-economic Factors on Fish Catch in Lower Ogun River, Isheri-olofin and Ihsasi, Ogun State, Nigeria

Nafisat Bolatito, Ikenweiwe, Adekule Adedoyin, Idowu, Nathanael Akinsafe, Bamidele, Otubusin Samuel and Eunice Oluremi, Fadipe*

Keywords:
Socio-economic, Fish-catch, Ogun River.

Abstract

This study was undertaken to investigate the effect of socio-economic factors on fish catch around the Lower Ogun River in Isheri-Olofin area (downstream) and Ishasi (upstream). Two hundred and fifty (250) respondents were randomly sampled. The data was analyzed using descriptive statistics and chi-square. It was found that a large percentage (74.4%) owned personal fishing craft. 18% of the respondents have not had any formal education, 38.4% had attended primary school, 42.4% had attended secondary school, and 1.2% had attended a tertiary institution. Majority (62%) of the respondents were married while 31.2 % were single. The result shows that 98.4% of the respondents constitutes male while the remaining 1.6% constitutes female. 70.4 % of the respondents fall within the age bracket of 15–40 years while the remaining 29.6 % are elderly with age range of 41-66 years. It was also discovered that there is a significance decline in the abundance of catch in the downstream compared to the upstream. There is significant difference between the two stations in all socioeconomic factors evaluated.

Department of Aquaculture and Fisheries Management University of Agriculture, PMB2240, Abeokuta, Ogun State, Nigeria
** Corresponding author's email: ikenweiwenb@unaab.edu.ng*

INTRODUCTION

The challenge to increase protein consumption in Nigeria appears to be more urgent now than ever (Mbanasor, 2002). Poor people are facing new barriers in both their production and returns on fish. Even by the standards of developing countries, artisanal fishers and fish workers are often among the poorest people and they generally operate on a small scale and use traditional fishing practices yet new technologies and environment requirement favour large scale capital intensive operation at the expense of traditional and small scale commercial fishing (Delgado *et al.,* 2003).

In Nigeria there are more than 6 million coastal and riverine artisanal fisher folks fishing the 46,300 km2 of maritime area and 125,470.82 km^2 of inland water bodies. They contribute 85% of domestic fish consumption in Nigerian (Fish for All Submit, 2005). About 300,000 coastal migrant fishermen, mostly Ghanaians, also depend on the fisheries resources as the main source of sustenance, assets and investment capital. Fishing supplies 75% of their animal protein intake, and more than 98% of the population of the fishing communities is dependent on fishing and fishery related activities (SFLP, 2002). There is an on-going quest for improved fishing techniques and gears to replace the low yielding traditional fishing methods. Fishermen are interested in new and improved fishing gears such as canoes, buoys, floats, nets, fish chemicals and mechanization with the use of an outboard engine (Watson *et al.,* 2006). The use of poor quality fishing materials limits the catch levels (Tietze *et al.,* 2005). Some of the synthetic nettings are, for instance, very expensive and poor in quality. Low tensile strength netting and slippage - prone single knots result in distorted or irregular mesh sizes especially in the gillnets. The dearth and high cost of fishing gear accessories are other problems and many fishermen often need to seek cheaper local options (AER, 2003). The buoyancy of the multi-various floats and the gravitational force of the cement sinkers are not quantified and the attitude of suspending the gear was more of the guess work than science, compromising on gear efficiency.

The Federal Bureau of Statistics FBS (1992) and Clark *et al.,* (2005) reported that non availability of credit scheme for small-scale capture fisheries militated against its capital-intensive expansion. Only a few financial institutions provide some credit without collateral for "small" loans. Small-scale fisheries are often considered too risky hence most banks do not include them in their credit loan scheme (Clark *et al.,* 2005).

Ogun state, being one of the eight coastal states in Nigeria has about 15 kilometers marine areas and numerous rivers, streams and inland waterways which support varied fishing activities (Adeokun *et al.,* 2006). The abject poverty of the fishermen also forced them to resort to illegal method of fishing with foul fishing gear which destroyed fish breeding grounds and afforded the harvest of immature fishes (Adeokun *et al.,* 2006).

Fish is a vital source of food for people. It is man's most important single source of high-quality protein, providing approximately 16% of the animal protein consumed by the world's population, according to (FAO, 1997). It is a particularly important protein source in regions where livestock is relatively scarce—fish supplies less than 10% of animal protein consumed in North America and Europe, but 17% in Africa, 26% in Asia and 22% in China (FAO, 2000). The FAO estimates that about one billion people worldwide rely on fish as their primary source of animal protein (FAO, 2000).

Fish also has substantial social and economic importance. The FAO estimates the value of fish traded internationally to be US$ 51 billion per annum (FAO, 2000). Over 36 million people are employed directly through fishing and aquaculture (FAO, 2000), and as many as 200 million people derive direct and indirect income from fish (Garcia and Newton, 1997). Consumption of food fish is increasing, having risen from 40 million tonnes in 1970 to 86 million tonnes in 1998 (FAO, 2000), and is expected to reach 110 million tonnes by 2010 (FAO, 1999). Increases in per capita consumption account for only a small portion; it is the growing human population in many countries in Asia, Africa and South America that is primarily responsible for this steadily growing demand for food fish. These data illustrate that a consistent source of fish is essential for the nutritional and financial health of a large segment of the world's population.

Today, fish is the only important food source that is still primarily gathered from the wild rather than farmed—with marine capture historically accounting for greater than 80% of the world's fish supply. Total landings from marine fisheries increased approximately 5-fold in the 40-year period from 1950 to 1990 (Mace, 1997). More recently, however, capture fisheries have not been able to keep pace with growing demand, and many marine fisheries have already been over-fished. In the period 1990–1997, fish consumption increased by 31% while the supply from marine capture fisheries increased by only 9% (FAO, 1999). This has intensified the pressure on the harvesters, which has translated into increased pressures on, and over-fishing of, many commercial fisheries. Nearly half of the known ocean fisheries are completely exploited (FAO, 1999), and 70% are in need of urgent management (MacLennan, 1995).

The general trend in fish supply suggested that domestic production is not increasing at the same rate with yearly increase in demand. The annual demand for fish is 40, 1128 tones against the supply level of 10,561 metric tons (Adekoya and Olunuga, 1999). The contribution of artisanal fisheries is poor due to low output, poor processing mechanism, ineffective distribution and marketing of fishing products and post harvest loss (Adeokun *et al.,* 2006). Post harvest losses according to Akande (1996), is estimated to be at least 40% of the total landings. Opele (2002) found out that these losses occurred as a result of difficulties in fish sorting from artisanal fishing nets, limitations of processing equipment, absence of cold storage on small fishing craft and poor water transport system.

The objective of the study is to determine the socio economic characteristics that could affect fish catch in the study area. Some of the socio economic characteristics of respondents that have been examined include age, sex, marital status, location, educational background, major occupation, minor occupation and source of credit.

MATERIALS AND METHODS

The study was conducted in Ogun State of Nigeria. Ogun State is situated within the tropic region and covers an area of about 16, 369, 37 square kilometers, and it occupies approximately 1.78% of the Country's total land area which is 923,778 square kilometers. It is one of the eight coastal (maritime states of the country, with a 15 kilometers coastline on its South-Eastern part in the Ogun water side Local Government Area. Ogun State is bordered on the South by Lagos State, on the East by Ondo State, in the North by Oyo and Osun States, and on the West by the Republic of Benin.

Essentially, the study was carried out in both Isheri-Olofin and Ishasi areas of lower course of Ogun River before its bifurcations into two tributaries. The entire study area lies between 30 16"E and 30 18"E Longitude, and Latitude of between 60 35"N and 60 38"N. The area is characterized by a typical rainforest climatic condition of rainy season from April to November and dry season from December through March. The dry season, which occurs between November and March of every year, is caused by a dry North-East trade wind coming from Sahara desert and causing harmattan. During these five months of dry season, the fishing operations are usually strenuous a bit due to the scorching of the Sun and the consequent drop in levels of water bodies. The rainy season is caused by the South-Western wind or the equatorial maritime air mass which brings warm and very moist air from the Atlantic Ocean.

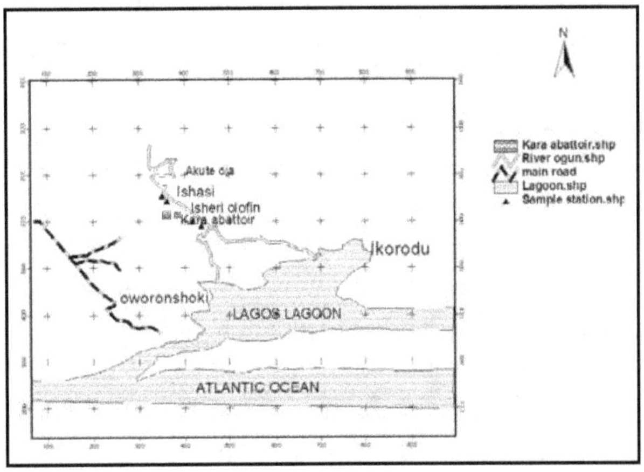

Figure 1: Map of Lower Ogun River

Artisanal fisheries are major activities in Lower Ogun River. At Isheri-Olofin, Lower Ogun River receives effluents from 'Kara' Abattoir which was established in 1984.

Data collection

The data was collected through field survey. A questionnaire was designed to collect a comprehensive profile of socioeconomic conditions of the fisher folk and their fishing and marketing characteristics. Five villages were randomly selected using fifty questionnaire per village. Since most members of the fishing communities are illiterate, a personal interview was conducted. 250 structured questionnaires were designed in order to elicit information from the fishermen and administered to randomly selected respondents from Isheri and Ishasi locations.

The questionnaires are structured in such a way to consists of both open and close ended questions. The open ended questions allow the respondent to express their own opinions about specific situations in order to fulfill the specific objectives of the study. Data collected include socio economic characteristics of the fisher folks in terms of sex, age, level of education, marital status, number of wives and household size, type of primary and secondary occupations, membership of cooperative society. Their fishing characteristics which include; types of fishing craft and gears used, fishing duration, possession of outboard engine, type of fish harvested, reason for fishing, method of disposal of catch, years of fishing experience. Their marketing characteristics which include; peak marketing period, type of labour, method of sale, quantity of fish caught per day, average selling price of fish, average monthly income, method of fish preservation.

Data Analysis

The primary data collected from the field was meticulously entered into the computer for processing and tabulation. Data from the study were analyzed using the following analytical tools:

(a) Descriptive Analysis such as percentages, tables, graphs and bar charts.

(b) Chi-square ($\chi 2$), which was used to test the parameters between the two stations.

RESULTS AND DISCUSSION

An average of 200 cows are slaughtered and butchered at the abattoir on daily basis. The effluents being discharged into the river chiefly contain the gut contents of the slaughtered and butchered cows. Data for this study were obtained from both primary and secondary sources. Primary data sources were those collected from the fishermen while the secondary data were obtained from research reports and also from literatures and other publications of relevance.

The followings are names of some fish species caught by the respondents:

1 *Synodontis nigrita* (Okokoniko)
2 *Heterobranchus longifilis* (Aso)
3 *Citharinus citharus* (Osu)
4 *Clarias anguillaris* (Aro)
5 *Hemichromis bimachulatus* (Epia)
6 *Hepsetus odoe* (Ija)
7 *Hydrocynus forskalii* (Akoko)
8 *Tilapia zilli* (Epia)
9 *Marcusenius senegalensis* (Afintin)

A total of 250 respondents were involved in the study of Isheri-olofin (130), and Ishasi (120) around lower Ogun River. At the end of the data collection the socio-economic characteristics, fishing characteristics and marketing characteristics of the respondents were tabulated in order to see the relationship in terms of frequencies and percentages. Some of the data collected were also displayed in graphical forms. The results of each objective of the study are displayed and explanations given as appropriate.

Table 1 below summarizes the different ages of the respondents. It is observed that the fishing profession is practiced by fisherfolks of different ages. That is, it is not limited to fisherfolks of certain age range. However, the result in Table 6 shows that 70.4 % of the respondents falls within the age bracket of 15–40 years while the remaining 29.6 % are elderly with age range of 41-66 years. This shows that most of the people who get involved in fishing are youth because they have the ability and strength to endure the rigor that is involved in fishing more than the elderly fishermen. It can be concluded that most of the marketers are in their economic active years.

It is indicative of the position of Oladoja (2005) and Adeokun *et al.,* (2002) that many of the fisherfolks who remain in the profession are there because they have difficulty in starting another profession rather than out of sustained interest.

Also, it shows that 98.4% of the respondents constitutes male while the remaining 1.6% con-

Table 1: Distribution of respondents' socio-economic characteristics

Socio-economic characteristics		
Gender	**Frequency**	**Percentage**
Male	246	98.4
Female	4	1.6
Total	**250**	**100**
Age		
15 - 24	78	31.2
25 - 34	69	27.6
35 - 44	47	18.8
45 – 54	35	14
55 – 64	12	7.6
Above 64	2	0.8
Total	**250**	**100**
Educational background		
No formal education	45	18
Primary education	96	38.4
Secondary education	106	42.4
Tertiary education	3	1.2
Total	**250**	**100**
Marital status		
Single	78	31.2
Married	155	62
Divorced	9	3.6
Widowed	8	3.2
Total	**250**	**100**
Household size		
1 – 5	46	18.4
6 – 10	158	63.2
Above 10	46	18.4
Total	**250**	**100**
Religion		
Christianity	135	54
Islam	76	30.4
Traditional	39	15.6
Total	**250**	**100**
Tribe		
Yoruba	144	57.6
Egun	55	22
Hausa	26	10.4
Igbo	25	10
Total	**250**	**100**
Major occupation		
Fishing	133	53.2
Artisan	19	7.6
Trading	27	10.8
Butchery	39	15.6
Transportation	1	0.4
Cattle rearing	9	3.6
Schooling	22	8.8
Total	**250**	**100**
Minor occupation		
Trading	24	9.6
Artisan	14	5.6
Herbalist	1	0.4
Transportation	11	4.4
Cattle rearing	6	2.4
Butchery	16	6.4
Barbing	3	1.2
Labour work	8	3.2
Fishing	117	46.8
None	50	20
Total	**250**	**100**
Credit source		
Personal savings	231	92.4
Cooperative	19	7.6
Total	**250**	**100**
Infrastructure facilities		
Hospital	63	25.2
Schools	56	22.4
Mobile toilet	19	7.6
Water supply	59	23.6
Electricity	10	4
Telecommunication	6	2.4
Community bank	37	14.8
Total	**250**	**100**
Water source		
Well	75	30
Pipe-borne	148	59.2
River	27	10.8
Total	**250**	**100**

stitutes female. This shows that on the general, male are more involved in fishing activities in the study areas than their female counterparts. This implies that both male and female participate in fishing activities, males being dominant. Female are therefore not left out in fishing. This is further supported by Williams and Awoyomi (1998) who observed that women in small-scale riverine fishing villages also perform other types of income-earning activities to supplement the household income. Such income sources were earned through sales of fisheries products and social services in fish distribution and marketing.

The result in Table 1 also reveals that majority (62%) of the respondents were married while 31.2 % were single. This might corroborate the stand that the marriage institution is still cherished and an indication of economic responsibilities of the respondents in caring for their dependents (Jibowu, 1992; Adeyemi *et al.,* 2002). The study also shows that very few (6.8%) of the respondents were either divorced or widowed. The result shows that more of the married men were involved in the fishing activities. This is justified by the fact that majority of the respondents have family members that they have to cater for, hence the more reason why they go into fishing.

The data in Table 1 illustrates that 18% of the respondents have not had any formal education, 38.4% had attended primary school, 42.4% had attended secondary school, and 1.2% had attended a tertiary institution. It can be concluded that most of the most of the respondents had secondary education, which is as a result of the presence of senior secondary schools in the study area. This finding substantiated the findings of Lawal and Idega (2004) who observed that the level of education attended by the respondents to a large extent determine the strategies which he/she may use to solve his/her marketing problems and to adopt new innovations without difficulties that will increase his profit as soon as they became available to him/her.

Table 1 shows the major occupation of the interviewed respondents. The study area is an area where there is diversification of trade just because of the presence of abattoir in the area. 53.2% of respondents are mainly into fishing,

8.8% are students which return to fishing after their school hours and during their holidays, 7.6% said fishing is a minor occupation while artisan is their major work, 10.8% said fishing was a minor occupation while trading was their major work, 15.6% said butchery was a major occupation, 0.4% said transporting was a major occupation, while 3.6% said cattle rearing was their major occupation. One can see from this result that the study areas are largely dominated by fisherfolks who are mainly into fishing profession.

Table 1 shows that 20% do not have a minor occupation, 9.6% said trading was a minor occupation, 5.6% said artisanship was a minor occupation, 46.8%said fishing was a minor occupation, 0.4% said herbalism was a minor occupation, 4.4% said transporting was a minor occupation, 2.4% said cattle rearing was a minor occupation, 6.4% said butchery was a minor occupation, 1.2% said barbing was a minor occupation, while 3.2% said labour work was a minor occupation.

Table 1 shows the responses of the respondents based on credit source. 92.4% said their credit source for fishing activities is through personal savings while 7.6% said it is through cooperatives. It can be said thus that those who source for income from co-operatives are likely to get better yield than their fellow counterparts because they readily buy equipments needed for fishing activities.

Table 2 above shows that most of the respondents are into commercial fishing. It is revealed that those into commercial fishing earn more income than the others.

Table 2 also shows that 77.6% said the reason for fishing is to make profit, while 22.4% said it is for home consumption. It is observed that not all the respondents are into fishing mainly because of profit making as a result there is decline in the fish caught in the area.

Table 2 above indicated that 25.2% said their method of disposal of catch is through middlemen, 52.8% said it is directly to the consumers, while 22% said they do not dispose it at all. Table 2 above shows that most of the respondents are into fishing mainly because of making profit. It is revealed that those who went for fishing early in the morning make more profit than those who went late in the evening.

From the result in Table 3, 58.8% of respondents said the peak marketing period is between April and June while 41.2% said it is between

Table 2: Distribution of respondents' fishing characteristics
Result on fishing characteristics

Type of fishing		
Commercial	171	68.4
Subsistence	79	31.6
Total	**250**	**100**
Reason for fishing		
For home consumption	56	22.4
To make profit	194	77.6
Total	**250**	**100**
Method of disposal of catch		
Through middlemen	63	25.2
Direct to consumers	132	52.8
None	55	22
Total	**250**	**100**
Years of fishing experience		
1 – 5 years	107	42.8
6 – 10 years	143	57.2
Total	**250**	**100**
Fishing period		
Morning	86	34.4
Evening	115	46
Both	49	19.6
Total	**250**	**100**
Number of fishing days		
1 - 4 days	18	7.2
5 - 8 days	232	92.8
Total	**250**	**100**
Migratory fisherfolks		
Yes	186	74.4
No	64	25.6
Total	**250**	**100**
Fishing duration		
4 – 6 months	100	40
7 – 9 months	102	40.8
Above 9 months	48	19.2
Total	**250**	**100**
Type of fishing gear used		
Cast net and spear	93	37.2
Hook and line	64	25.6
Lift net and drag net	40	16
Gill net and Gura trap	53	21.2
Total	**250**	**100**
Type of fishing craft used		
Dug out canoe	123	49.2
Ghana made canoe	50	20
Out board engine	13	5.2
None	64	25.6
Total	**250**	**100**
Age of craft		
1-4 years	127	50.8
5-8 years	59	23.6
None	64	25.6
Total	**250**	**100**
Ownership of craft		
Bought	162	64.8
Inherited	30	12
No canoe	58	23.2
Total	**250**	**100**

July and September of each year.

The result in Table 3 also shows that 51.6% of the respondents use family labour as the type of labour used in fishing activities and as a result of this, little or no expenses is incurred during fishing activities. 18% use hired labour and 30.4% use self labour. It is also revealed from the study that only those that uses out board engine only spend money on the fueling.

Table 3: Distribution of respondents marketing characteristics
Result on marketing characteristics

Peak marketing period		
April – June	147	58.8
July – September	103	41.2
Total	**250**	**100**
Time of Sale of Catch		
After harvest	163	65.2
After storage	87	34.8
Total	**250**	**100**
Type of labour used		
Family labour	129	51.6
Hired labour	45	18
Self labour	76	30.4
Total	**250**	**100**
Method of selling fish		
In baskets	164	65.6
Singly	26	10.4
None	60	24
Total	**250**	**100**
Average monthly income		
₦1000 – 20000	109	43.6
₦21000-40000	125	50
₦41000-60000	13	5.2
Above ₦60000	3	1.2
Total	**250**	**100**
Membership of cooperatives		
Yes	167	66.8
No	83	33.2
Total	**250**	**100**
Benefit derived		
Taking loans	98	39.2
Access to middlemen	1	0.4
Control of price	69	27.6
Not a member	82	32.8
Total	**250**	**100**
Problems faced in fishing		
Storms	35	14
Post harvest losses	19	7.6
Lack of capital	33	13.2
Harsh weather	23	9.2
Over exploitation	27	10.8
Dangerous aquatic animal	16	6.4
No modern equipments	27	10.8
Flooding	28	11.2
Water borne disease	24	9.6
Inadequate storage facilities	16	6.4
Poverty	2	0.8
Total	**250**	**100**
Method of preservation		
Smoking	116	46.4
Sun drying	70	28
Icing	20	8
Salting	44	17.6
Total	**250**	**100**

Table 3 reveals that 65.6% of the respondents sell their produce in baskets, 24% sell singly while the remaining 10.4% do not sell their catch at all. Those that do not sell are those that go into fishing simply for home consumption.

The result in Table 3 indicates that 66.8% of the respondents belong to cooperative association while 33.2% does not belong to any fishermen association. The larger percentage recorded from those that belong to an association is largely due to the benefits they derive from such an association.

From the chi-square analysis, it was discovered that there is no significant difference (p=0.00 and 0.002) between the two stations in their type of fishing. There is significant relationship between the two stations in their reasons for fishing (p=0.00). There is significant difference between the two stations in their fishing period (p=0.00). There is significant difference between the two stations in their migratory fishing activity (p = 0.00). There is no significant difference between the two stations in their fishing duration per annum.

There is no significant difference between the two stations in their fishing gear and mesh size. There is significant difference between the two stations in their average monthly income (p = 0.00). There is no significant difference between the two stations in the type of labour used for their fishing activities. There is significant difference between the two stations in the size of labour required for their fishing activities. There is no significant difference between the two stations in the number of fish caught per day. There is significant difference between the two stations in the selling price of fish (p=0.00). There is no significant difference in the preservation method between the two stations.

CONCLUSION

It can be deduce that the various activities around the water body are the reasons why there is a decline of fish catch in study area. A toilet is located directly towards the edge of the water body which releases dark and brown water into

Table 4: Chi Square analysis of the two stations

	Chi-square	df	p
Pearson Chi-square	0.6321759	1	.42656
M-L Chi-square	0.6332872	1	.42615
Yates Chi-square	0.4342131	1	.50993
Fisher exact, one-tailed			.25516
two-tailed			.49629
McNemar Chi-square (A/D)	20.66116	1	.00001
(B/C)	12.4031	1	.00043
Pearson Chi-square	13.02492	3	.00458
M-L Chi-square	13.32676	3	.00398
Pearson Chi-square	31.86423	16	.01042
M-L Chi-square	36.79567	16	.00224

the water. Blood from slaughtered cows are allowed to run freely and the remnants are washed directly into the water body. Also, along the side of the river are those processing the blood which is used as blood meal for producing livestock feed. All the waste from these activities are washed directly into the water thereby polluting the entire water body.

More so, further observation shows that the respondents were using fishing gears with small mesh sizes to catch fish which also affect the amount of fish catch in the study area. It was also observed that the number of fishing effort is high. This is as a result of the large population around the water body.

The study has shown that majority of the respondents had their fishing gear and craft either by inheritance or purchase; this will therefore enable them to adopt better production practices if a government service is extended to the people. Funding was also found to be through personal savings by the majority of the respondents. However, since a large percentage of the respondents are members of one association or the other, assistance in terms of funding could be better channeled through the associations. The socio-cultural characteristics showed that no fewer than four ethnic groups inhabited the study area as a result of the abattoir in the area. Other migrants groups were also found living together harmoniously. However, the there is a continuous increase in the load of debris of human and abattoir waste being deposited in the water body.

RECOMMENDATION

From the study carried out so far on the socioeconomic lives of fisherfolks in Isheri and Ishasi area of Lower Ogun River, South West, Nigeria, the followings are therefore recommended:

1. Further research should be carried out to determine the physico-chemical parameters of the water body.
2. Associations existing among the respondents should be strengthened by the government and non-governmental organizations.
3. Government and corporate bodies should come to the aid of people living in these communities in order to solve the numerous problems encountered in the study areas.

Finally, considering the ecological and socioeconomic importance of fish, it is imperative to put in place measures to adequately regulate the discharge of abattoir effluents into the natural water bodies.

REFERENCES

1- Adekoya, B.B. and A.O. Olunuga, (1999). A proposal on farm Adaptive Research trial on cage culture in Ogun State, Nigeria OGADEP, Adeokuta. Paper presented to the preliminary meeting on fish cage culture, Ogun and Oyo States held at OGADEP Abeokuta on the 27th July 1999. pp: 8-11

2- Adeokun O.A., F.O. Adereti and A.I. Opele (2006). Factors Influencing Adoption of Fisheries Innovations by Artisanal Fishermen in Coastal Areas of Ogun State, Nigeria. Journal of Applied Sciences Research, 2(11): 966-971.

3- AER, (2003). Concerted Action: Co-ordination of Research in Fishery Economics and Promotion of Common Methods for Economic Assessment in EU Fisheries. Economic performance of selected European fishing fleets. Annual Report. 289 p.

4- Aguero, M., (1987) Economic Consequences of Excessive effort, in Indo Pacific Fishery Commission Symposium on the Exploitation and Management of Marine Fishery Resources in Southeast Asia, IPFC/87/Symp/IV/WP.3,16-169, 1987.

5- Ajayi OY (2001). Economics of Fish Farming in Remo and Ijebu Division of Ogun State, Nigeria. B. Agric. Dissertation, Unpublished. Department of Agricultural Economics and Farm Management, Ogun State University.

6- Akande, G.R., (1996). Post-harvest processing in fisheries". A paper presented at training for officers of UNDP assisted programme on artisanal fisheries development Ogun State at Federal College of fisheries marine technology Lagos. Jan. 15th – 5th Feb. 1996. pp: 7.

7- Akanni, K.A. (2008). Catch Levels and Capital Investment of Artisanal Fishermen in Lagos State, Nigeria. Turkish Journal of Fisheries and Aquatic Sciences 8: 361-368.

8- Akinyemi, O. (1998). Sustainable Management of Nigeria Fisheries in the 21st Century. Faculty of Agriculture and Forestry. Ibadan: University of Ibadan.

9- ALCOM (1994) Background Report on Fish Farming in Morogoro, Tanzania. ALCOM, Morogoro, P: 37.

10- Anthonio, O.R. and Akinwumi, J.A. (1991). Supply and distribution of fish in Ibadan, Nigeria. Geog. Journal, 14 (2): 16.

11- Central Bank Of Nigeria (CBN), 2000. Statistical Bulletin, Vol: 10.

12- Clark, C.W., Munro, G. and Sumaila, U.R. 2005. Subsidies, buybacks, and sustainable fisheries. Journal of Environmental Economics and Management, 50: 47- 53.

13- Cleaver K.M, Creiber, G.A. (1994). Reversing the Spiral, Population, and Agriculture and Environment nexus in sub Saharan Africa. World Bank, Washington. D.C.

14-VDelgado, C.L., Wada, N., Rosegrant, M. W., Meijer, S. and Ahmed, M. (2003); The Future of Fish: Issues and Trends to 2020, International Food Policy Research Institute, Washington DC, USA, 19: 1-22.

15- Economic Survey, (2000-01). Government of Pakistan, Finance Division, Economic Advisor's Wing, Islamabad–Pakistan

16- Elliot, O. (1993) Nigerian Coastal Fisheries Resources in Proceedings of the National Conference on Conservation of Aquatic Resources Organized by the National Advisory Committee on Conservation of Renewable Resources in Cooperation with the Federal Department of Fisheries (FDF), Abuja and the Nigerian Institute for Oceanography and Marine Research (NIOMR), Lagos, Nigeria Egborge, A. B. M., O. J. Omoloyin, A. Olojede, and S. A. Manu, eds, 1993.

17- FAO (1991). Fish for Food and Employment. Food and Agricultural Organization, Rome.

18- FAO (1995). Yearbook Production. Vol 45. Rome: FAO

19- FAO, (2009). World Review of Fisheries and Aquaculture section of the leading scientific consensus report produced in 2009 by the Food & Agriculture Organization (FAO):"The State of World Fisheries and Aquaculture.

20- FAO, World Bank (1996). Soil fertility initiative. Food and Agriculture organization of the United Nations, Rome.

21- FBS, (1992). Fishery Baseline Survey. Lagos State Agricultural Development Project, LSADP 34 – 79

22- FDF (2005). Report of Presidential Committee on Fisheries and Aquaculture Development.Volume one: Consolidated Report. Federal Department of Fisheries, Federal Ministry of Agriculture and rural Development. September (2005).

23- Fish for All Summit (2005) Presentation of Nigeria's Fisheries Resources, Abuja, Nigeria 22-25 August 2005

24- Haakonsen M. J. (1992) West African Artisanal Fisheries in Fishing For Development: Small – Scale Fisheries in Africa, Tredten, I. and B. Hersoug, eds. The Scandinavian Institute of African Studies, Uppsala, Motala, Grafisca AB Motala.

25- Hague M (1992). The Economic and Feasibility of Aquaculture in Northern Tanzania. B.A. dissertation, University of Stirling, Scotland p.40

26- Ijaz, K., (1995). Sociology of Maternity and Child Health: some observations in the context of Pakistan, Department of Rural Sociology, University of Agriculture, Faisalabad–Pakistan

27- Jinadu O. O. (1994). Small-scale Fisheries In Lagos State, Nigeria: Economic Sustainable Yield Determination. Federal College of Fisheries and Marine Technology, Wilmot Point, Victoria Island, Lagos, Nigeria.

28- Kusemiju, K. (1993) Development, Utilization and Conservation of Aquatic Resources in Nigeria, in Proceedings of the National Conference on Conservation of Aquatic Resources, eds. Egborge A. B.

M.; O. J. Omoloyin, A. Olojede, and S. A.Manu, Organised by National Advisory Committee on Conservation of Renewable Resources In Cooperation with Federal Department of Fisheries, Abuja and the Nigerian Institute for Oceanography and Marine Research, Lagos pp: 313.

29- Manyong, V.M., J. Smith, K. Weber, S.S. Jagtab and B. Oyewole, (1996). Macro Characterization of Agricultural System in West Africa; An overview in Resource and Crop management Research. Monograph. 21: 9-12.

30- Matata P. Z., O. C. Ajay, P. A. Oduol and A. Agumya (2010). Socio-economic Factors Influencing Adoption of Improved Fallow Practices Among Smallholder Farmers in Western Tanzania. African Journal of Agricultural Research 5 (8): 818-823.

31- Mbanasor J.A. (2002), Resource Use Pattern among Poultry Enterprises in Abia State Nigeria. Nigeria Journal of Animal Production, 29 (1): 64-70.

32- NEPAD (2001). The New Partnership for Africa's Development. Printed by McDarnells Press, Abuja, Nigeria for NEPAD, Nigeria. 59 pp.

33- OchaiI, S., (1995). An Integrated production consumption Analysis of small scale Farming Households in Kogi State. Unpublished Ph.D. Dissertation Department of Agricultural Economics, University of Ibadan, Nigeria.

34- Ogunfiditimi, T.O., (1983). A new approach to food Distribution in Nigeria Nutrition and food policy and Strategic Studies, pp: 105-108.

35- Ogunsumi L. O. (2007). Socio-economic Conditions of Peasant Farmers: The Case of Agricultural Technologies' Sustainability in Southwest Nigeria. World Journal of Agricultural Sciences, 3 (5): 678-684.

36- Ogunsumi, L.O., (2004). Analysis of Sustained use of Agricultural Technologies on Farmers' Productivity in Southwest Nigeria, Unpublished Ph.D. Thesis, Federal University of Technology, Akure, Nigeria.

37- Ojo, S.O. Fagbenro O.A. (2004). Poverty Reduction Strategy in Nigeria –Improving productivity and Technical Efficiency in Artisanal Fisheries in Niger Delta Region. Paper presented at the 12th Bi-annual conference of the International Institute of Fisheries Economics and Trade (IIFET), Tokyo, Japan.

38- Olagunju, F. I., Adesiyan I. O. and Ezekiel A.A. (2007). Economic Viability of Cat Fish Production in Oyo State, Nigeria. J. Hum. Ecol., 21(2): 121-124 (2007)

39- Olayide, S.O., (1980). Characteristic, problems and significance of farmers. Nigerian small farmers-Problems and Prospects in Integrated Rural development. Edited by S. O. Olayide *et al.,* Centre for Agricultural and Rural Development (CARD). University of Ibadan Nigeria, pp: 1-15.

40- Opele, A.I., (2002). "Factors influencing Adoption of innovations in artisanal fisheries in Ogun State" An unpublished M.SC. Thesis, Department of Agricultural Extension and Rural Sociology Olabisi Onabanjo University, Ago Iwoye, Nigeria, pp: 35.

41- Ovie S. I. and Raji A. (2006). Fisheries co-management in Nigeria: an analysis of the underling policy process. WorldFish Center, Africa and West Asia Programme PO Box 1261 Maadi 11728 Cairo – Egypt December 2006

42- Owolabi, O.O. (2002). Wildlife contribution to meat protein consumption in the rural community in south-western Nigeria. Proceeding of the 27th annual Conference Nigeria Society for Animal Production (NSAP) March 17 -21, Akure.

43- Pauly, D. (1984) Fish population dynamics in tropical waters: a manual for use with programmable calculators, ICLARM Studies and Reviews 8, 325, (1984).

44- Pauly, D. (1990) On Malthusian Overfishing, NAGA ICLARM Quarterly 3-4, January (1990).

45- Pauly, D. (2006). Major trends in small-scale fisheries, with emphasis on developing countries and some implications for the social sciences. Maritime Studies (MAST), 4: 7-18

46- Sandhu T. A., Chaudhry A., Akbar N. and I. Ahmad (2005). Effect of Socio-Economic Factors on the Female Education in Rural Areas of Faisalabad (Pakistan). Journal of Agriculture and Social Sciences 1813–2235/2005/01–1–41–42 http://www.ijabjass.org.

47- Slang, V.C. (1973). Comparison on the Economic Potential of Agricultural Land, Animal Husbandry and Oceans Fisheries, the case of Aiwa Agriculture. F.A.O Technical Conference on Aquaculture, Kyoto, Japan.

48- Stephen, K.H., (1993). Educational Psychology. The study of educational growth. The John Hopkins Univ. New York.

49- Sustainable Fisheries Livelihoods Programme (2002) Contribution of Fisheries Research to the Improvement of Livelihoods in West African Communities, Case Study: Nigeria, SFPL/DFID-FAO 59p

50- Tietze, U., Lash, R., Thomsen, B. and Rihan, D. (2005). Economic Performance and Fishing Efficiency of Marine Capture Fisheries. FAO Fisheries Technical Paper. No 482. Rome, 68 pp.

51- Vanlauwe, B., Giller K.E. (2006). Popular myths around soil fertility management in sub-Saharan Africa. Agriculture, Ecosystems & Environment. 116: 34- 46.

52- Watson, R., Revenga, C. and Kura, Y. (2006). Fishing Gear Associated with Global Marine Catches: Database Development. Fisheries Research 79: 97-100.

53- Weischet, A.O. and Cariedes C., (1993). 'Research Transaction Cost Economics the Governance of Contractual Relations; Journal of Law and Economics 22: 233-261.

54- Wetengere K. (2009). Socio-economic Factors Critical for Adoption of Fish Farming Technology: The Case of Selected villages in Eastern Tanzania. International Journal of Fisheries and Aquaculture. 1 (3): 028-037.

24

Factors Affecting the Technical Efficiency Level of Inshore Fisheries in Kuala Terengganu, Malaysia

Nurul Aisyah[1], Nalini Arumugam[1], Mohd. Ariff Hussein[2] and Ismail Latiff[2]*

Abstract

The objective of this study is to determine factors affecting the technical efficiency of the inshore fisheries in Kuala Terengganu. Data for the study was collected from a survey conducted between June and August 2007 where 100 fishermen in 14 villages were chosen by stratified sampling. Data envelopment analysis (DEA) and Tobit analysis were employed to determine the technical efficiency level and factors influencing technical efficiency among the fishermen. Results of the study show that, most fishing units exhibit a low degree of technical efficiency. This implies that either fishing inputs were used inefficiently or insufficient inputs were used in fishing activities. The mean technical efficiency for the sample was estimated to be 55% for the peak season and 40% for the non peak season. About 37% and 62% of the fishermen had less than 40% level of technical efficiency in peak season and non peak season respectively. Management variables (planning, staffing and controlling) and demographic variables (size of horsepower, size of family and formal education) exert positive effects on technical efficiency of inshore fisheries in Kuala Terengganu. These findings suggest that there is much room for improvement in efficiency among a large segment of the inshore fishermen. With appropriate training and using more advanced technologies, fishermen' level of technical efficiency can be raised.

Keywords:
Fishermen, Technical efficiency, Data envelopment analysis.

[1] *Faculty of Agriculture and Biotechnology, Universiti Sultan Zainal Abidin, Kampus Gong Badak, 21300 Kuala Terengganu, Terengganu. Malaysia.*

[2] *Department of Agribusiness and Information Technology, Universiti Putra Malaysia, 43400 Serdang, Selangor Darul Ehsan. Malaysia.*

* *Corresponding author's email: nalini@unisza.edu.my*

INTRODUCTION

From ancient times, fishing has been a major source of food for humanity and a provider of employment and economic benefits to those engaged in this activity. The marine capture fisheries in Malaysia and in Terengganu have developed from an inshore traditional fishery to its present mix of inshore traditional and commercial, and deep-sea fishery sub-sectors. Although the bulk of the marine fish landings come from the inshore commercial and the deep sea fishery sub-sectors, the inshore traditional fishery is no less important. A number of factors are believed to have contributed to the inefficiency of the inshore traditional fishery such as vessel size, age of fishermen and vessel horsepower.

In the year 2009, the fisheries sector contributed 3.4% to the Malaysia GDP. Terengganu is one of the Malaysian state which is situated in north-eastern Penisular Malaysia and, is bordered in the east by the South China Sea. In Terengganu, production from marine captures fisheries in the year 2007 contributed 81,007 tonnes (5.86%) of the nation's fish production valued at RM 384million. It also provides employment to 8,651 (5,884 are local workers and 2767 are foreign workers) fishermen who work on 2,422 units of licensed fishing vessels. Total landings in Terengganu decreased by 16.50% from 111,394 million in 2006 to 81,007 million tonnes in 2007. The decline of capture fisheries in Terengganu is a symptom of many complex problems that have no easy solutions. In general, problems and constraints faced by small scale fisheries in most countries in the region are similar, only varying from one country or village to another in terms of local importance. In contrast to large-scale commercial fisheries, inshore fisheries are owner-operated and labor intensive, employing rudimentary technologies. Inshore fisheries harvest the sea from comparatively small vessels, powered by sail, paddles or outboard motors of limited power, have limited fishing range and generally deploy passive fishing gears that are set and later retrieved (Squire and Hj Omar, 1998). In Terengganu, about 1909 of the fishermen working on licensed vessels used traditional fishing gear and the dominant gear type used is hooks and lines.

In the Eight Malaysia Plan (2001-2005), marine fisheries achieved only 0.6% in average annual growth rate from targeted 5.9%. It means the fisheries subsector recorded a positive but slow growth rate. The measurement of efficiency has long been a subject of study for many economists and hence is of great interest to policy planners. It is important as a first step in identifying the process for resource saving and how to increase productivity. In an economy, where resources are scarce and opportunities of new technologies are lacking, efficiency study will be able to show the possibility of raising productivity by improving efficiency without increasing the resource base or developing new technologies. A firm is technically efficient if it produces a higher level of output as compared to another firm at the same level of input, Yotopoulos and Lau (1973).

The Fisheries Comprehensive Licensing Policy (FCLP) divides Malaysian fishing waters into four zones:

Zone A: 0-5 miles from shore, reserved for traditional fisheries;

Zone B: 5-12 miles from shore, for commercial fisheries that uses gear such as trawls and purse-seines below 40 GRT (Gross Registered Ton);

Zone C: 12-30 miles from shore, for commercial fisheries that uses boats above 40 GRT;

Zone C2: 30 miles from the shore and beyond, for commercial fisheries that uses boats 70 GRT and above.

This is to protect the juveniles of fish that are concentrated in the inshore waters from the intensive fishing pressure of these commercial gears. This study was focused on fishermen from Zone A only which is the vessel can go to 0-5 miles from shore and this zone only reserved for traditional fisheries such as hook and lines.

MATERIALS AND METHODS

Source of data

The data collection was carried out from June to July 2007. The data collected were based on fishing activities from January until December 2006. 14 villages in Kuala Terengganu were selected for this study. Survey respondents consisted of 100 fishermen engaged in inshore fisheries in Kuala Terengganu were chosen

recommendation for performance improvement (Fukuyama, 1993; Grabowski *et al.,* 1994). Non-parametric analysis also avoids the problem arising from multicollinearity among variables (Elyasiani and Mehdian, 1993).

Model specification

The methodology used in this study is a two-stage process commonly used in the literature. The efficiency scores are estimated in the first stage and then the scores are used as dependent variables in the second stage. In the first stage, the data envelopment analysis (Charnes *et al.*, 1978) or DEA is used to estimate the level of technical efficiency. For the second stage, Tobit regression is used to find the factors that influence the technical inefficiency.

In establishing the model specification, data envelopment analysis was adopted and tested. DEA is multi-factor productivity analysis model for measuring the relative efficiencies of a homogenous set of decision making units (DMUs). The general form of the model is:

$$\text{Efficiency} = \frac{\text{weighted sum of output}}{\text{weighted sum of inputs}} \quad (1)$$

Assuming that there are n DMUs, each with **m** inputs and **s** outputs, the relative efficiency score of a test DMU p is obtained by solving the following model proposed by Charnes *et al.*, (1978):

$$\text{Maximize} \quad Z_0, \frac{\sum_{k=1}^{s} v_k y_{kp}}{\sum_{j=1}^{m} u_j x_{jp}}$$

$$\text{Subject to} \quad \frac{\sum_{k=1}^{s} v_k y_{ki}}{\sum_{j=1}^{m} u_j x_{ji}} \leq 1 \forall i$$

$$v_{kv}, \quad u_j \geq 0 \quad \forall k, j, \qquad (2)$$

Where
$k = 1$ to $\boldsymbol{s}$,
$j = 1$ to $\boldsymbol{m}$,
$i = 1$ to $\boldsymbol{n}$,
y_{ki} = amount of output $\boldsymbol{k}$ produced by fishermen $\boldsymbol{i}$,
x_{ji} = amount of input j utilized by fishermen $\boldsymbol{i}$,
v_k = weight given to output (catch per kg) k,
u_j = weight given to input (GRT, no of worker, distance from shore and expenditure per trip) j.

The fractional program shown as (2) can be converted to a linear program as shown in (3).

$$\text{Maximize} \quad Z_0, \sum_{k=1}^{s} v_k y_{kp}$$

$$\text{Subject to} \quad \sum_{j=1}^{m} u_j x_{jp} = 1$$

$$\sum_{k=1}^{s} v_k y_{ki} - \sum_{j=1}^{m} u_j x_{ji} \leq 0 \qquad \forall i$$

$$v_k, u_j \geq 0 \qquad \forall k, j$$

(3)

The above problem is run n times in identifying the relative efficiency scores of all the DMUs. Each DMU selects input and output weights that maximize its efficiency score. In general, a DMU is considered to be efficient if it obtains a score of 1 and a score of less than 1 implies that it is inefficient.

Variables used in the DEA Model divided into two which is the dependent variable and independent variable. The dependent variable is total output (catch) in kilogram per month. While, for independent variables are vessel capacity measured by Gross Registered Tonnage (GRT), number of crew employed per vessel for fishing trip, including captain, input usage (diesel/petrol, lubricant and oil, ice, food and miscellaneous variable inputs) and distance traveled measured by nautical miles from shore line.

For the second stage, Tobit regression was used to determine which factors influence efficiency, using a log-linear form of the model. Tobit regression is an alternative to ordinary least squares regression (OLS) and is employed when the dependent variable is bounded from below or above or both, with positive probability pileup at the interval ends, either by being censored or by being corner solutions.

As DEA is a non-parametric method, no econometric tests are available to gauge the appropriateness of the estimated frontier and efficiency scores. In a probit model the variable of theoretical interest, y*, is unobserved; what is observed is a dummy variable, y, which takes on a value of 1 if y*i is greater than 0, and 0 otherwise. In contrast, devised what became known as the Tobit (Tobin's probit) or censored normal regression model for situations in which y is observed for values greater than 0 but is not observed (that is, is censored) for values of zero or less.

The standard Tobit model is defined as

$$y^*_i = x_i\beta + \epsilon_i$$

$$y_i = y^*_i \quad \text{if} \quad y^*_i > 0$$

$$y_i = 0 \quad \text{if} \quad y^*_i \leq 0 \qquad (1)$$

where y*i is the latent dependent variable, yi is the observed dependent variable, xi is the vector of the independent variables, β is the vector of coefficients, and the Єi's are assumed to be independently normally distributed: Єi ~ N(0,σ) (and therefore yi ~ N(xiβ,σ)). Note that observed 0's on the dependent variable can mean either a "true" 0 or censored data. At least some of the observations must be censored data, or yi would always equal y*i and the true model would be linear regression, not Tobit.

Tobit is chosen by assuming that the concentration of the dependent variable clusters toward the left limit (i.e., zero) and because it does not only explain the value of the dependent variable or the probability of limit (e.g. point of technical efficiency) and non-limit (e.g. points of technical inefficiency) responses, but also the size (e.g. value) of non-limit responses (Tobin, 1958). Tobit regression is an alternative to ordinary least squares regression (OLS) and is employed when the dependent variable is bounded from below or above or both, with positive probability pileup at the interval ends, either by being censored or by being corner solutions. In the former case (censored) observations outside the limiting interval are recorded as the border values. That is if the range is given by the interval [a;b], observed y < a is recorded as y = a, and likewise observed y > b is recorded as y = b. In the latter case (corner solutions) the observations are by nature limited from below or above or both with a positive probability at the 'corners' (interval ends). DEA scores are limited to the interval [0;1] and accordingly only has a positive probability to attain one of the corner values. As such the two-limit tobit model, which is generally used to model censored or corner solution data limited both from below and above, is necessarily a mis-specification when applied to DEA scores, as this method requires a positive probability to attain both corner values.

When the dependent variable y is limited to the interval [0;1] it may be described by the model:

$$y_i^* = \sum_{k=1}^{n} \beta_k x_k + u_i, \qquad (1)$$

$$y_i = \left[\frac{1+sign\,(y*_i)}{2}\right].\min(1, y_i^*),$$

$$\text{sign}\,(y_i^*) = \begin{cases} 1; & y_i^* \geq 0, \\ -1; & y_i^* < 0, \end{cases}$$

where u_i ~ N (0, σ) are independent and identically normal distributed (iid.) residuals of the observations and x = (x_i, ..., x_n) is the vector of explanatory variables. Thus

$$y = \begin{cases} y^*; 0 \leq y^* \leq 1 \\ 0;\ y^* < 0; \\ 1; 1 < y^* \end{cases}; \quad y^* = \sum \beta_k x_k + u. \qquad (2)$$

The probability that a recorded y is equal to 0 is given by

$$P\,(y = 0) = F(-\Sigma\beta_k x_k |\ 0, \sigma)$$

$$= \frac{1}{\sqrt{2\pi\sigma^2}} \int_{\infty}^{-\Sigma\beta_k x_k} e^{-t^2/(2\sigma^2)}\, dt \qquad (3)$$

Given the basic iid. assumption for the residuals

u. F ($x|$ μ, σ) is the density function. Likewise:

$$P(y=1) = F(-(1-\Sigma\beta_k x_k)|\ 0, \sigma)$$

$$= \frac{1}{\sqrt{2\pi\sigma^2}} \int_{-\infty}^{-(1-\Sigma\beta_k x_k)} e^{-t^2/(2\sigma^2)} dt. \tag{4}$$

Finally when the recorded y is between 0 and 1 the probability to observe y is given by

$$P(y_i|0 < y_i < 1) = f\left(y_i - \sum \beta_k x_k \left|0, \sigma\right.\right)$$

$$= \frac{1}{\sqrt{2\pi\sigma^2}} \int_{-\infty}^{-(1-\Sigma\beta_k x_k)} e^{-t^2/(2\sigma^2)} dt. \tag{5}$$

where f($x|$ μ, σ) is the N (μ,σ) frequency function. Thus the combined likelihood function for the recorded censored dataset is given by

$$L = \prod_{y_i=0} P(y_i = 0) \prod_{y=1_i} P(y_i = 1) \prod_{0<y_i<1} P(y_i|0 < y_i < 1) \tag{6}$$

This shows why the two-limit tobit model is mis-specified when applied to DEA scores, as the scores only have a positive pileup at the right hand side of the interval [0;1]. As none of the scores are equal to zero the first multiplication in (6) will be left out.

RESULTS

Data Envelopment Analysis

The technical efficiency represents the degree of success to produce maximum output from given levels of inputs. One percent technical efficiency means that the fishermen could have produced one percent more output from existing level of inputs. Range 80% and above are indicated as the most efficient level of technical efficiency. While 40% and below are categorized as a very low efficiency. For the peak season, the technical efficiency for individual fishermen ranged from 0.073 to 1 with a mean of 0.5469. The most striking result obtained by the study is the variation of technical efficiency by season. Most vessel are technically inefficient (below 0.4) during peak season. In this season, there are two regimes, low efficiency (below 0.4) or very high efficiency (above 0.8) figure 1.

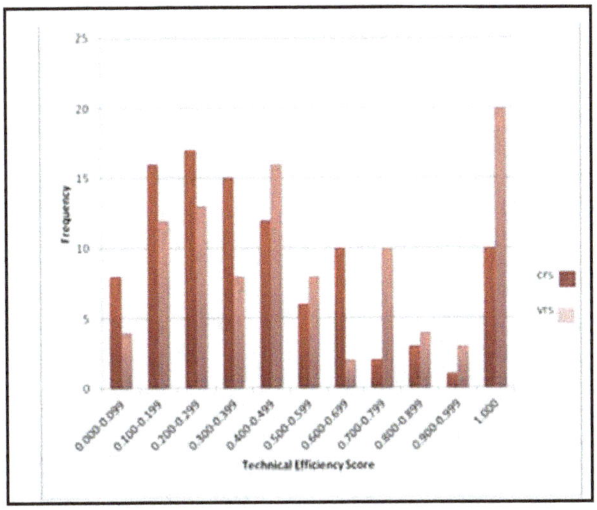

Figure 1: Frequency distribution of technical efficiency index for peak season.

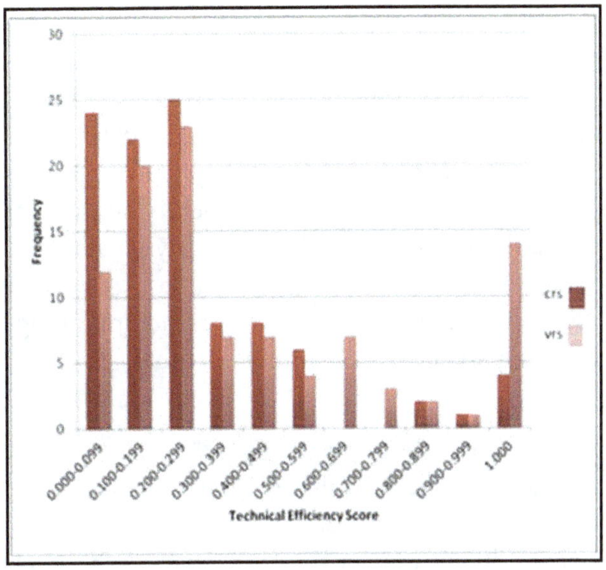

Figure 2: Frequency distribution of technical efficiency index for non peak season.

During the non peak season, there are also two efficiency regimes that are slightly different from each other. The lower level regime (below 0.4) dominates the higher level regime (above 0.8) giving a very low overall efficiency (figure 2). Technical efficiency for each individual fisherman ranged from 0.025 to 1 with a mean of 0.401.

Table 1: Estimated technical inefficiency function for the peak and non peak season

	Peak Season			Non Peak Season		
Variables	Coefficient	Std. Error	Prob.	Coefficient	Std. Error	Prob.
Intercept	0.241384	0.589132	0.6820	0.309043	0.607469	0.6109
Age of fishermen	-0.001428	0.003739	0.7025	0.005306	0.004276	0.2146
Horsepower	-0.530796	0.186765	0.0045	-0.320788	0.073901	0.0303
Trip	0.004026	0.005539	0.4673	0.000522	0.005873	0.9292
Household	0.001221	0.011453	0.9151	-0.031021	0.012403	0.0344
Ownership	0.039246	0.116423	0.7360	0.108585	0.116797	0.3525
Education	-0.040293	0.092198	0.0421	-0.130203	0.103580	0.2087
Management variables						
Planning	-0.028901	0.012684	0.0227	-0.036081	0.014199	0.0110
Organizing	-0.001886	0.012711	0.8821	-0.014900	0.014267	0.2963
Staffing	-0.004620	0.016805	0.7834	-0.037948	0.018743	0.0429
Directing	-0.005902	0.018307	0.7472	-0.013732	0.019389	0.4788
Controlling	-0.008175	0.015851	0.6060	-0.035009	0.016630	0.0353

Tobit Regression

Table 1 provides the estimated technical inefficiency function, where the dependent variable is technical inefficiency as opposed to technical efficiency. Thus, a negative sign indicates a decrease in technical inefficiency or an increase in technical efficiency. The statistically significant variables in the technical inefficiency function are education, horsepower and planning for a peak season and horsepower, household, planning, staffing and controlling for a non peak season.

CONCLUSION

The empirical results of the study show that 20% of the fishermen had an impressive efficiency index greater than 90% during the peak season. On the other hand 39% of them had an efficiency index of less than 40%. During the non peak season, the number of fishermen declined from 20% to 15% for the most efficient vessels efficiency score and about 63% of the fishermen surveyed had less than 40%. The mean technical efficiency for the sample were estimated to be 54.6% for the peak season and 40.1% for the non peak season implying that more effort can still be done to increase the efficiency level. For the technical inefficiency test, there were 8 inefficiency variables shown to have significant impact against technical efficiency. They are: formal education, horsepower engine and planning for the peak season and size of family, horsepower engine, planning, staffing and controlling for non peak season.

The findings of this study suggest that there exists room for improvement in efficiency among a sizeable proportion of the fishermen studied. With appropriate training and introduction of more advanced technologies fishermen' efficiency level can be raised. Also higher level of engine power to steam in different fishing grounds and thus conducts more fishing operations than smaller horsepower in a given trip. Hence fishermen should use bigger horsepower engine to increase their catch. This can be achieved through loans, subsidies and incentives from the government.

REFERENCES

1- Aigner, D.J., Lovell, C. A. K. and Schmidt, P. (1977). Formulation and Estimation of Stochastic Frontier Production Function Models. Journal of Econometrics, 6: 21-37.

2- Bravo-Ureta, B. and Pinheiro, A. (1993). Efficiency Analysis of Developing Country Agriculture: A Review of the Frontier Function Literature, Agricultural and Resource Economics Review, Vol. 22, No. 1, pp. 88-101.

3- Charnes, A., Cooper, W. W. & Rhodes, E. (1978). Measuring the Efficiency of Decision Making Units. European Journal of Operational Research, 2, 429-444.

4- Elyasiani, E. and Mehdian, S. (1993). Measuring Technical and Scale Inefficiencies in the Beer Industry: Non-parametric and Parametric Evidence, The Quarterly Review of Economics and Finance, Vol. 33 No. 4, pp. 383-408.

5- Farrell, M.J. (1957). The Measurement of Productive Efficiency. Journal of the Royal Statistical Society Series A, vol. 120, pp.253-90.

6- Fukuyama, H. (1993). Technical and Scale Efficiency of Japanese Commercial Banks: A Non-Parametric Approach, Applied Economics, Vol. 25, pp. 1101-12.

7- Grabowski, R., Rangan, N. and Rezvanian, R. (1994). The Effect of Deregulation on the Efficiency of US Banking Firms, Journal of Economics and Business, Vol. 46, pp. 39-54.

8- Hellriegel, D., Jackson, S. and Slocum, J.W. (2002) Management: a Competence-based Approach. 9 ed. Cincinnati: South-Western College Pub.

9- Kirkley, J., Squires, D. and Strand, I. (1998). Characterizing Managerial Skill and Technical Efficiency in a Fishery, Journal of Productivity Analysis, 9: 145-160.

10- Kirkley, J., Paul C.M. and Squires, D. (2002). Capacity and Capacity Utilization in Common-Pool Resource Industries: Definition, Measurement, and a Comparison of Approaches. Environmental and Resource Economics, 22(1,2): 71–97.

11- Lewin, A.Y., Morey, R.C. and Cook, T.J. (1982). Evaluating the Administrative Efficiency of Vourts, OMEGA, 10: 401-11.

12- Meeusen. W. and van den Broeck, J. (1977). Efficiency Estimation from Cobb-Douglas Production Functions with Composed Error. International Economic Review, Vol. 18, pp. 435-444.

13- Miller, J.L. and Adam Jr, E.E. (1996). Slack and Performance in Health Care Delivery, International Journal of Quality and Reliability Management, 13(8): 63-74.

14- Resti, A. (2000). Efficiency Measurement for Multi-Product Industries: A Comparison of Classic and Recent Techniques Based on Simulated Data. European Journal of Operation Research, 121:559-78.

15- Rosko, M. D. (1990). Measuring Technical Efficiency in Health Care Organizations. Journal of Medical Systems, Vol. 14 No. 5, pp. 307-21.

16- Schultz, T.W. (1964). Transforming Traditional Agriculture. New Haven: Yale University Press.

17- Sherman, H. (1984). Hospital Efficiency Measurement and Evaluation: Empirical Test of a New Technique, Medical Care, 22 (10): 922-38

18- Soteriou, A.C., Karahanna, E., Papanastasiou, C. and Diakourakis, M.S. (1998). Using DEA to Evaluate the Efficiency of Secondary Schools: The Case of Cyprus, International Journal of Educational Management, 12 (2): 65-73.

19- Squires, D., Grafton, Q. Alam, F. and Ishak (1998). Where the Land Meets the Sea: Integrated Sustainable Fisheries Development and Artisanal Fishing. Discussion Paper No. 98-26, Department of Economics, University of California, San Diego.

20- Squires, D., Grafton, Q., Alam, M.A. & Omar, I.H. (2002). Technical Efficiency in the Malaysian Gill Net Artisanal Fishery. Economic and Environment Network, Sydney: The Australian National University.

21- Thanassoulis, E., Dyson, R.G. and Foster, M.J. (1987). Relative Efficiency Assessments Using Data Envelopment Analysis: An Application to Data on Rates Departments, Journal of the Operational Research Society, 38 (5): 397-411.

22- Tomkins, C. & Green, R. (1988). An Experiment in the Use of Data Envelopment Analysis for Evaluating the Efficiency of UK University Departments of Accounting, Financial Accountability & Management, 4 (2): 147-64.

23- Youtopoulos, P.A. & Lau, L.J. (1973). A Test of Relative Economic Efficiency: Some Further Results. American Economic Review 49 (1-3): 476-495.

25

Applying CVM for Economic Valuation of Drinking Water in Iran

*Morteza Tahami Pour and Mohammad Kavoosi Kalashami**

Abstract

Economic valuation of water is useful in the administration and management of water. Population growth and urbanization caused municipal water demand increase in Iran. Limited water resource supply and urban water network capacity raised complexity in water resources management. Present condition suggests using economic value of water as a criterion for allocating policies and feasibility study of urban water supply projects. This study use contingent valuation method for determining economic value of drinking water in Kohkiloye & Boyerahmad province. Required data set were obtained from 177 questionnaires by applying stratified random sampling in 2011 year. From 136 investigated urban households 111 ones are willing to pay more for qualified drinking water. Also, from 41 investigated rural households only 3 ones are willing to pay more for qualified drinking water. Results indicated that economic value of drinking water is 6877 Rial per cubic meter.

Keywords:
Economic Value of Water, Willingness to Pay, Municipal Water Consumption, Contingent Valuation Method, Kohkiloye & Boyerahmad Province

[1] *Ph.D Student of Agriculture Economics, Department of Agriculture Economic, University of Tehran, Iran.*
* *Corresponding author's email: mkavoosi@ut.ac.ir*

INTRODUCTION

Iran is one of the arid and semi-arid countries of the world with average precipitation of 251 mm per year. The total renewable water resources of Iran is 130 billion cubic meters, out of which 92 percentages is used for agriculture, 6 percentages for domestic use and services and 2 percentages for industrial uses (Assadollahi, 2009). Rapid population growth and low irrigation efficiency in agricultural sector of Iran have increased the demand for water resources. Therefore, rational management for water supply and demand and optimum use of the available water resources is necessary. Managing water as an economic good is an important way of achieving efficient and equitable use, and of encouraging conservation and protection of water resources. Water resources provide variety of products and services for human being include food, drinking water, hygiene, entertainment and hydropower. Water demand amount and variety increase beside water supply limitations in municipal regions cause competition among users and impose pressure on drinking water networks. Widespread network of water in cities, high cost of drinking water supply and important role of water in sustainable development raised complexity in water resource management in most cities of Iran. Water allocation and using water economic value as a criterion for allocating is one of the mentioned complexity in water resource management (Turner *et al.,* 2004). The value of water in alternative uses is important for the rational allocation of water as a scarce resource, whether by regulatory or economic means. Water economic value could be used as a proper framework for water pricing and water allocating policies. Also, benefit-cost analysis of water supply projects mainly uses water economic value for economic benefit determination. Most of development documents in Iran insist on estimating economic value of water in different uses and applying this amount as a criterion for allocating. Fourth development program of Iran noted economic value of water should be calculated in water basins. Also, in Fifth development program of Iran using water economic value for allocating and supplying water resource has been mentioned. Many studies estimate economic value of water in agriculture usage applying parametric and non-parametric methods like production function and mathematical programming (Hussain and Young, 1985, Thompson, 1988, Chakravorty and Roumasset, 1991, He and Tyner, 2004). Against irrigation water, drinking water considered as a final commodity in economical modeling and its economic value determination approach completely different from the situation in which water is a production input like in agriculture and industry (Lehtonen *et al.,* 2003). When water considered as a final commodity, contingent valuation method could properly used for its economic value determination. Many studies use CVM for estimating and calculation economic value of drinking water (Gnedenko and Gorbunova, 1998, Gnedenk *et al.,* 1999, Farlofi *et al.,* 2007, Guha, 2007).

According to the importance of economic water valuation in Kohkiloye and Boyerahmad province, present study estimated economic value of drinking water in municipal and rural districts of mentioned province based on willingness to pay of households for proper quality of drinking water. Estimated WTP for drinking water could be used in benefit-cost analysis of urban water supply projects in Kohkiloye and Boyerahmad province.

MATERIALS AND METHODS

Several approaches applied for economic valuation of water which categorize into two main groups include economic valuation of water as a production input and economic valuation of water as an economic commodity. In agriculture and industry, water used as an input and approaches like production function or mathematical programming applied for economic valuation of water. In municipal usage water treated as a final commodity and water consumers gained utility by using water. Hence, in this case approaches like contingent valuation method, choice modeling and water market transactions have been used for estimating economic value of water (Young, 2005). CVM studies are very popular among mentioned approaches used for economic valuation of drinking water, present study apply CVM for achieving to the foresaid goals.

CVM usually used as a standard and flexible tool for measuring non-use values and non-market use values of natural resources and environment (Hanemann *et al.,* 1991 and Hanemann, 1994). CVM determined individuals WTP in hypothetical scenarios (Lee, 1997). In this approach a hypothetical market for drinking water with proper quality established and some proposed price suggested to individuals for acquiring mentioned commodity. Individual's answer to the proposed prices or bids has been formed based on maximizing utility. As shown below, accepting proposed price means that utility gained by accepting is more than the utility of denying proposed price.

$$U(1,Y-A;S)+\varepsilon_1 \geq U(O,Y;S)+\varepsilon_0 \quad (1)$$

In above equation U is indirect utility, Y is individual income, A is proposed price or bid and S is social-economical characteristics. Utility difference is as below:

$$\Delta U = U(1,Y-A;S)-U(O,Y;S)+(\varepsilon_1-\varepsilon_0) \quad (2)$$

According to the model characteristics logit or probit functional form has been used for estimating valuation function. Considering Pi as the probability of accepting proposed price (A) by individual, logit functional form could be formed as below:

$$P_1 = F_\eta(\Delta U) = \frac{1}{1+\exp(-\Delta u)} = \frac{1}{1+\exp\{-(\alpha-\beta A+\gamma Y+\theta S)\}} \quad (3)$$

In which $F_\eta(\Delta U)$ is cumulative distribution function. θ, γ and β are regression coefficients and it is expected that θ>0, γ>0 and β≤0. After estimating above logit function it is possible to calculate expected WTP using integral.

Logit regression model coefficients have been determined using maximum likelihood estimator (Lehtonen and *et al.,* 2003). Integral of $F_\eta(\Delta U)$ between 0 to infinity has been calculated based on below equation:

$$E(WTP) = \int_0^\infty F_\eta(\Delta U)dA = \int_0^\infty \left(\frac{1}{1+\exp\{-\alpha^* + \beta A\}}\right)dA \quad (4)$$

In above equation E(WTP) is expected willingness to pay and α* is adjusted constant which is constructed as below:

$$[\alpha^* = (\alpha + \gamma Y + \theta S)] \quad (5)$$

One of the main advantages of logit estimation is that it is possible to investigate change in variables amount on the probability of accepting proposed price by individual i. The probability of bid acceptance by individual defined as below:

$$P_i = F(X_i^* \lambda) = \frac{1}{1+\exp^{X_i^* \lambda}} \quad (6)$$

In which, X_i^* is the vector of variables and λ is the vector of coefficients. In order to evaluate the effect of each variable quantity change on the probability of bid acceptance, the derivation of above equation has been calculated (Maddala, 1991):

$$\frac{\partial P_i}{\partial X_{ik}} = \frac{\exp^{X_i^* \lambda}}{(1+\exp^{X_i^* \lambda})^2} \lambda_k \quad (7)$$

This equation generates marginal effect of each variable which is very useful in policy making analysis of model results. Requested data set were obtained through a survey using questionnaires. Mainly, information like education, age, individual satisfaction of drinking water quality and quantity, monthly income and expenditure, water consumption quantity and willingness to pay for drinking water has been asked from each individual by interviewing. Present study used stratified random sampling method for determining the sample size. From 177 sample size 136 and 41 questionnaires gathered from urban and rural districts of Kohkiloye & Boyerahmad province, respectively. Explanatory variables which were used in logit model for estimating economic value of drinking water could be summarized in below table 1.

RESULTS AND DISCUSSION

Investigating sample individuals' satisfaction of drinking water quality showed that 7 percentages of rural individuals and 40 percentages of urban individuals satisfied with present condition of drinking water quality. On the other hand, 93 percentages of investigated rural individuals and 60 percentages of investigated urban

Table 1

Explanatory variables	Description
Region	Dummy variable (1= urban district, 0= rural district)
Water quality	Dummy variable (1=satisfaction from drinking water quality, 0=dissatisfaction from drinking water quality)
Water quantity	Dummy variable (1=satisfaction from drinking water quantity, 0=dissatisfaction from drinking water quantity)
Monthly income	In 1000 Rials
Monthly water consumption	In cubic meter
Family size	-
Proposed price or bid	In 1000 Rials
Family supervisor gender	Dummy variable (1=male, 0=female)
Family supervisor age	-
Family supervisor education	-

individuals dissatisfied with drinking water quality. Reasons of dissatisfaction could be summarized as kidney diseases occurrence, parasite diseases occurrence and existence of pollution in rural drinking water.

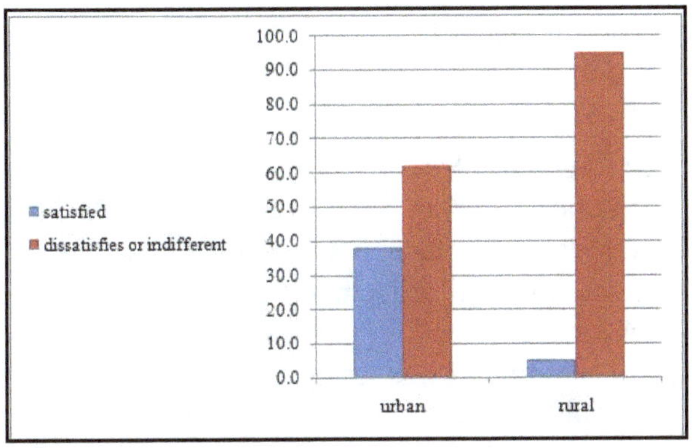

Figure 1: Comparing satisfaction level of drinking water quality in urban and rural districts (percentages).

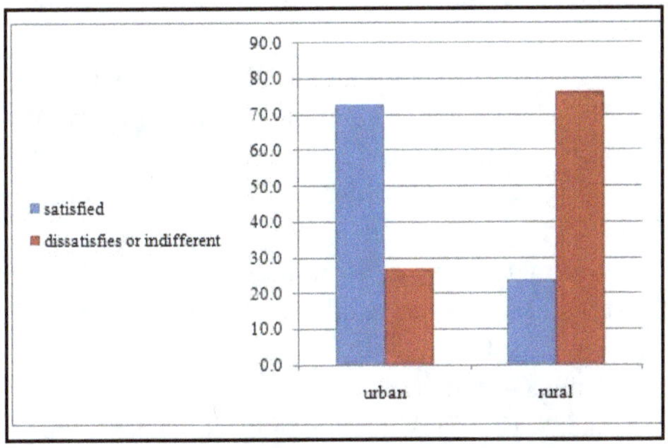

Figure 2: Comparing satisfaction level of drinking water supply in urban and rural districts (percentages).

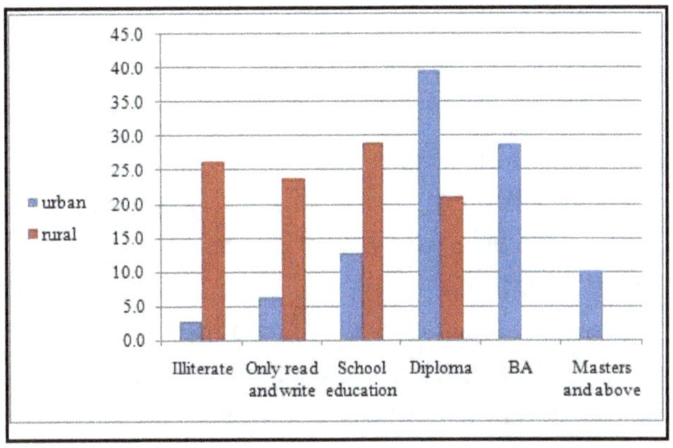

Figure 3: Comparing education level of sample individuals in urban and rural districts (percentages).

Investigating sample individuals' satisfaction of drinking water quantity showed 17 percentages in rural districts and 69 percentages in urban districts satisfied by present condition of drinking water quantity or supply. Also, 83 percentages of investigated people in rural districts and 31 percentages in urban districts dissatisfied from drinking water quantity or supply. Reasons of dissatisfaction include water quantity, dissection, and network insufficient pressure. Hence, households forced to use tankers or reserve drinking water.

Investigating sample individuals' education level showed that frequency of diploma degree

Table 2: Comparing willingness to pay in urban and rural districts.

District	Description	Willing to pay more	Will not pay more
Urban districts	Frequency	111	25
	percentages	81	19
Rural districts	Frequency	38	3
	percentages	93	7

Source: research findings.

Table 3: Logit model estimation result.

Variables	Coefficients	S.D.	t-statistics	Marginal Effect
Bid	-0.107	0.042	-2.503	-0.024
Monthly income	0.00064	0.00036	1.76	0.00014
Family supervisor education	0.085	0.041	2.061	0.0192
Region	-1.113	0.679	-1.637	-0.25
Constant	1.834	0.639	2.868	-

Log of likelihood function = -86.5
Percentage of Right Predictions = 0.77
Source: Research findings.

is more in urban districts. Education level frequency of sample is shown in figure 3.

From 136 investigated urban households 111 ones are willing to pay more for qualified drinking water. Also, from 41 investigated rural households only 3 ones are willing to pay more for qualified drinking water.

After estimating different models with explanatory variables, below model select as the best estimations.

Results revealed that bid, monthly income, family supervisor education level and dummy of region variables had direct and statistically significant effects on probability of bid acceptance. Marginal effect value of monthly income variable showed that one unit increase (1000 Rials increase) in monthly income 0.0192 unit increase the probability of bid acceptance. Also, marginal effect of bid showed that 1000 Rials increase in bid amount 0.024 unit decrease the probability of bid acceptance by sample individuals. One level increase in family supervisor education 0.0192 unit increase the mentioned probability amount.

Percentages of right prediction in logit model were 77 percentages. So, 77 percentages of individuals' responses could be simulated by model. Using model coefficients, the amount of α^* calculated ($\alpha^* = 2.171$). Expected quantity of WTP for drinking water consumption equals 213187 Rials per month. Considering average drinking water consumption for each household as 31 cubic meters, value of a cubic meter drinking water equals 6877 Rials.

CONCLUSION

Results revealed that drinking water consumers were willing to pay 6877 Rials per cubic meter in Kohkiloye & Boyerahmad province. This price could be used as a good framework for allocating planning of drinking water. Also, for calculating economic benefits of new drinking water supply projects this value could be used. Applying CVM approach for economic valuation of drinking water provides good view for managers and planners about consumers' demand of drinking water.

REFERENCES

1- Assadollahi, S.A. (2009). Groundwater Resources Management in Iran. Secretary General, IRNCID and Deputy of Protection and Exploitation of Iran Water Resources Management Company.

2- Chakravorty, V. and J. Roumasset. (1991). Efficient Spatial Allocation Irrigation Water. American journal of Agricultural Economics. 73: 165-173.

3- Farlofi, S., Mabugu, R., & Ntshingila, S. (2007). Domestic Water Use and Aalues in Swaziland: Contingent Valuation Analysis, Agrekon Magazine. (Abstract)

4- Gnedenk, E., Gorbunova, Z., & Safonov, G. (1999). Contingent Valuation of Drinking Water Quality in Samara City, Moscow State University, zone "I", room 75, Moscow, 117234, Vorobievy Gory.

5- Gnedenko E.D., & Gorbunova, Z.V. (1998). A Contingent Valuation Study of Projects Improving Drinking Water Quality", Modern Toxicological Problems. (Abstract).

6- Guha, S. (2007). Valuation of Clean Water Supply by Willingness to Pay Method in Developing Nation, a Case Study in Calculate, India. Jabalpur University.

7- Hanemann, W. M. (1994). Valuing the Environment through Contingent Valuation. Journal of Economic Perspectives. 8(4): 19-43.

8- Hanemann, W. M., Loonis, J., & Kanninen, B. (1991). Statistical Efficiency of Double-bounded Dichotomous Choice Contingent Valuation. American Journal of Agricultural Economics, 73(4): 1255-1263.

9- He, L., & Tyner, W. (2004). Improving Irrigation

Water Allocation Efficiency Using Alternative Policy option in Egypt, http://econpapers.hhs.se

10- Hussain, R.Z., & Young, R.A. (1985). Estimates of the Economic Value Productivity of Irrigation Water in Pakistan from Farm Surveys, Water Resources Bulletin. 26(6): 1021-1027.

11- Lee, C. (1997). Valuation of Nature-based Tourism Resources Using Dichotomous Choice Contingent Valuation method, Tourism Management. 18(8): 587-591.

12- Lehtonen, E., Kuluvainen, J., Pouta, E., Rekola, M., & Li, C. (2003). Non-market Benefits of Forest Conservation in Southern Finland. Environmental Science and Policy. 6: 195-204.

13- Maddala, G.S. 1991. Introduction of Econometrics. 2nd Edition, Macmillan, New York.

14- Thompson, C.D. (1988). Choice of Flexible Functional Forms: Review and Appraisal, Western Journal of Agricultural Economics. 13: 169-183.

15- Turner, J.C., Douglas, C.L., Hallum, C.R., Krausman, P.R., & Ramey, R.R. (2004). Determination of Critical Habitat for the Endangered Nelson's bighorn sheep in southern California. Wildlife Society Bulletin, 32: 427-448.

16- Young, R.A. (2005). Determining the Economic Value of Water. Concepts and Methods. Washington, DC: Resources for the Future.

The Economic and Welfare Effects of Different Irrigation Water Pricing Methods, Case Study of Khomein Plain in Markazi Province of Iran

Gholamreza Zamanian [1], Mehdi Jafari[2] and Shahram Saeedian [3]*

Abstract

The scarcity of water resources and supply resources limitation, have caused an increasing gap between water supply and demand specially in recent decades in almost all regions of the globe. One of the best known solutions proposed by the economists is using the different water pricing approaches thereby obtaining the optimal allocation and social justice. To this purpose, this paper uses the positive Mathematical Programming (PMP) and Econometric Mathematical Programming (EMP) in a comparative analysis to study the economic and welfare impacts of alternative water pricing approaches in the agricultural sector during agricultural period 2011/2012 in Khomein plain of Markazi province in Iran. Results show that the EMP can be a better alternative approach instead of PMP to better analyze of agricultural policies. According to the final outcomes, it is suggested to apply the block tariff in place of volumetric pricing method to reach the optimal allocation and promoting the water efficiency in the price range of 198 to 853 Rials.

Keywords:
Economic and welfare effects, Water pricing, Mathematical programming, Khomein plain

[1] Assistant Professor in Economics, University of Sistan and Baluchestan, Iran.
[2] Ph.D Student of Agricultural Economics, University of Sistan and Baluchestan, Iran.
[3] Graduate student of Agricultural Economics, University of Sistan and Baluchestan, Iran.
* Corresponding author's email: m.jafari365@gmail.com

INTRODUCTION

During recent decades, as for to population growth and improving life standards, water demand faced a dazzling speed. To deal with this, primarily, the strategy of discovering and exploiting new water resources came up. Increasing healthy and reliable water resources, producing more food and electricity and rural economic development were some of the benefits of this policy. It has to be mentioned that, nowadays, more than 70 percent of irrigation water is provided from fresh water and the increasing growth of this demand can be easily predicted (Jafari 2013). Extending the exploitation of non-renewable water resources has been one of the main approaches to provide with this increscent water demand in Iran. However, continuing the expansionist policies which were followed previously, due to impairment losses of species, ecosystems and water resources pollution are not possible any more. Additionally, the difficulty of finding new water resources and also the externalities of constructing huge water projects has increased the marginal cost of water extracting. To solve these problems, water management pattern with brand new policies like as concentrating on productivity improvement, managing the water demand and reallocating water between consumers as more suitable solutions are changing. A substantial number of studies show that governments, in order to reach the optimum allocation and rising water productivity used some policies like decentralization of irrigation water management, pricing systems, water laws and commercial plans (see Dinar and Maria, 2005; Johansson *et al.*, 2002; Tiwari and Dinar, 2002; Tsure, 2004; Roe, 2005; Veettil, 2011 for surveys). What emerges from these studies is that the relationship between variables and different existent characteristics in agricultural environment as the special irrigation type, water laws, structural frameworks and alternative cropping systems can affect the results significantly. The results of Liao *et al.* (2007), Frija *et al.* (2008), Herrera *et al.* (2004), Speelman *et al.* (2010 and 2011) mentioned that farmers willingness to pay could be influenced by environmental conditions and when water laws are not defined properly, it leads to inefficiency of water pricing systems, non-optimum water allocation, increasing trade-off costs and expenditures and finally inappropriate evaluation of water resources (Fragoso and Marques, 2013). More than 60 percent of Iran, Markazy province and particularly Khomein region have a dry and semi-dry climate. Khomein as a flat and talented agricultural region as for 240 mm annual rainfall and also increasing population growth as well as extending agricultural activities, faces rising water demand and contrary to its shortage. With considering the substantial relationship between water resources stock and rainfall, surveys show that despite the moderate and extreme drought happened during 2008 to 2011 in khomein region, annual exploitation from underground water resources increased 13.1 million m3 on average (Mosayebi and Maleki, 2012) which led to completely drying of 164 deep and shallow wells, 172 Ghanats, 57 natural fountains, 38 rivers and 13 soiled-dams. And also 349 deep and shallow wells, 79 Ghanats, 21 natural fountains, five rivers and 5 soiled-dams have 1 to 10 liters in second water that are so likely to get dried in the near future (Agricultural organization of Markazi province, 2012). Water has been regarded as a free commodity in Iran, historically. The act of pricing this scarce input and increasing the current prices encounters many problems. Currently water pricing in Iranian agricultural sector is done on the basis of "Justly Distribution of Water" law and regarding the underlying crop. As in this system, pricing is not based on water consumption volume; there is not enough motivation to efficient and economic allocation of water and its marginal return is often higher than the price and providing and distributing costs. The extension limitation of water resources and weak management companies with huge water losses make the applying of water demand-side policies as complementary inputs taxes or product taxes unavoidable. These policies have been investigated by different researchers in Iran. Hossain zad (2004) and Asadi *et al.* (2007) showed that as for low elasticity of water demand in agricultural sector of Iran, increasing the price of this input decreases the water demand slightly. So, the water price has to be increased substantially or alternative policies are to be introduced. But, it is to be noted that efficiency improvement and water allocation without suitable economic policies and instruments is not conceivable. Moreover, the result of a policy or its impact depends highly on the farmers' reaction to the applied policies. The farmers' reaction is depended on farm condition, individual attitudes and characteristics. It is not possible to examine alternative policies in laboratory conditions and the policy maker is seeking to get a good intuition about policy implications in agricultural sector and farmers' reaction to the policies. To this end, this paper is going to study the different water pricing methods impacts on water demand, water allocation among irrigated agricultural crops, farmers' revenue, costs and other inputs demand in Khomein plain. The rest of the article is organized as follows: In part twowe discuss he analytical framework. Part three, elaborates the data and empirical models. Part four presents the empirical results and discussion and finally in part five concludes.

Analytical framework

The water demand and supply

Generally, the demand for Irrigation water is come from the market demand of agricultural products. Suppose a farm with n products and an input of water, the profit is defined as:

$$\pi = \sum_{j=1}^{n} \left[p_j f_j(q_j) - w q_j \right]$$

$j = 1, 2, ..., n$ (1)

Where $fj(qj)=yj$ is an ascending and strictly concave function, j is production yield, pj indicates the market price of j-th product and water price is shown by w. Essential prerequisite to maximize the profit is as

$$\frac{\partial \pi}{\partial q_i} = 0 \text{ or } f'(q_j(w)) = \frac{w}{p} \Rightarrow q_j(w) = f_j'^{-1}(w/p) \quad (2)$$

Where $qj(w)$ shows the amount of entering water with price w. In other words, the water demand function of farmer is

$$q(w) = \sum_{j=1}^{n} q_j(w) = \sum_{j=1}^{n} f_j'^{-1}(w/p) \quad (3)$$

The individual water demand is specified by $qj(w)$ and the aggregative water demand for all farmers is the sum of individual demands presented as

$$q(w) = \sum_{i}^{k} q_i(w) = \sum_{i=1}^{k}\sum_{j=1}^{n} f_{ij}'^{-1}(w/p) \quad (4)$$

Water demand could be measured with considering it as a free commodity and also with supposing it limited in x liter. Here, the thing which seems important is that we have to know the farmers willingness to pay for Δ unit more water. When they use water at x level, their revenue is $p\times f(x)$. Expectedly, the additional income from using Δ unit more water is $p[f(x+\Delta)-f(x)]$. The additional income $pf(x)$ which is due to the little amount Δ, is indeed the maximum price that farmers are willing to pay for consuming additional units of irrigation water. This price is called Shadow price of water and its value is positive if the water constraint is binding. In other words, the problem of allocating the water between products can be solved by maximizing profit condition as:

$$\pi(p,x) = \max \sum_{j=1}^{n} p_j f_j(q_j)$$

$$s\,t \sum_{j=1}^{n} q_j \leq x \quad (5)$$

Which its lagrangian form is as

$$\ell = p_j f_j(q_j) - \lambda \left[x - \sum_{j=1}^{n} q_j \right] \quad (6)$$

Where λ here is a coefficient of constraining factor of water, and shows the shadow price of it. This strategy can be applied for more inputs, variables, constraints, infinite puechased inputs and crops that use the water in their production process. In words, a combination of non-linear production functions with linear programming can be combined into a non-linear programming frame. Also, in both of these cases, the Irrigation water demand function could be obtained subject to maximizing the profit at different water levels which allows to obtain different allocative amounts of qj with shadow price of water λ. In the usual approach, the irrigation water demand can be extracted by regression analyzing of the observed information of water price and quantity. However, due to some problems like as unavailability of information and the variability of water price in small scales, this method causes imprecise estimations (Tsure, 2005).

Water pricing

The intersection between the non-decreasing marginal cost function and the descending slope derived from demand function determines the marginal cost of water. This happens In the great irrigation projects, when the average cost function is decreasing and the marginal cost curve placed under the average cost $(w^* < AC(w^*))$. Therefore, the real profit of supplier does not meet the fixed costs and in order to continue the activity in long-run, subsidies have got to be given to the suppliers. In the long-run, financing the suppliers' costs increases in order to cover their costs which often results in decreasing the average pricing costs. In this case, water price is set up following the exploited demand function and average cost. By the way, despite the farmer is able to return the overall water cost, but this policy is not efficient in the average pricing cost because it does not ensure the maximization of producer and farmers welfare. As it was mentioned by Tsure (2005), determining the water price by moving through the marginal cost curve toward the average cost could provide producers with positive profit and simultaneously decreases the farmers' profit. Since this decreasing profit is greater than that increasing profit, so the total welfare will experience a down fall by taking this method of pricing (Tsure, 2005). Hence, according to Tsure and Dinar (1997), water pricing on the basis of marginal cost can give the optimal water allocation but implementing this method requires some prohibitive operations like monitoring and management and collecting exact data. Thus, alternative water pricing methods are applied across the world including volumetric approach under which, the water costs are measured directly by estimating the water volume consumpted; input-output approach that irrigation water valuing is done based on products or inputs (except water) used in the production process ; regional method that water is priced on the basis of irrigation methods used in the region. Usually the differences in irrigation costs come as for the kind and amount of irrigation, irrigation method and irrigation season in a special region; blocking method in which variable volumetric tariffs are used proportional to an specified level of water consumption; two component tariff which usually comprises the pricing method based on marginal cost and annual fixed costs for water right (which has different values in each region depending on irrigation method); the last approach is the tax method in which water costs payments are considered according to the added value of the sown area which is caused by the irrigation water. Each of these water pricing methods, leads to different levels of welfare and net benefits and choosing one of them is based on the implementation costs which vary from one region to another as for the climatic issues, demographic, social structure, water rights, time and economic conditions. Thus, the pricing method is considered which has the most benefit. Without considering the implementation costs, one of the efficient approaches is the volumetric method. Tsure and Dinar (1997) compared the results of the volumetric and regional pricing methods. Results showed that if 7.5 percent of the outcome from water would used for operational expenditures, the regional pricing method has a much better return in comparison to the other approaches. Most notably, the supply and demand function and also the pricing methods based as the theoretical base of this paper are derived from Tsure (2000, 2005), Dinar (2000), Dinar and Maria (2005), Dinar and Mody (2004).

Positive mathematical programming (PMP) models

Recently, there appears an increasing interest to apply sort of generalized mathematical programming in agricultural sector. Heckely and Britz (2005) ratiocinate this interest by some reasons. First, an expanded range of political tools in addition to supportive policies based on pricing are come up. Also, as for to developing the multipurpose agriculture which is so important, it is more likely that with existing technical constraints, most of the old mathematical programming models give incoherent results. After presenting the positive mathematical programming by Howitt (1995) for calibration, it was applied in agriculture widely. Positive mathematical programming (PMP) was developed to overcome the difficulties of normative mathematical programming (Howitt, 1998). At most one concave profit function and the MC parameter as well in the non-linear variable cost function are used for PMP models. Therefore, this model is able to reproduce the observed situation and evaluate the policies and to suggest more reliable policies. The main model which was presented by Howitt (1995) had two components. The first component was a linear model with calibration constraints in order to build the dual value of resources and constraints and in the second component given obtained dual, calibration parameters (including MC as the coefficient of concave cost function(in the short-run as the coefficient of the non-linear profit function)) are estimated to maximize the model given the linear constraints. The main idea hidden in this method is using the dual values for non-linear calibration of the objective function in order to obtain the simple and exact basic situation observed data). Sabuhi *et al.* (2006) put it well"In order to specify the non-linear object function, each type of non-linear function which can set the marginal cost of preferential activities to their related prices in the level of observed activities, is possible to be utilized. So, in this study, for analyzing the policies we use the quadratic cost function which its general presentation is outlined in Heckely and Britz (2005), Henry *et al.* (2007) surveys. As it was shown in equations (1) and (11), these models show the maximum sum of farmers' surpluse.

$$max\ \pi = g\acute{m}l \quad (7)$$

$$\text{s.t}$$
$$Al \le b[\lambda]$$
$$l \le l^0 + \varepsilon[\rho]$$
$$l \ge 0$$

Where π indicates the profit function in the short-run which is corresponding to the gross yield of farm in the short-run. *n* and *gm* present the vector of gross yield corresponding to each activity and the non-negative variable of sown area of each product, respectively; shows the technical coefficients matrix; *b* is the $m\times 1$ vector of available inputs (like land, water, labor and chemical fertilizer); indicates the $m\times 1$ vector of shadow prices for each input; ρ and l^0 present the $n\times 1$ vector of observed sown area for each product in the base year and the corresponding shadow price of it, respectively; ε is the littlest number as the calibration constraint which is used to prevent the linear dependency.

When ρj the the is determined then in the second step, using the PMP approach, the variables of non-linear cost function $C^v(l^0)$ are estimated in which the marginal cost of the activity $MC^v(l^0)$ has formed from two components: the known costs of activity *(c)* and the unknown marginal cost which are given below

$$MC^v = \frac{\partial c^v(l^0)}{\partial l} = d + Ql^0 = c + \rho \quad (8)$$

Where Q and *d* are a $n\times 1$ positive, determined and symmetric vector of linear coefficients and the quadratic matrix of variable. To simplify, the diagonal elements of matrix Q as for the standard estimation approach from $qjj = \rho j/l_j^0$ for determining the quadratic function of costs are placed in the model below

$$\max \pi = gm'l - \frac{1}{2}l'Ql \quad 0$$
$$s\,t \qquad Al \le b[\lambda], \qquad l \ge 0$$

The econometric mathematical programming (EMP) models

According to Heckely and Britz (2005), PMP approach faces some important limitations for instance the calibration constraints have to have the zero degree of freedom while this issue needs much of data or a so flexible functional form to cover all constraints. The other limitation is that different approaches to estimate the calibration parameters lead to considerable differences in simulation behavior. Buysse *et al.,* (2007) state that to obtain the more realistic simulation behavior, econometric programming models which can estimate the objective function and constraints given the external information are suitable alternatives for PMP models. The main axiom of this strategy using the lagrangian model is presented like below

$$\ell = gm'l - \frac{1}{2}l'Ql + \lambda(b - ul) \quad (10)$$

If the land is the only fixed resource, then $A=u$ and *u* is a $n\times 1$ vector of the sum of them. The first optimization condition is

$$\frac{\partial \ell}{\partial l} = gm - Ql - \lambda u = 0$$
$$\frac{\partial \ell}{\partial \lambda} = b - ul = 0 \qquad (11)$$

Thus, the unknown parameters λ and Q can be estimated using some of econometric measures. In the case that the observed data are less than the parameters that has to be estimated, we face the *III-posed* situation. In this case, using the generalized Maximum Entropy (GME) the aforementioned situation can be solved (Golan *et al.*, 1996). Similar to Heckely and Wolf (2003) combining information related to land demand elasticity extracted from the sample, we can have a better estimation. The simple structure of the GME model constrained to the optimization conditions used in the programming model is

$$\max_{w_t, w^e, Q, l, \lambda} H(w_t, w^e) = -\sum_{t=1}^{T} w_t' \ln(w_t) - w^{e'} \ln(w^e)$$

$$s\,t \qquad (12)$$

$$gm_t^0 - \lambda_t u - Q(l_t^0 - \varepsilon_t) = 0$$
$$u'(l_t^0 - \varepsilon_t) = b_t^0 \qquad (13)$$

$$\varepsilon_t = Vw_t = \sum_{s=1}^{2} \sigma_{jts} w_{jts} \qquad (14)$$

$$diag[E] = V^e w^e = diag\left[(Q^{-1} - Q^{-1}u (u'Q^{-1}u)^{-1} uQ^{-1})(\frac{\overline{gm^0}}{l^0})\right] \qquad (15)$$

$$Q = LL', \qquad L = 0 \ \forall\ i \geq j \qquad (16)$$

$$\sum_{s=1}^{2} w_{jts} = 1, \qquad \sum_{s=1}^{2} w_{js}^{e} = 1 \qquad (17)$$

Where H is the entropy variable, w_t and w^e are the probabilities values as for to the error and the estimated elasticity E; gm_t^0 and l_t^0 are vector of production marginal yield and the production physical amount for each observation t, respectively; λ is the shadow price of fixed resources (like land); Q is the symmetric positive and determined matrix of production marginal cost coefficients; V and V^ε are the known matrix of errors and supportive values of elasticity. Equation (12) indicates the maximum entropy; (13) is the first optimization condition; (14) and (15) allow to calculate the error term (ε_t) and elasticity (E) as for to the second optimization condition that the variable cost function has to be non-descending. (16) is included to ensure the concavity of variable cost function and being positive and determined of Q; (17) makes sure that the sum of the probabilities and elasticity are equal to unity.

The stochastic errors of each observation (ε_t) have zero mean and a standard deviation of σ_{jts}. To apply the GME approach, it was necessary to carry out re-parameterization of the error term as expected values of a probability distribution (Vw_t). This is calculated based on known values of standard deviation, which are spread by two support points (the n×n×2 V matrix). Incorporation of out of sample information through the use of priors on elasticities allows us to obtain more accurate estimates for the Q matrix. In our case the elasticity estimates (E) are given by the product between the n×n Jacobian matrix of the land demand functions $\{Q^{-1}\text{-}Q^{-1}u(\acute{u}Q^{-1}u)^{-1}u^{-1}\}$ and the mean of observed gross profit divided by the mean of observed land allocation to crop $(\frac{\overline{gm^0}}{l^0})$. As for the error estimates, the elasticities (E) also have to be re-parameterized as the expected values of a probability distribution (we). In this case, for the central value of prior elasticities two support points were also considered and the values of standard deviations are bounded in the $n \times n \times 2V^e$ matrix (Fragoso and Marques, 2009).

After estimating the σ, Q, ε^t, w^e, wtvalues, the values are placed in the defined programming frame and it will be used to simulate the water pricing policies.

It is to be mentioned that in order to survey the different impacts of either of two EMP and PMP models on cropping pattern in this paper, after estimating the two models, we investigate the resulted diversity pattern. Genarally, for measuring the diversity of determined optimal cropping plans, regardless the different definitions that are presented for cropping diversity, we can measure it using two indexes including sown area and gross income. There exist Numerous indexes for calculating the diversity of a cropping plan. Shanon and bor, Simpson, Herfindal, Entropy and corrected concentration index are some of the most famouse indexes that are used to this end (Karbasi *et al.*, 2010). In this paper, we use Entropy Diversity index as it is used for large-scales like our case. This index is measured according to the following equation (Chang and Mishra, 2008):

$$EI = \sum_{i=1}^{i=n} \frac{X_i}{\sum_{i=1}^{i=n} X_i} \log\left(\frac{X_i}{\sum_{i=1}^{i=n} X_i}\right)^{-1} \quad (18)$$

Where X_i indicates the sown area of the activity. In this equation, if the EI is greater than zero, the cropping diversity is high and if it is equal to zero or less than it, there is no cropping diversity.

Data

In the current study, Khomein plain accommodated 7543 farmers, was used as the underlying statistical population and in order to collect the data related to the quantity of inputs consumption required for producing crops which are included as: water, labor, machinery, chemical fertilizer and maure and herbicide were gathered through a three-stage stratified sampling and given the Cochran-Orcut formulation according to number of farmers in rural districts and villages over 2011-2012 agricul-

Table 1: Descriptive statistics of variables

variable	mean	SD	Min	max
Yield				
Irrigated wheat	3113.7	1840.5	200	8000
Dry wheat	1157	909	125	4000
Irrigated barely	2615	1421	200	7000
Dry barely	1411	852	2000	8000
Dry pea	370	250	150	1350
Bean	2568	889	1000	5000
Potato	19470	12890	3333	32000
Onion	5000	0	5000	5000
Alfalfa	8696	5658	750	20000
Irrigated corn	40000	0	40000	40000
Inputs				
Labor(rial)	360163.8	453615.4	19845.24	6775431
Chemical Fertilizer phospat(kg/ha)	136.2	33.6	63.5	171.9
Chemical Fertilizer azot(kg/ha)	199	68.8	83.3	318.8
Animal fertilizer(kg/ha)	5527	6208	0	14428
Herbicide(kg/ha)	0.683	0.589	0	1.7
Machinery (rial)	63269.5	74904.9	4328.1	984926.1
Irrigated wheat	4933.3	56.6	4820	5010
Dry wheat	4773.6	169.41	4560	4890
Irrigated barely	5944	198.8	5370	5948
Dry barely	5944	0	5944	5944
Dry pea	5850	0	5850	5850
Bean	37100	2325	36400	42000
Potato	2100	0	2100	2100
Onion	6000	0	6000	6000
Alfalfa	6500	3250	3180	7000
Irrigated corn	1000	0	1000	1000
Labor(hour)	5945.9	1164.6	4926.1	7215
Chemical fertilizer(kg)	118.4	84.7	68.9	216.4
Animal fertilizer(kg)	9.8	1.03	8.7	10.8
Herbicide(kg)	5627.6	1708.2	4926	7575
Machinery (hour)	6566.5	651.8	5911.3	7215

Source: Research findings

tural year, 36, 41, 30, 32, 39, 45 and 27 questionnaires were distributed among Chahar cheshme, Khoram dasht, Ashena khor, Hamze loo, Rastagh, Salehan and Gale zan (Totally 250 questionnaires) rural districts, respectively. Farmers of each village were chosen by systematic sampling method so that on the basis of farmers number and sample size related to each rural district, the sample of each village was determined. Also, the information related to the sown area and the production quantity of the under study region was derived from the Markazi province's Jihad-e-Agriculture organization data bank The descriptive statistics are reported in the Table 1.

RESULTS

The results are presented in two sections. The first section compares the results from PMP and EMP approaches using observed data aimed to choose a model which is able to explain the farmers' behavior in the best way. The second section is related to survey the alternative water pricing policies on water consumption, irrigated land area, farm profit and total welfare.

Results of EMP and PMP

Results of measuring the Entropy index for the PMP and the EMP models are showing that changing the cropping pattern is on the basis of reproducing the PMP model and also decreasing the diversity of cropping pattern is based on the EMP model reproduction. In the next step, as for which model predicts the farmers' behavior, we compare the results exploited from the PMP and EMP models with observed data. Also in the second section of the results, evaluation of the impact of alternative water pricing policies on water consumption, the irrigated area, farm profit and total welfare is discussed.

Irrigation water demand

Given that Evaluating the irrigation water demand and irrigated land area is not only beneficial for choosing the best policy analysis model but it can be used for creating pricing scenario assumptions which are essential for simulation. In order to survey and compare the EMP and PMP models in Figures 1 and 2 respectively, we evaluate the water demand quantity and the percentage of irrigated land versus the shadow price in either of the models.

As shown in Figure 1, the PMP pattern has a more flexible curve than EMP which indicates that the water constraint has a more considerable effect in PMP model for simulation of products substitution. This result is more obvious in figure 2 where the sown area is stated as a function of shadow price. Although, the EMP curvature is more than PMP, but it is a more suitable model to predict farmers' behavior regarding policy changes. Evaluating the irrigation water

Table 2: Comparative results of PMP and EMP models in comparison to the base year.

Activity	Sown area in the base year(Hectare)	Models PMP	EMP	EMP-base
Irrigated wheat	8531	8531	8533	0.02
Dry wheat	9422	9422	9450	0.3
Irrigated barely	2708	2708	2700	-0.3
Dry barely	24	24	0	-100
Dry pea	151	151	157	0.03
Bean	3496	3496	3500	0.11
Potato	114	114	110	-3.51
Onion	266	266	293	10.5
Alfalfa	1494	1494	1498	0.27
Irrigated corn	35	35	0	-100
Total sown area	26241	26241	26241	0
Entropy Index	0.6576	0.6576	0.6572	-34
Water consumption('000 M3)	23558	23558	20106	-14.8
Dual value of land (Rial /hectare)	71685	71685	68385	-4.8

Source: Research findings

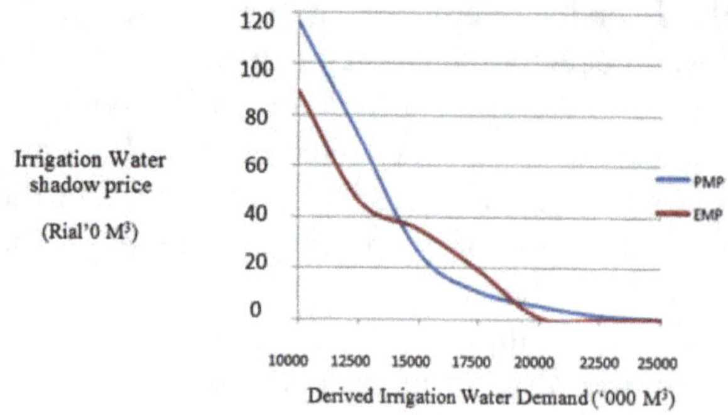

Figure 1: Derived Irrigation Water Demand from PMP and EMP models

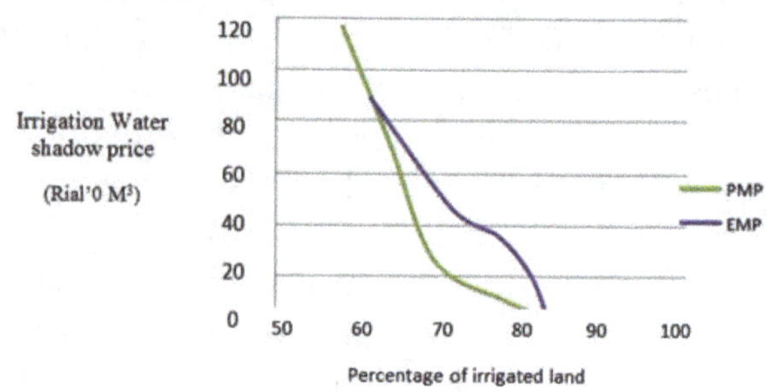

Figure 2: Percentage of irrigated land area from PMP and EMP models

demand and irrigated land area is not only beneficial to choose the best policy analysis model but it can be used for creating pricing scenario assumptions which are essential for simulation.

In the first part of demand curve i.e., 17500 to 25000 M3, The shadow price of water is in the range of 0 to 198 rials which have an elasticity of 0.008. In this price range, the changing percentage in water consumption is much less than the price changing percentage. In the second part of water demand curve i.e., 15000 to 17500 M3, price is in the range of 198.2 to 382 and the elasticity increases to 0.017 which means that more changing is expected from farmers regarding the price changes. In the third part and the availability range of 12500 to 15000 M3, price is between 382 and 420.3 and the elasticity comes out as 0.049. Interestingly, in the last part of demand curve, the elasticity increases to 0.108 which is the largest change in the consumption regarding to the price changes.

As for to the aforementioned results and the objective of this paper the simulation of the irrigation water pricing policies was done regarding the volumetric and the block tariff. In the volumetric tariff, simulation was done for 198, 382, 420.3 and 853.3 rials as optimal prices for each cubic meter of irrigation water. For the block method, the water costs were divided to three parts and in each part 50, 100 and 150 percent of water costs coverage was simulated.

Evaluation of the irrigation water pricing policies

In this section, the effect of water pricing policies on water consumption in all over the region,

Table 3: The water demand elasticity resulted from water demand for each shadow price.

Number	Water Demand	Water shadow price	Demand elasticity
1	17500-25000	0-198	0.008
2	15000-17500	198.2-382	0.017
3	12500-15000	382-420.3	0.049
4	10000-12500	420.3-853	0.108

Source: Research findings

Table 4:The economic effects of alternative irrigation water pricing policies using the EMP model

Policy	Welfare Changes	Sown area	Water consumption	Total Profit
Tariff 1	-24%	-16%	-8%	-1.5%
Tariff 2	-33%	-20%	-14%	-4%
Tariff 3	-42%	-38%	-19%	-25%
Tariff 4	-52.1%	-50%	-23%	-40%
Block Pricing	-30%	-17%	-21%	-26%

Source: Research findings

gross profit and total welfare of the society is tested. The final out comes are featured in the Table 4.

As featured in Table 4, total welfare (supplier plus consumer welfare) under the volumetric tariff of 198 rials, had the least reduction of 24% and in the second place, block tariff with 30 percent reduction proportional to the base year shows the least reduction in the total welfare. The distance between block and volumetric methods even exceeds 20 percent (for Tariff 4 case) which shows that generally, the block tariff provides a more satisfying total welfare level in comparison to the volumetric tariff. In the volumetric tariff of 198 rials, the water saving is 8% and in the block tariff, it reaches to 21%. As it is seen, the difference between these two policies in upper tariffs is negligible. The less reduction of farm profit is related to the tariffs of 198 and 382 rials for each cubic meter of volumetric method which shows 1.5% and 4% reduction, respectively. Considering the results, it can be said that in the lower tariffs, the block and volumetric tariffs effect on farm profit is almost identical but in the higher levels of pricing tariffs, the profit reduction in block tariff is less than volumetric one so that this difference in 420.3 and 853.3 rials for each cubic meter of water has 11% and 26% more reduction of profit. The sown area in the tariff of 198 rials for volumetric tariff is 84% and under block tariff is 83%. On the hand, attention has to be paid that increasing water price results in reducing the irrigated land areas thereby reducing the water consumption. Therefore, for this level of pricing tariff, block tariff allows to have a 3 percent water saving for 1 percent of reduction in the sown areas. Furthermore, placing 382, 420.3 and 853.3 tariffs will reduce the irrigated lands to 80, 62 and less than 50 percent, respectively.

CONCLUSION

In a brief summarizing, given the comparative results of the two models including Positive Mathematical Programming (PMP) and Econometric Mathematical programming (EMP) in reproducing the observed values and also the water demand and irrigated water amounts, it is understood that econometric mathematical programming model is more suitable and it is suggested to us this approach to better analyze the simulation of the effects of agricultural policies on farmers behavior. On the other hand, the simulation results show that pricing policies in irrigation sector are extremely affected by the local, structural and institutional situation. Also the pricing policies often are seeking objectives as economic efficiency, reducing costs, justice and resources conservation which are opposite to each other. The simulation analysis of alternative irrigation water pricing policies indicates that the block pricing policy is considerably capable to influence the allocation, efficiency improvement and water saving with taking into account of farmers' profit and the total welfare of suppliers and consumers. So, in order to sustaine the water resources and management and influencial reduction in irrigation water demand, it needs to incease the water price significantly but this plan will face serious reactions by beneficiaries of surface water and and also the policy makers and administrative authorities related to water issue. Given the afformantioned issues, there appears that for properly managing the water demand, they should go forward with accurate planning and scheduling the water price increasing (so that, the average water price approaches to long-run marginal cost companied with reforming the economic structure of the country) and consolidated which are likely to improve the irrigation water demand management. In this line, management and planning the water resources distribution and also correcting the water rules and presenting a suitable pattern of determining the rate of water price as block tarrif which is appropriate for the khomein region or other similar plains are the most influencial policies to reconstruct a progressive irrigation water managent system.

REFERENCES

1- Agricultural Static Yearbook, (2012), Agricultural Organization of Markazi Province.

2-Asadi, H., Soltani, Gh., & torkamaani, J.(2007). Irrigation water pricing in Iran: A case study on land downstream of Taleghan dam. Agricultural Economic and Development (Special agricultural policy), 15: 58.

3- Buysse, J., Van Huylenbroeck, G., & Lauwers, L. (2007). Normative, positive and econometric mathematical programming as tools for incorporation of multi-functionality in agricultural policy modelling. Agriculture, Ecosystems and Environment, 120, 70-81.

4- Chang, H., & Mishra. P.L. (2008). Impact off-farm labor supply on food expenditures of the farm household. Journal of Food Policy, 51, 619-624.

5- Dinar, A., & Maria-Saleth, R. (2005). Issues in Water Pricing Reforms: From Getting Correct Prices to Setting Appropriate Institutions. In Folmer, H. and Tietenberg, T. (Eds.). International Yearbook of Environmental and Resource Economics New Horizons in Environmental Economics, Edward Elgar Publishing, Cheltenham, UK.

6- Dinar, A., & Mody, J. (2004). Irrigation water management policies: Pricing and allocation principles and implementation experiences. Natural Resources Forum, 28, 112-122.

7- Fragoso, R., & Marques, C. (2009). Avaliação económica de tarifas de água no uso agrícola: um estudo de caso no Sul de Portugal. Revista de Economia e Sociologia Rural, 47(3), 699-718.

8- Frija, A., Chebil, A., Speelman, S., & Van Huylenbroeck, G. (2008). Effect of changes in the institutional structure of irrigation water property rights on the willingness to pay of farmers for water: case of Tunisia. XIIth Congress of European Association of Agricultural Economics. EAAE, Ghent, Belgium.

9- Golan, A., Judge, G., & Miller, D. (1996). Maximum Entropy Econometrics, Wiley, Chichester United Kingdom.

10- Heckelei, T., & Britz, W. (2005). Models based on positive mathematical programming: state of art and further extensions. In: Arfini, F. (ed.) Modelling Agricultural Policies: State of Art, New Challenges. Monte Universita Parma, Parma, 48-74.

11- Heckelei, T., & Wolff, H. (2003). Estimation of constrained optimization models for agricultural supply analysis based on generalized maximum entropy. European Review of Agricultural Economics, 30(1), 27-50.

12- Henry de Frahan, Buysse, B., Polomé, J., Fernagut, B., Harmignie, O., Lauwers, L., Van Huylenbroeck, G., & Van Meensel, J. (2007). Positive mathematical programming for agriculture and environmental policy analysis: review and practice. In: Weintraub, A., Bjorndal, T., Epstein, R., & Romero, C. (eds.), Management of Natural Resources: A Handbook of Operations Research Models, Algorithms and Implementations. Kluwer Academic Publihers.

13- Herrera, P.A., van Huylenbroeck, G., & Espinel, R.L. (2004). An application of the contingent valuation method to assess the efficiency of the institutional structure of irrigation property rights: the case of the peninsula of Santa Elena, Ecuador. International Journal of Water Resources Development, 20: 537-551.

14- Howitt, R. (1995). Positive mathematical programming. American Journal of Agricultural Economics, 77(2), 329-342.

15- Hussein zad, J. (2004). The role of agricultural water demand management policy of the precious, Iranian Water Resources Management Conference, University of Tabriz.

16- Jafari, M. (2013). Investigating the effect of drought and increasing price of agricultural inputs on cropping patterns of Khomein plain. The Dissertation of M.Sc. in Agricultural Economics Engineering, University of Sistan and Baluchestan.

17- Johansson, R.C., Tsur, Y., Roe, T.L., Doukkali, R., & Dinar, A. (2002). Pricing irrigation water: a review of theory and practice. Water Police, 4, 173-199.

18- Karbasi, A., & Falsafizade, N. (2010). Factors affecting the diversity of agricultural crops, Journal of Agricultural Economics Research, Vol 4.

19- Liao, Y.S., Giordano, M., & Fraiture, C. (2007). An empirical analysis of the impacts of irrigation pricing reforms in China. Water Policy, 9, 45-60.

20- Mills, G. (1984). Optimization in Economic Analysis. George Allen & Unwin Ltd, London.

21- Mosayebi, M., & Maliki, M. (2012). Evaluate the status of water resources in terms of droughts and wet periods (Case Study: Watershed Khomeini turret - Central Province), First National Conference on Wilderness, Wilderness International Research Center, Tehran University.

22- Ray, I. (2007). Get the prices right: a model of water prices and irrigation efficiency in Maharashtra, India. In Molle, F., Berkoff, J. (eds.), Irrigation water pricing: The gap between theory and practice. CAB International, Oxfordshire, UK.

23- Roe, T., Dinar, A., Tsur, Y., & Diao, X. (2005). Feedback links between economy wide and farm-level policies: With application to irrigation water management in Morocco. Journal of Policy Modeling, 27, 905-928.

24- Rogers, P., Bhatia, R., & Huber, A. (1998). Water as a social and economic good: How to put the principle into practice. Global Water Partnership/ Swedish International Development Cooperation Agency, Stockholm, Sweden.

25- Rogers, P., Silva, R., & Bhatia, R. (2002). Water

is an economic good: How to use prices to promote equity, efficiency, and sustainability. Water Policy, 4, 1-17.

26- Speelman, S., Farolfi, S., Frija, A., D'Haese, M., & D'haese, L. (2010a). The impact of the water rights system on smallholder irrigators' willingness to pay water in Limpopo province, South Africa. Environment and Development Economics, 15, 465-483.

27- Speelman, S., Frija, A., Buysse, J., & Van-Huylenbroeck, G. (2011). The importance of irrigation water rights: lessons from South Africa and Tunisia. Water Policy, 13(5), 1-14.

28- Tiwari, D., & Dinar, A. (2002). Balancing future food demand and water supply: The role of economic incentives in irrigated agriculture. Quarterly Journal of International Agriculture, 41(1-2), 77-97.

29- Tsur, Y., & Dinar, A. (1997). On the relative efficiency of alternative methods for pricing irrigation water and their implementation. World Bank Economic Review, 11, 243-262.

30- Tsur, Y., Roe, T.L., Doukkali, R.M., & Dinar, A. (2004). Pricing irrigation water: Principles and cases from developing countries. RFF Press, Washington, DC, USA

31- Tsure, Y. (2005). Economic aspects of irrigation water pricing. Canadian Water Resources Journal, 30(1), 31-46.

32- Veettil, P.C., Speelman, S., Frija, A., Buysse, J., &Van-Huylenbroeck, G. (2011). Complementarity between water pricing, water rights and local water governance: A Bayesian analysis of choice behaviour of farmers in the Krishna river basin, India. Ecological Economics, 70, 1756-1766.

33- Ward, F.A. (2009). Economics in integrated water management. Environmental Modelling and Software, 24, 948-958.

Permissions

All chapters in this book were first published in IJAMAD, by Islamic Azad University; hereby published with permission under the Creative Commons Attribution License or equivalent. Every chapter published in this book has been scrutinized by our experts. Their significance has been extensively debated. The topics covered herein carry significant findings which will fuel the growth of the discipline. They may even be implemented as practical applications or may be referred to as a beginning point for another development.

The contributors of this book come from diverse backgrounds, making this book a truly international effort. This book will bring forth new frontiers with its revolutionizing research information and detailed analysis of the nascent developments around the world.

We would like to thank all the contributing authors for lending their expertise to make the book truly unique. They have played a crucial role in the development of this book. Without their invaluable contributions this book wouldn't have been possible. They have made vital efforts to compile up to date information on the varied aspects of this subject to make this book a valuable addition to the collection of many professionals and students.

This book was conceptualized with the vision of imparting up-to-date information and advanced data in this field. To ensure the same, a matchless editorial board was set up. Every individual on the board went through rigorous rounds of assessment to prove their worth. After which they invested a large part of their time researching and compiling the most relevant data for our readers.

The editorial board has been involved in producing this book since its inception. They have spent rigorous hours researching and exploring the diverse topics which have resulted in the successful publishing of this book. They have passed on their knowledge of decades through this book. To expedite this challenging task, the publisher supported the team at every step. A small team of assistant editors was also appointed to further simplify the editing procedure and attain best results for the readers.

Apart from the editorial board, the designing team has also invested a significant amount of their time in understanding the subject and creating the most relevant covers. They scrutinized every image to scout for the most suitable representation of the subject and create an appropriate cover for the book.

The publishing team has been an ardent support to the editorial, designing and production team. Their endless efforts to recruit the best for this project, has resulted in the accomplishment of this book. They are a veteran in the field of academics and their pool of knowledge is as vast as their experience in printing. Their expertise and guidance has proved useful at every step. Their uncompromising quality standards have made this book an exceptional effort. Their encouragement from time to time has been an inspiration for everyone.

The publisher and the editorial board hope that this book will prove to be a valuable piece of knowledge for researchers, students, practitioners and scholars across the globe.

List of Contributors

Fakhrodin Ghasemi Sahebi
M.Sc., Irrigation and Drainage, Razi University, Kermanshah, Iran

Masoud Hekmat
Department of Agricultural Management, Shoushtar Branch, Islamic Azad University, Shoushtar, Iran

Eman Pourkhiz
Young Researchers Club, Ghaemshahr Branch, Islamic Azad University, Iran

Vali Borimnejad
Associate Professor, Department of Agricultural Economics, Islamic Azad University, Karaj Branch, Karaj, Iran

Fatemeh Salimian
MSc Student, Department of Agricultural Economics, Islamic Azad University, Karaj Branch, Karaj, Iran

Mohammad Reza Bakhshi
Associate professor of Agronomy Department, Faculty of Agriculture, University of Birjand

Behnaz Nejati
Former Graduate Student Department of Geography, Faculty of Literature and Humanities, University of Birjand

Mofid Shateri
Associate professor of Department of Geography, Faculty of Literature and Humanities, University of Birjand

Nasrin Ohadi
Master of Agricultural Economics, University of Sistan and Baluchestan, Iran

Jaleh Kurki Nejad
PhD student of Agricultural economics, Faculty Member, University of Payam-Noor Sirjan, Iran

Masoud Samian
Ph.D. Student, Agricultural Extension and Education Department, College of Agriculture, Bu-Ali Sina University, Hamedan, Iran

Karim Naderi Mahdei, Hamid Balali and Reza Movahedi
Assistant Professor Agricultural Extension and Education Department, College of Agriculture, Bu-Ali Sina University, Hamedan, Iran

Heshmatollah Saadi
Associate Professor, Agricultural Extension and Education Department, College of Agriculture, Bu-Ali Sina University, Hamedan, Iran

Soleiman Rasouliazar and Saeid Fealy
Department of Agricultural Management, Mahabad Branch, Islamic Azad University, Mahabad, Iran

Hassan Afrakhteh
Professor in the Department of Rural Planning, Kharazmi University, Tehran, Iran

Maryam Armand and Fatemeh Askari Bozayeh
PhD Student of Rural Planning, Kharazmi University, Tehran, Iran

Morteza Tahamipour
Assistant Professor, Department of Economics, University of Shahid Beheshti

Mohammad Kavoosi Kalashami
Assistant Professor, Department of Agricultural Economics, Faculty of Agricultural Sciences, University of Guilan

Amirhossein Chizari
Assistant Professor of Agricultural Economics, Department of Economics and Agricultural Development, University of Tehran

Habib Shahbazi
Assistant professor, Department of Agricultural Economics, Sayyed Jamaleddin Asadabadi University (SJAU), Hamedan, Iran

Hamidreza Alipour
Department of Management, Rasht Branch, Islamic Azad University, Rasht, Iran

Akram Abbasifar
Research Assistant, Consumer and Producer Protection Organization, Ministry of Industry, Mine and Business, Tehran, Iran

Zahra Ghaffari Moghadam
Institute of Agricultural Center at the University of Zabol

Mostafa Mardani, Saman Ziaee, Elham Kalbali and Samira Soltani
Department of Agricultural Economics, University of Zabol, Bonjar Street, Zabol City, Sistan and Balochestan Province, Iran

Mohammad Taleghani
Associate Professor, Faculty of Management and Accounting, Department of Industrial Management, Rasht Branch, Islamic Azad University, Rasht, Iran

Hayat Lionboui and Tarik Benabdelouahab
National Institute of Agricultural Research, Research Center of Tadla, Beni Mellal, Morocco

Aziz Hasib and Abdelali Boulli
Sultan Moulay Slimane University, Faculty of Science and Technology, Beni Mellal, Morocco

Hamidreza Mirzaei Khalil and Mahdi Esfandiari
Department of Agricultural Economics, Shahid Bahonar University of Kerman, Kerman, Iran

Javad Shahraki and Morteza Yaghoubi
Department of Economics, University of Sistan and Baluchestan, Zahedan, Iran

Mahmoud Mohammadghasemi
PhD Student of Agricultural Economics, University of Sistan and Baluchestan
Social and Economic Research Department, Sistan Agriculture and Natural Resources Research and Education Center, AREEO, Zabol, Iran

Javad Shahraki
Associate Professor of Economics, University of Sistan and Baluchestan

Mahmood Sabouhi Sabouni
Professor of Agricultural Economics, University of Ferdowsi Mashhad

Zahra Hosseinpur
M.Sc. Student in Rural Development, Faculty of Agricultural Sciences, University of Guilan, Guilan, Iran

Mohammad Hossein Menhaj and Mohammad Kavoosi-Kalashami
Assistant Professor, Department of Agricultural Economics, Faculty of Agricultural Sciences, University of Guilan, Guilan, Iran

Kolsoum Azizi Mizab
MSc. in Agricultural Management, Tabriz Branch, Islamic Azad University, Tabriz, Iran

Azadeh Falsafian
Assistant Professor, Department of Agricultural Management, Extension and Education, Tabriz Branch, Islamic Azad University, Tabriz, Iran

Maryam Shadpour-Rashti
MSc. Student, Irrigation and Drainage, Water Engineering Department, Faculty of Agricultural Sciences, University of Guilan, Rasht, Iran

Mohammadreza Khaledian
Water Engineering Department, Faculty of Agricultural Sciences, University of Guilan, Rasht, Iran, and and Department of Water Engineering and Environment, Caspian Sea Basin Research Center

Mohammadhassan Biglouei
Water Engineering Department, Faculty of Agricultural Sciences, University of Guilan, Rasht, Iran

Mojtaba Rezaei
Rice Research Institute of Iran, Agricultural Research, Education and Extension Organization (AREEO), Rasht, Iran

Javad Shahraki
Assistant Professor of Agricultural Economics, University of Sistan and Baluchestan

Neda AliAhmadi
Master of Agricultural Economics in MA great of the University of Sistan and Baluchestan

Seyed Mahdi Hosseyni
Ph.D Student at University of Sistan and Baluchestan, Iran

Javad Shahraki
Assistant Professor Economics, University of Sistan and Baluchestan, Iran

Seyed Abolghasem Mortazavi and Reza Hezareh
Faculty Member and MSc Student, Department of Agricultural Economics, University of Tarbiat Modares, Tehran, Iran

Sina Ahmadi Kaliji
Former MSc student, Department of Agricultural economics, Gorgan University of Agricultural Sciences and Natural Resources, Gorgan, Iran

Samira Shayan Mehr
MSc Student in Agricultural Economics, University of Kordestan, Iran

Govinda Bhandari
Environment Professionals' Training and Research Institute Pvt. Ltd. Kathmandu, Nepal

Nafisat Bolatito, Ikenweiwe, Adekule Adedoyin, Idowu, Nathanael Akinsafe, Bamidele, Otubusin Samuel and Eunice Oluremi, Fadipe
Department of Aquaculture and Fisheries Management University of Agriculture, PMB2240, Abeokuta, Ogun State, Nigeria

Nurul Aisyah and Nalini Arumugam
Faculty of Agriculture and Biotechnology, Universiti Sultan Zainal Abidin, Kampus Gong Badak, 21300 Kuala Terengganu, Terengganu, Malaysia

Mohd. Ariff Hussein and Ismail Latiff
Department of Agribusiness and Information Technology, Universiti Putra Malaysia, 43400 Serdang, Selangor Darul Ehsan, Malaysia

Morteza Tahami Pour and Mohammad Kavoosi Kalashami
Department of Agriculture Economic, University of Tehran, Iran

Gholamreza Zamanian
Assistant Professor in Economics, University of Sistan and Baluchestan, Iran

Mehdi Jafari
Agricultural Economics, University of Sistan and Baluchestan, Iran

Shahram Saeedian
Graduate student of Agricultural Economics, University of Sistan and Baluchestan, Iran

Mina Salehnia
Department of Agricultural Economics, Faculty of Agriculture, University of Tabriz, Iran

Baballah Hayati and Mohammad Ghahremanzadeh
Associate Professors, Department of Agricultural Economics, Faculty of Agriculture, University of Tabriz, Iran

Morteza Molaei
Assistant Professor, Department of Agricultural Economics, Faculty of Agriculture, University of Urmia, Iran

Index

www.ingramcontent.com/pod-product-compliance
Lightning Source LLC
LaVergne TN
LVHW060940220926
848485LV00015B/1147